Selected Topics on
Continuous-Time Controlled
Markov Chains and Markov Games

ICP Advanced Texts in Mathematics

ISSN 1753-657X

Series Editor: Dennis Barden *(Univ. of Cambridge, UK)*

ICP Advanced Texts in Mathematics – Vol. 5

Selected Topics on Continuous-Time Controlled Markov Chains and Markov Games

Tomás Prieto-Rumeau

Universidad Nacional de Educación a Distancia, Spain

Onésimo Hernández-Lerma

CINVESTAV-IPN, Mexico

Imperial College Press

Published by

Imperial College Press
57 Shelton Street
Covent Garden
London WC2H 9HE

Distributed by

World Scientific Publishing Co. Pte. Ltd.
5 Toh Tuck Link, Singapore 596224
USA office: 27 Warren Street, Suite 401-402, Hackensack, NJ 07601
UK office: 57 Shelton Street, Covent Garden, London WC2H 9HE

British Library Cataloguing-in-Publication Data
A catalogue record for this book is available from the British Library.

ICP Advanced Texts in Mathematics — Vol. 5
**SELECTED TOPICS ON CONTINUOUS-TIME CONTROLLED
MARKOV CHAINS AND MARKOV GAMES**
Copyright © 2012 by Imperial College Press

ISBN-13 978-1-84816-848-0
ISBN-10 1-84816-848-9

Typeset by Stallion Press
Email: enquiries@stallionpress.com

Printed in Singapore.

To Guadalupe and my parents, Emilio and Michèle.

To Marina, Gerardo, Adrián, Claudia, and Lucas.

Preface

This book concerns continuous-time controlled Markov chains and Markov games. The former, which are also known as continuous-time Markov decision processes, form a class of stochastic control problems in which a single decision-maker wishes to optimize a given objective function. In contrast, in a Markov game there are two or more decision-makers (or players, or controllers) each one trying to optimize his/her own objective function.

The main features of the control and game models studied in the book are that the time variable is *continuous*, the state space is *denumerable*, and the control (or action) sets are *Borel spaces*. Moreover, the transition and reward rates of the dynamical system may be *unbounded*. Controlled Markov chains and Markov games have many important applications in areas such as telecommunication networks, population and epidemic models, engineering, operations research, etc. Some of these applications are illustrated in this book.

We note that most of the material presented here is quite recent: it has been published in the last six years, and it appears in book form for the first time.

One of the main goals of this book is to study the so-called advanced optimality criteria for controlled Markov chains (e.g., bias, overtaking, sensitive discount, and Blackwell optimality), which are refinements of the basic criteria, namely, discounted and average reward optimality. To make this a self-contained book, we also give the main results on the existence of controlled Markov chains and the basic optimality criteria. For the corresponding technical details — some of which have been skipped here — the reader can consult Guo and Hernández-Lerma's *Continuous-Time Markov Decision Processes: Theory and Applications* [52].

A particular emphasis is made regarding the application of the results presented in the book. One of our main concerns is to propose assumptions on the control and game models that are easily verifiable (and verified) in practice. Furthermore, we study an algorithm to solve a certain class of control models, and establish some approximation results that allow us to give precise numerical approximations of the solutions to some problems of practical interest.

Hence, the book has an adequate balance between, on the one hand, theoretical results and, on the other hand, applications and computational issues. It is worth mentioning that the latter were, somehow, missing in the literature on continuous-time controlled Markov chains.

Finally, the topic of zero-sum two-person continuous-time Markov games, for both the basic — discounted and average payoff — and some "advanced" optimality criteria — bias and overtaking equilibria — appears for the first time in book form.

This book is mainly addressed to researchers in the fields of stochastic control and stochastic games. Indeed, it provides an extensive, rigorous, and up-to-date analysis of continuous-time controlled Markov chains and Markov games. It is also addressed to advanced undergraduate and beginning graduate students because the reader is not supposed to have a high mathematical background. In fact, a working knowledge of calculus, linear algebra, probability, and continuous-time Markov chains (at the level of, say, Chapter 4 of R. Durrett's book *Essentials of Stochastic Processes* [31]) should suffice to understand the material herein. As already mentioned, the reader interested in the theoretical foundations of controlled Markov chains can consult [52].

We have carefully written this book, with great dedication and commitment. We apologize, however, for any errors and omissions it might contain.

The authors, April 2011

Contents

Chapter 1

Introduction

1.1. Preliminary examples

Before giving a formal definition of control and game models, we propose two motivating examples.

1.1.1. A controlled population system

We describe next a *controlled population system* inspired by the models in [52, Example 7.2] and [136, Sec. IV]. (In Sec. 9.4 below we will consider a generalization of this controlled population system.) We call it a controlled system because there is a *controller* (also known as a *decision-maker*) who observes a *random dynamical system*, and takes *actions* so as to optimize the system's behavior according to a given *optimality criterion*.

The state space The state variable, denoted by i, is the population size, which takes values in the *state space*

$$S = \{0, 1, 2, \ldots\}.$$

We suppose that the population system is observed *continuously in time*, at times labeled $t \geq 0$. The time horizon may be finite (that is, we observe the state variable on a time horizon $0 \leq t \leq T$, for some finite time $T > 0$) or infinite (which means that the population system is observed at all times $t \geq 0$). We will denote by $x(t) \in S$ the random state of the system at time $t \geq 0$. We will refer to $\{x(t)\}_{t \geq 0}$ as the *state process*.

  The sources of variation of the population are described next.

The birth rate The population is subject to a natural birth rate, denoted by $\lambda > 0$. This means that each individual of the population can give birth

1

to a new individual with a transition probability rate which equals λ. More precisely, suppose that the population size is $i \in S$ at time $t \geq 0$, and let $P_i^+[t, t+\delta]$ denote the probability that a new individual is born on the time interval $[t, t + \delta]$. Then we have that

$$\lim_{\delta \downarrow 0} \frac{P_i^+[t, t+\delta]}{\delta} = \lambda i \quad \forall \, t \geq 0. \tag{1.1}$$

Note that, in (1.1), the "individual" birth rate λ is multiplied by the population size i.

The death rate   The population is also subject to a natural death rate $\mu > 0$. So, if the population size is i at time $t \geq 0$, and $P_i^-[t, t+\delta]$ denotes the probability that an individual dies on the time interval $[t, t + \delta]$, then we have

$$\lim_{\delta \downarrow 0} \frac{P_i^-[t, t+\delta]}{\delta} = \mu i \quad \forall \, t \geq 0. \tag{1.2}$$

The immigration rate   At this point, note that the birth and death rates described above are *not controlled*, meaning that the decision-maker cannot modify them. That is why we called them the *natural* birth and death rates. On the other hand, the immigration rate, defined next, may be controlled by the decision-maker.

Put $A = [a_1, a_2] \subset \mathbb{R}^+$, and let $a \in A$ be the controlled immigration rate. The interpretation is that the decision-maker can encourage or discourage immigration by following suitable immigration policies. Hence, when the decision-maker *chooses* the control $a \in A$, the probability $P_i^a[t, t+\delta]$ that an immigrant individual arrives, on the time interval $[t, t+\delta]$, at the population under study when its size is i at time $t \geq 0$ verifies that

$$\lim_{\delta \downarrow 0} \frac{P_i^a[t, t+\delta]}{\delta} = a \quad \forall \, t \geq 0. \tag{1.3}$$

We assume that the controller takes actions continuously in time. So, let $a(t)$ in A, for $t \geq 0$, denote the controller's action at time t.

The catastrophe rate   In addition, we suppose that the population is subject to "catastrophes". The rate $b \in B = [b_1, b_2] \subset \mathbb{R}^+$ at which catastrophes occur is controlled by the decision-maker (for instance, by using adequate medical policies or implementing fire prevention programs, the controller can decrease the catastrophe rate b).

Moreover, the catastrophe is supposed to have a random size. This means that, if a catastrophe occurs when the population size is $i \in S$, then the probability that $1 \leq k \leq i$ individuals die in the catastrophe is $\gamma_i(k)$. We suppose that $\gamma_i(k) > 0$ and that

$$\sum_{1 \leq k \leq i} \gamma_i(k) = 1.$$

Consequently, the transition rate from a state $i > 0$ to a state $0 \leq j < i$ corresponding to a catastrophe under the action $b \in B$ is

$$b \cdot \gamma_i(i - j). \tag{1.4}$$

More explicitly, the rate b corresponds to the catastrophe, and then, *conditional on the catastrophe*, $i - j$ individuals perish with probability $\gamma_i(i-j)$. The new state of the system is thus j.

We denote by $b(t) \in B$, for $t \geq 0$, the action chosen by the controller at time $t \geq 0$.

The action set As seen in the previous paragraphs, the controller chooses his/her actions in the set $A \times B$. We will refer to $A \times B$ as the *action set*.

The transition rate matrix Our previous discussion on the dynamics of the system can be summarized by the transition rate matrix $[q_{ij}(a, b)]_{i,j \in S}$. Here, $q_{ij}(a, b)$ denotes the transition rate from the state $i \in S$ (row) to the state $j \in S$ (column) when the controller chooses the actions $a \in A$ and $b \in B$. This transition rate matrix is

$$\begin{pmatrix} -a & a & 0 & 0 & 0 & 0 & \cdots \\ \mu + b & -(\mu + \lambda) - a - b & \lambda + a & 0 & 0 & 0 & \cdots \\ b\gamma_2(2) & 2\mu + b\gamma_2(1) & -2(\mu + \lambda) - a - b & 2\lambda + a & 0 & 0 & \cdots \\ b\gamma_3(3) & b\gamma_3(2) & 3\mu + b\gamma_3(1) & -3(\mu + \lambda) - a - b & 3\lambda + a & 0 & \cdots \\ \vdots & \vdots & & & & \vdots & \ddots \end{pmatrix}.$$

The matrix $[q_{ij}(a, b)]_{i,j \in S}$ above is constructed as follows. The transition rate from state i to $i + 1$ is obtained by summing the corresponding transition rates in (1.1) and (1.3). For the transition rate from i to $i - 1$ we proceed similarly, and we sum (1.2) and (1.4) for $j = i - 1$. Finally, the transition rate from i to j, for $0 \leq j < i - 1$, is given by (1.4). The diagonal terms $q_{ii}(a, b)$ are such that the rows of the transition rate matrix sum to zero. This technical requirement comes from the fact that the transition probabilities sum to one (by rows), and so the corresponding derivatives must sum to zero. Further details on this issue are given in Chapter 2.

Policies   A *control policy* or simply a *policy* is a "rule" that prescribes the actions chosen by the controller. Typically, a policy is a function of the form

$$\varphi(t, i) = (a, b) \in A \times B,$$

which is given the following interpretation: the controller observes the state of the system $x(t) = i \in S$ at time $t \geq 0$, and then he/she takes the actions $a(t) = a \in A$ and $b(t) = b \in B$. The process $\{x(t), a(t), b(t)\}_{t \geq 0}$, which describes the evolution of the system and the controller's actions, is called the *state-action process*.

The above defined policies are called *Markovian* because they depend only on the current state of the system, say $x(t)$, and the time variable t. In general, however, although we will not consider them in this book, policies can be *history-dependent*. This means that the actions $a(t)$ and $b(t)$ may depend on the *history* $\{x(s), a(s), b(s)\}_{0 \leq s \leq t}$ of the state-action process up to time t. More specific classes of policies (such as *randomized* or *stationary* policies) will be introduced in Chapter 2.

The reward rates   We suppose that the controller earns rewards (or incurs costs) continuously in time. Typically, the reward rates depend on both the state of the system and the actions.

For this controlled population system, suppose that there is a reward rate function $R(i)$ depending on the population size $i \in S$. Usually, R will be an increasing function of i. In some particular cases, R will be a linear function.

In addition, we assume that there is a cost rate $C_1(i, a)$ associated with the action $a \in A$ when the population size is $i \in S$. (The function C_1 captures, e.g., the cost of the immigration policy, but also the benefits of having a larger working population.) Similarly, the cost rate for controlling the catastrophe rate $b \in B$ when the population size is $i \in S$ is denoted by $C_2(i, b)$.

Hence, if the decision-maker selects the actions $(a(t), b(t)) \in A \times B$ when the state of the system is $x(t) \in S$, at time $t \geq 0$, then he/she obtains an infinitesimal net reward

$$(R(x(t)) - C_1(x(t), a(t)) - C_2(x(t), b(t))) \cdot \delta$$

on the "small" time interval $[t, t + \delta]$.

The optimality criterion   The optimality criterion is concerned with the performance evaluation of the policies. As an illustration, if the controller wants to maximize his/her total expected reward on the finite horizon $[0, T]$, then he/she will consider

$$\varphi \mapsto E^{\varphi} \left[\int_0^T [R(x(t)) - C_1(x(t), a(t)) - C_2(x(t), b(t))] dt \right] \qquad (1.5)$$

for each policy φ. In (1.5), E^{φ} denotes expectation under the policy φ. Hence, the *finite horizon control problem* consists in finding a policy with the maximal total expected reward (1.5).

Suppose now that there is a depreciation rate $\alpha > 0$ (related to the inflation rate), and that the controller wants to maximize his/her total expected rewards brought to their present value. The *discounted optimality criterion* consists in finding a policy φ that maximizes

$$E^{\varphi} \left[\int_0^{\infty} e^{-\alpha t} [R(x(t)) - C_1(x(t), a(t)) - C_2(x(t), b(t))] dt \right].$$

Furthermore, we can assume that the controller has a given budget, say θ, for the (discounted) expenses on the immigration policy and the catastrophe prevention programs. In this case, the controller has to find the policy that maximizes the expected discounted reward

$$E^{\varphi} \left[\int_0^{\infty} e^{-\alpha t} R(x(t)) dt \right]$$

within the class of policies that satisfy the constraint

$$E^{\varphi} \left[\int_0^{\infty} e^{-\alpha t} [C_1(x(t), a(t)) + C_2(x(t), b(t))] dt \right] \leq \theta.$$

This is a *constrained control model* similar to those that we will study in Chapter 8.

Conclusions   Finally, we summarize the main elements of the controlled population system described above.

- The state space. (As in the example in this section, the control models studied in this book have denumerable state space.)
- The action set.
- The transition and reward rates.
- The optimality criterion.

1.1.2.   *A prey-predator game model*

In the control model described in Sec. 1.1.1, there was a single controller handling the stochastic dynamical system. In a game model, we suppose that there are *several players*. Our next example is a simplified version of the Kolmogorov prey-predator model, which is based on the Lotka–Volterra equation; see, e.g., [19, 39].

The state space   We assume that there are two interacting species in a given environment: species 1 is the prey, while species 2 is the predator. The bidimensional state variable (i, j) stands for the total population i and j of the prey and predator species, respectively. So, the state space is

$$S = \{0, 1, 2, \ldots\} \times \{0, 1, 2, \ldots\}.$$

The state process $x(t) = (i(t), j(t))$, for $t \geq 0$, gives the (random) size of the two populations at time $t \geq 0$. As in the control model in Sec. 1.1.1, the state variable is observed continuously at times $t \geq 0$.

The action set   The prey species takes actions $a \in A \subset \mathbb{R}$. The action a models the struggle for survival of the prey (for instance, moving to safer areas under the hunting pressure of the predator). The predator species takes actions $b \in B \subset \mathbb{R}$, which model its hunting intensity (e.g., involving more individuals in hunting). Hence, the action set for species 1 is A, while B is the action set for species 2.

   Therefore, this model is a *two-player* game in which the prey and predator species compete.

Strategies of the players   The players use Markov policies (as defined in Sec. 1.1.1). More precisely, the players observe the state of the system $(i(t), j(t))$, and then they independently choose their actions $a(t) \in A$ and $b(t) \in B$. Hence, the policies of the players are given by functions

$$\phi : [0, \infty) \times S \to A \quad \text{and} \quad \psi : [0, \infty) \times S \to B,$$

for species 1 and 2, respectively, which depend on the state of the system and the time $t \geq 0$.

   A usual convention in game models is that the players' policies are referred to as *strategies*, rather than policies.

The transition rates   We assume that the two species have natural birth and death rates. Namely, the birth and death rates of the prey species are $\lambda_1 > 0$ and $\mu_1 > 0$, respectively, while the corresponding birth and death rates of the predator species are $\lambda_2 > 0$ and $\mu_2 > 0$.

In addition, we suppose that the hunting rate is of the form

$$F(a, b) \cdot i \cdot j$$

when the populations have respective sizes $i \geq 0$ and $j \geq 0$, and the species take actions $a \in A$ and $b \in B$. The function F accounts for the interaction between the actions of the species. On the other hand, the $i \cdot j$ term models the fact that the hunting rate grows with both the prey and the predator population.

Finally, taking into account that the prey is the predator's food supply, it is natural to assume that hunting has an effect on the predator's birth rate. This effect is modeled by a term of the form $G(a, b) \cdot i \cdot j$, interpreted as the hunting rate defined above.

Hence, the transition rates of this prey-predator population system are as follows. If the system is at state $(i, j) \in S$ and the species take the actions $(a, b) \in A \times B$, then transitions to the next states can occur:

- $(i + 1, j)$ with transition rate $\lambda_1 i$,
- $(i - 1, j)$ with transition rate $(\mu_1 + F(a, b)j) \cdot i$,
- $(i, j + 1)$ with transition rate $(\lambda_2 + G(a, b)i) \cdot j$,
- $(i, j - 1)$ with transition rate $\mu_2 j$.

The above transition rates are inspired by the Lotka–Volterra differential equations, which are of the form

$$i'(t) = (c - Fj(t)) \cdot i(t),$$
$$j'(t) = (d + Gi(t)) \cdot j(t),$$

for "continuous" state variables $i(t)$ and $j(t)$, and constants c, d, F, and G. (For a differential game based on the Lotka–Volterra equations, the interested reader can consult [18].)

The payoff rates and the optimality criterion Each species has its own payoff rate function. Suppose, for instance, that the species want to maximize their long-run expected average population. Then the payoff rates r_1 and r_2 for the species 1 and 2 are

$$r_1(i, j, a, b) = i \quad \text{and} \quad r_2(i, j, a, b) = j,$$

for $(i, j) \in S$ and $(a, b) \in A \times B$, respectively. The goal of the prey species is to maximize the payoff function

$$V_1(\phi, \psi) = \lim_{T \to \infty} \frac{1}{T} E^{\phi, \psi} \left[\int_0^T i(t) dt \right],$$

while the predator species maximizes the payoff function

$$V_2(\phi, \psi) = \lim_{T \to \infty} \frac{1}{T} E^{\phi, \psi} \left[\int_0^T j(t)dt \right],$$

where $E^{\phi, \psi}$ denotes expectation under the pair of strategies (ϕ, ψ). This is called an *average payoff* game because the payoff functions are given by long-run averages.

Zero-sum games   A two-player game is called a *zero-sum* game if $V_1 + V_2 \equiv 0$, where V_i, for $i = 1, 2$, denotes the payoff of player i. In this case, $V_1 = -V_2$ and so the players have opposite criteria. As an illustration, suppose that the predator species wants to maximize the long-run average number of hunted individuals, that is,

$$V_2(\phi, \psi) = \lim_{T \to \infty} \frac{1}{T} E^{\phi, \psi} \left[\int_0^T [F(a(t), b(t)) \cdot i(t) \cdot j(t)]dt \right]. \qquad (1.6)$$

On the other hand, the prey species wants to minimize the long-run average number of hunted individuals (1.6) or, equivalently, maximize

$$V_1(\phi, \psi) = -\lim_{T \to \infty} \frac{1}{T} E^{\phi, \psi} \left[\int_0^T [F(a(t), b(t)) \cdot i(t) \cdot j(t)]dt \right].$$

This is indeed a *zero-sum* game because the payoff functions verify that $V_1 + V_2 \equiv 0$.

Noncooperative equilibria   Usually, in game models, both players cannot maximize their payoff functions simultaneously. Then, optimal strategies correspond to an equilibrium, namely, a *noncooperative* or *Nash* equilibrium. For the case of the zero-sum game, we say that a pair of strategies (ϕ^*, ψ^*) is a Nash equilibrium if

$$V_2(\phi^*, \psi) \leq V_2(\phi^*, \psi^*) \leq V_2(\phi, \psi^*)$$

for every pair of strategies (ϕ, ψ). The interpretation is that the players cannot improve their payoffs by unilaterally changing their strategies. Observe that the pair of strategies (ϕ^*, ψ^*) is a *saddle point* of the payoff function $(\phi, \psi) \mapsto V_2(\phi, \psi)$.

We note that to ensure the existence of such noncooperative equilibria, generally one has to consider *randomized strategies*. These will be defined in Chapter 10.

Conclusions The main elements of the prey-predator game model are thus:

- The state space. (Note that the state space is a bidimensional denumerable set.)
- The action sets of the players.
- The transition rates of the system and the payoff rates of the players.
- The optimality criterion.

The common feature with the control model in Sec. 1.1.1 is that the state space is denumerable. These are indeed the continuous-time control and game models we are concerned with in this book.

1.2. Overview of the book

After the motivating examples described in Sec. 1.1, we next give a more general description of the control and game models we will deal with in this book, together with some bibliographic references.

We are mostly concerned with continuous-time infinite horizon denumerable state controlled Markov chains (CMCs), also known as Markov decision processes (MDPs), or, more generally, as controlled Markov processes. The two final chapters are devoted to the study of zero-sum Markov games. We usually (but not exclusively) analyze these problems by means of dynamic programming techniques.

Control models A controlled Markov chain is a stochastic dynamical system that is "controlled" by a decision-maker or controller. The controller takes actions that modify the behavior of the stochastic system. Besides, the controller receives a reward (or incurs a cost) for handling the system. Therefore, the goal of the decision-maker is to find the optimal actions that yield the maximum reward (or minimum cost) with respect to a suitably defined optimality criterion. These control problems are termed Markovian because, typically, the state process behaves as a Markov process, that is, the transitions of the system depend on its current state (and not on the history of the process) and the current action taken by the controller.

We suppose that the time variable is continuous (as opposed to discrete-time models), that is, we assume that the controller observes the state of the system and can take actions at each time $t \geq 0$ of the time horizon. In contrast, in a discrete-time setting, observations and actions occur only at certain discrete times, for instance, $t = 0, 1, 2, \ldots$ It is important to mention that, since the controller does observe the state of the system, we are not faced with a partially observable (or incomplete information) control model. Moreover, we assume that the observed states take values in a denumerable set (the state space), while the controller chooses his/her actions in a fairly general (Borel) space, called the action (or control) space. This allows us to consider action spaces that can be "discrete" (that is, finite or countably infinite), "continuous" (say, an interval in $\mathbb{R}$), or even infinite-dimensional (for instance, in Sec. 2.5, we will consider an action space that is an infinite-dimensional metric space, while in Chapter 11 we will deal with an action space that is a family of probability measures). We will also assume that the rewards are earned continuously in time.

Control problems can be roughly classified according to the nature of the above three elements: the state space, the action space, and the time variable. Here, as was already mentioned, we shall consider a denumerable state space, Borel action space, continuous-time CMC.

The continuous-time case   Continuous-time CMCs were introduced by Bellman [11, Chapter 11] for the finite horizon case, while Howard [77, Chapter 8] dealt with infinite horizon problems with *finitely* many states. Other authors, for instance, Miller [112, 113], Rykov [144], and Veinott [161] made a systematic analysis of the discounted, average, and total expected reward optimality criteria. Veinott [161] introduced the sensitive discount optimality criteria. At these early stages, the various authors usually considered finite state and finite action CMCs, and most of the results for continuous-time models were deduced from their discrete-time counterparts (as suggested by Howard [77, p. 113] and later used by Veinott [161, p. 1654]). This approach was subsequently known as the *uniformization technique*, also used for denumerable state CMCs with *bounded* transition and reward rates. However, for most models of practical interest, such as queueing systems and population models, among others (see also the motivating examples in Sec. 1.1), the transition and reward rates are typically *unbounded*. In this case, the uniformization technique cannot be used. Hence, the interest in continuous-time denumerable state space CMCs mainly arises when considering *unbounded* transition and reward rates. This is the case we deal with in this book.

Summing up, continuous-time CMCs were studied from the early developments of dynamic programming, in parallel with the discrete-time case. Later, however, discrete-time models received much more attention, whereas continuous-time MDPs remained less developed. For instance, it is worth mentioning that the first book exclusively dealing with continuous-time CMCs is the recent book by Guo and Hernández-Lerma [52]. Nevertheless, in recent decades more attention has been paid to continuous-time models. This is because they accurately model situations of practical interest arising in, for instance, telecommunications, control of populations, and mathematical finance, to name just a few fields of application.

Optimality criteria Regarding the various optimality criteria, the basic ones are the infinite horizon discounted reward criterion and the long-run average reward optimality criterion. The latter requires, however, some results on the total reward optimality criterion for finite horizon problems. Moreover, some refinements of the discounted and average optimality criteria, such as bias, variance, overtaking, sensitive discount, and Blackwell optimality have been introduced in the literature. Our main concern here is to study these so-called "advanced" or refined optimality criteria. For average reward problems, we will also deal with pathwise optimality (as opposed to "expected" optimality). Constrained optimality criteria are analyzed in Chapter 8 of this book.

Stochastic games Stochastic games were introduced by Shapley [150] in 1953. His formulation is similar to that of control models, but now for two players. As noted by Shapley himself, control models can be seen as one-player games. More precisely, the game model consists of a stochastic dynamical system observed by two players. Then, at each stage, both players take an action and, depending on their actions, they receive a reward or incur a cost. Then the system makes a random transition according to the players' actions. In noncooperative games, each player tries to maximize or minimize his/her reward or cost, according to the corresponding optimality criterion. A Markov game is a stochastic game in which the state's transition probability function depends only on the current state of the system and the current actions of the players.

As for the control case, a game model depends on the nature of the time variable, the state space of the system, and the action spaces of the players. Here we consider the particular case of zero-sum games. The usual optimality criteria are the total expected payoff (for finite horizon problems), and the discounted and average payoff optimality criteria. Refined optimality criteria (such as bias and overtaking equilibria, as in control models) can also be analyzed.

One of the main concerns when dealing with zero-sum stochastic games is to prove the existence of the value function and a Nash equilibrium (also referred to as a noncooperative equilibrium or a saddle point). The existence of the value function and the Nash equilibrium, for different optimality criteria, has been proved by Parthasarathy [120], Federgruen [33], Mertens and Neyman [109, 110], and Sobel [152], among others. Perhaps surprisingly, the existence of Nash equilibria for a general nonzero-sum stochastic game with *uncountable* state space remains an *open problem*.

In this book, we deal with a two-person zero-sum continuous-time Markov game. We assume that the state space is denumerable, and also that the action space of each player is a Borel space. We deal with the basic optimality criteria (in analogy to control models): discounted and average optimality. Further, we study refinements of the average payoff criterion such as bias and overtaking equilibria.

1.3. Contents

This book is divided into eleven chapters, including this introduction. Chapters 2 to 9 are devoted to the study of controlled Markov chains (CMCs), while Chapters 10 and 11 deal with two-person zero-sum Markov games. Each chapter begins with an introduction which, in particular, gives bibliographic references on the topic studied therein. Also, each chapter ends with a section of conclusions.

In Chapter 2 we introduce the control model we will be dealing with. We also study the existence of a CMC, and we propose some results on ergodicity, which are interesting for both controlled and uncontrolled Markov chains.

Chapter 3 is devoted to the study of the basic optimality criteria (which include the total expected reward criterion for finite horizon problems, and the discounted and average reward criteria for infinite horizon models). A recurrent issue throughout this book is the vanishing discount approach to average reward optimality. In this chapter, we give our first results on this topic.

Further issues on discounted and average reward optimality are given in Chapter 4. These include the policy iteration algorithm and approximation results. The approximation theorems turn out to be useful tools for computationally solving discounted and average reward CMCs.

Our analysis of the refined optimality criteria begins in Chapter 5 with the bias, overtaking, and variance criteria. These criteria are refinements of the average reward optimality criterion in the sense that, within the class of average optimal policies, we find subclasses of "better" policies. See Sec. 5.1 for details.

On the other hand, the sensitive discount optimality criteria, described in Chapter 6, deal with asymptotic discount optimality as the discount rate goes to zero. In this chapter, we also establish further results on the vanishing discount approach to average optimality.

In Chapter 7 we study the Blackwell optimality criterion, which is the strongest criterion analyzed in this book. Blackwell optimality is obtained as the "limit" of sensitive discount criteria.

Constrained CMCs, for both discounted and average reward problems, are studied in Chapter 8. We use the Lagrange multipliers approach to constrained problems. We also explore the vanishing discount approach to average constrained models, and it turns out that the results do *not* entirely coincide with the unconstrained case.

Our analysis of control models ends in Chapter 9 with some applications. We study a controlled queueing system, two controlled population systems, and an epidemic process. We give conditions ensuring that there indeed exist optimal policies for the optimality criteria studied so far, and we also provide some illustrative computational results based on the approximations introduced in Chapter 4.

The two final chapters of this book deal with zero-sum Markov games. In Chapter 10, we study the basic optimality criteria, namely, discounted and average payoffs. The corresponding optimality equations, as well as the existence of optimal strategies, are also analyzed.

Finally, as for control models in Chapter 5, we study bias and overtaking equilibria for Markov games in Chapter 11. We explain why further refined optimality criteria (such as, for instance, sensitive discount optimality) do not make sense for a game model.

The book ends with an exhaustive list of the notation we use here, the bibliography, and an index including the main concepts and terminology introduced throughout.

1.4. Notation

We next introduce some notation that will be used in the text.

Given a metric space $\mathcal{X}$, the σ-field of Borel subsets of $\mathcal{X}$ is denoted by $\mathbb{B}(\mathcal{X})$. A *Polish space* is a complete and separable metric space. A Borel subset of a Polish space is called a *Borel space*. The set of real numbers is denoted by $\mathbb{R}$, while $\mathbb{R}^+$ is the set of nonnegative real numbers. Measurability (of sets, functions, etc.) is always understood with respect to the corresponding Borel σ-field.

Let $(\Omega, \mathcal{F})$ be a probability space, and P a probability measure on $(\Omega, \mathcal{F})$. The abbreviation P-a.s. is used for a property that holds "P almost surely", that is, a property that holds on a set of P-probability one. A sequence $\{Y_n\}_{n \geq 1}$ of random variables on $(\Omega, \mathcal{F})$ *converges in probability* to a random variable Y if

$$\lim_{n \to \infty} P\{|Y_n - Y| > \varepsilon\} = 0 \quad \forall \, \varepsilon > 0.$$

We will write $Y_n \xrightarrow{p} Y$. We say that the probability measure P is a *Dirac* or *unit measure* if there exists $\omega_0 \in \Omega$ such that $\{\omega_0\} \in \mathcal{F}$ and $P\{\omega_0\} = 1$.

If $\{P_n\}_{n \geq 1}$ and P are probability measures on a Borel space $(\mathcal{X}, \mathbb{B}(\mathcal{X}))$, we say that $\{P_n\}$ converges weakly to P, which is denoted by $P_n \xrightarrow{w} P$, if

$$\lim_{n \to \infty} \int_{\mathcal{X}} f(x) P_n(dx) = \int_{\mathcal{X}} f(x) P(dx)$$

for every bounded and continuous function $f : \mathcal{X} \to \mathbb{R}$. A well-known result is that if $\mathcal{X}$ is compact then, given an arbitrary sequence $\{P_n\}_{n \geq 1}$ of probability measures on $(\mathcal{X}, \mathbb{B}(\mathcal{X}))$, there exist a subsequence $\{n'\}$ of $\{n\}$ and a probability measure P on $(\mathcal{X}, \mathbb{B}(\mathcal{X}))$ with $P_{n'} \xrightarrow{w} P$.

The *indicator function* of a set $B \subseteq \mathcal{X}$ is denoted by $\mathbf{I}_B$, that is, $\mathbf{I}_B(x) = 1$ if $x \in B$ and $\mathbf{I}_B(x) = 0$ if $x \notin B$. The constant real-valued function on $\mathcal{X}$ that equals one is denoted by $\mathbf{1}$.

Given two integers i and j, the *Kronecker delta* δ_{ij} equals 1 if $i = j$, and equals 0 if $i \neq j$.

We will also use the standard *Landau notation*: o and O. More precisely, given sequences $\{A_n\}_{n \geq 1}$ and $\{B_n\}_{n \geq 1}$ of positive numbers, we will write

$$B_n = o(A_n) \quad \text{and} \quad B_n = O(A_n)$$

if

$$\lim_{n \to \infty} B_n/A_n = 0 \quad \text{and} \quad \limsup_{n \to \infty} B_n/A_n < \infty,$$

respectively. Furthermore, we write $A_n \sim B_n$ if $A_n = O(B_n)$ and $B_n = O(A_n)$. For a positive function f, notations such as $O(f(t))$ or $o(f(t))$ as $t \to 0$ or $t \to \infty$, are defined similarly.

Finally, the symbols $:=$ and $=:$ indicate an equality by definition.

Chapter 2

Controlled Markov Chains

2.1. Introduction

In this chapter we introduce the continuous-time controlled Markov chain
(CMC) model that we will study in the subsequent chapters. This is the
standard denumerable state space, Borel action space, continuous-time control model studied by several authors; see, for instance, [48, 52, 53].

We begin with the definition of the CMC model in Sec. 2.2, and then
we describe the behavior of the associated stochastic dynamical system. In
this section, we also introduce the class Φ of admissible Markov policies,
which include, as particular cases, the stationary policies.

Section 2.3 is devoted to the study of the *existence* of CMCs. That is,
for each control policy $\varphi \in \Phi$, we give conditions ensuring the existence of
the corresponding controlled Markov process $\{x^\varphi(t)\}_{t\geq 0}$, which describes
the evolution of the dynamical system under the policy φ. More precisely,
we propose *Lyapunov-like* conditions, also known as *drift* conditions, on the
system's transition rates. Here, we do not prove the existence results on the
controlled Markov process. For proofs, the interested reader is referred to
[44] or [52, Appendix C].

In this book, one of the strongest conditions that we will impose on the
control model is *uniform exponential ergodicity* in the class of stationary
policies. This issue is studied in Sec. 2.4. The uniform exponential ergodicity property is required when dealing with the average reward and related
optimality criteria. We give two sufficient conditions yielding uniform exponential ergodicity. The first one consists of the monotonicity conditions
proposed by Lund, Meyn, and Tweedie [106], and later used in, e.g., [45, 52];
our second sufficient condition is a strengthened Lyapunov condition introduced by Prieto-Rumeau and Hernández-Lerma in [131, Theorem 7.2]; see

also [129, Theorem 2.5]. This is generalized here in Theorem 2.11, and its proof is given in Sec. 2.5.

To conclude, in Sec. 2.6 we make some comments and give some bibliographic references on CMCs.

2.2. The control model

Model definition   The basic control model we will deal with is

$$\mathcal{M} := \{S, A, (A(i)), (q_{ij}(a)), (r(i, a))\},$$

which consists of the following elements:

- The *state space* S, which is assumed to be a denumerable set (with the discrete topology). Without loss of generality, we suppose that S is the set of nonnegative integers:

$$S := \{0, 1, 2, \ldots\}.$$

- The *action space* A, assumed to be a Polish space. The set of actions available at state $i \in S$ is $A(i) \subseteq A$, which is a Borel subset of A. The family of feasible state-action pairs is

$$K := \{(i, a) \in S \times A : a \in A(i)\}.$$

- The *transition rates* $q_{ij}(a) \in \mathbb{R}$, for $i, j \in S$ and $a \in A(i)$. We suppose that the transition rates verify the following conditions:

  (i) For each i and j in S, the function $a \mapsto q_{ij}(a)$ is measurable on $A(i)$.
  (ii) For each $i \in S$ and $a \in A(i)$, we have $q_{ij}(a) \geq 0$ whenever $j \neq i$. Also, the transition rates are *conservative*, that is,

$$-q_{ii}(a) = \sum_{j \neq i} q_{ij}(a) < \infty \quad \forall\, (i, a) \in K.$$

  (iii) The transition rates are *stable*, i.e., for each $i \in S$

$$q(i) := \sup_{a \in A(i)} \{-q_{ii}(a)\} \qquad (2.1)$$

  is finite.

- The *reward rate* function $r : K \to \mathbb{R}$, where $a \mapsto r(i, a)$ is supposed to be measurable on $A(i)$ for each $i \in S$.

The continuous-time CMC model $\mathcal{M}$ represents a system that roughly behaves as follows. (Further details are given in Sec. 2.3.) Suppose that the system is at state $i \in S$ at time $t \geq 0$, and that the controller chooses the action $a \in A(i)$. Then, on the time interval $[t, t+dt]$, for small dt, the controller receives a reward $r(i,a)dt$ and, moreover, either the system makes a transition to the state $j \neq i$ with probability $q_{ij}(a)dt + \mathrm{o}(dt)$, or remains at state i with probability $1 + q_{ii}(a)dt + \mathrm{o}(dt)$.

Control policies For technical reasons we will consider "randomized" control policies defined as follows.

Let Φ be the family of real-valued functions

$$\varphi = \{\varphi_t(B|i) : i \in S, \ B \in \mathbb{B}(A(i)), \ t \geq 0\},$$

such that $t \mapsto \varphi_t(B|i)$ is measurable on $[0, \infty)$ for every $i \in S$ and $B \in \mathbb{B}(A(i))$, and, in addition, $B \mapsto \varphi_t(B|i)$ is a probability measure on $A(i)$ for every $t \geq 0$ and $i \in S$. We say that $\varphi \in \Phi$ is a *randomized Markov policy* or, in short, a Markov policy.

The interpretation of a Markov policy $\varphi \in \Phi$ is that if the system is at state $i \in S$ at time $t \geq 0$, then the controller randomly chooses an action in $A(i)$ according to the probability measure $\varphi_t(\cdot|i)$. Such a policy is called "Markov" because at each time $t \geq 0$ it only depends on the time t itself and the current state of the system. This in turn means that the corresponding state process is a continuous-time Markov chain; the transition rates are as in (2.2) below.

If $\varphi \in \Phi$ is such that the probability measure $\varphi_t(\cdot|i)$ is a Dirac measure for every $t \geq 0$ and $i \in S$, then φ is called a *nonstationary deterministic policy*. Such a policy can be identified with a measurable function $\varphi : [0, \infty) \times S \to A$ such that $\varphi(t, i)$ is in $A(i)$ for every $t \geq 0$ and $i \in S$. We define the transition rates associated with $\varphi \in \Phi$ as

$$q_{ij}(t, \varphi) := \int_{A(i)} q_{ij}(a)\varphi_t(da|i) \tag{2.2}$$

for $i, j \in S$ and $t \geq 0$. Our assumptions above on the transition rates (in particular, the conservativeness and the stability properties) ensure that $q_{ij}(t, \varphi)$ is well defined, finite, and measurable in $t \geq 0$. By dominated convergence, the transition rates (2.2) are also conservative, i.e.,

$$-q_{ii}(t, \varphi) = \sum_{j \neq i} q_{ij}(t, \varphi) \quad \forall i \in S, \ t \geq 0, \ \varphi \in \Phi, \tag{2.3}$$

and stable, that is,

$$-q_{ii}(t, \varphi) \leq q(i) < \infty \quad \forall i \in S, \ t \geq 0, \ \varphi \in \Phi. \tag{2.4}$$

Stationary policies   In the family of randomized Markov policies Φ, we consider two relevant classes.

We say that a Markov policy $\varphi = \{\varphi_t(B|i)\}$ is *stationary* if $\varphi_t(B|i) := \varphi(B|i)$ does not depend on $t \geq 0$. We will write $\varphi \in \Phi_s$. For a stationary policy φ the transition rates (2.2) become

$$q_{ij}(\varphi) := \int_{A(i)} q_{ij}(a)\varphi(da|i) \quad \forall i, j \in S.$$

Let $\mathbb{F}$ be the family of functions $f : S \to A$ such that $f(i)$ is in $A(i)$ for every $i \in S$. If f is in $\mathbb{F}$, then f can be identified with the randomized stationary policy φ such that $\varphi(\cdot|i)$ is the Dirac measure concentrated at $f(i)$ for every $i \in S$. We say that $f \in \mathbb{F}$ is a *deterministic stationary policy*. The corresponding transition rates are $q_{ij}(f) := q_{ij}(f(i))$, for $i, j \in S$.

The inclusions $\mathbb{F} \subseteq \Phi_s \subseteq \Phi$ are obvious.

Convergence of stationary policies   In the set of deterministic stationary policies $\mathbb{F}$, we will consider the topology of componentwise convergence. That is, given $\{f_n\}_{n\geq 1}$ and f in $\mathbb{F}$, we have $f_n \to f$ if and only if $f_n(i) \to f(i)$ for every $i \in S$. In the family of randomized stationary policies Φ_s, we will consider the topology of the componentwise weak convergence. That is, given $\{\varphi_n\}_{n\geq 1}$ and φ in Φ_s, we have $\varphi_n \to \varphi$ if the probability measures $\varphi_n(\cdot|i)$ converge weakly to $\varphi(\cdot|i)$ for all $i \in S$; equivalently, using the notation introduced in Sec. 1.4,

$$\varphi_n(\cdot|i) \xrightarrow{w} \varphi(\cdot|i) \quad \forall i \in S. \tag{2.5}$$

If the action sets $A(i)$ are compact for every $i \in S$, then $\mathbb{F}$ is metrizable and compact with the topology of componentwise convergence. The proof of this fact is simple: consider the metric

$$\mathbf{d}(f, f') := \sum_{i \in S} \frac{d(f(i), f'(i))}{2^i |A(i)|}$$

on $\mathbb{F}$, where d is the metric on the action space A, and $|A(i)|$ is the (finite) diameter of the compact set $A(i)$. This metric on $\mathbb{F}$ corresponds to the topology of the componentwise convergence, and it is also easily seen that $\mathbb{F}$ is compact. Similarly, if the action sets $A(i)$ are compact for each $i \in S$, then Φ_s is a compact metric space with the topology of componentwise weak convergence; see, e.g., Theorems A.22 and A.25 (Prohorov's theorem) in [38].

2.3. Existence of controlled Markov chains

Now we address the issue of the existence of a state Markov process for each control policy $\varphi \in \Phi$.

The family of matrices $Q^\varphi(t) = [q_{ij}(t, \varphi)]_{i,j \in S}$, for $t \geq 0$, is usually known as a *nonhomogeneous Q^φ-matrix*. By (2.3) and (2.4), the Q^φ-matrix is conservative and stable.

We say that the family $\{P_{ij}(s,t)\}_{i,j \in S}$, for $0 \leq s \leq t$, is a *nonhomogeneous transition function* if the following properties are verified for each $i, j \in S$ and $t \geq s \geq 0$:

(i) $P_{ij}(s,t) \geq 0$ and $\sum_{j \in S} P_{ij}(s,t) \leq 1$.

(ii) Given $u \in [s,t]$, the *Chapman–Kolmogorov equation* holds:

$$P_{ij}(s,t) = \sum_{k \in S} P_{ik}(s,u)P_{kj}(u,t). \tag{2.6}$$

(iii) $\lim_{t \downarrow s} P_{ij}(s,t) = P_{ij}(s,s)$ and $P_{ij}(s,s) = \delta_{ij}$, where δ denotes the Kronecker delta.

If, in addition, $\sum_{j \in S} P_{ij}(s,t) = 1$ for every $i \in S$ and $t \geq s \geq 0$, then the transition function is said to be *regular*.

We say that the so-defined transition function is a Q^φ-*process* for the Q^φ-matrix $[q_{ij}(s,\varphi)]$ if

$$\lim_{t \downarrow s} \frac{P_{ij}(s,t) - \delta_{ij}}{t - s} = q_{ij}(s,\varphi)$$

for every $i, j \in S$ and $s \geq 0$. In this case, the transition function $\{P_{ij}(s,t)\}$ will be written $\{P_{ij}^\varphi(s,t)\}$ for every $i, j \in S$ and $0 \leq s \leq t$.

By [167, Theorem 3], for every $\varphi \in \Phi$ there exists a Q^φ-process which, however, might be neither regular nor unique. To ensure regularity, we impose Assumption 2.1 below, where we use the following definition: a monotone nondecreasing function $w : S \to [1, \infty)$ such that $\lim_{i \to \infty} w(i) = \infty$ will be referred to as a *Lyapunov function* on S. Such Lyapunov functions are also known as moment or norm-like [111] functions.

Assumption 2.1. There exist a Lyapunov function w on S and constants $c \in \mathbb{R}$ and $b \geq 0$ such that, for every $(i, a) \in K$,

$$\sum_{j \in S} q_{ij}(a)w(j) \leq -cw(i) + b. \tag{2.7}$$

The inequality (2.7) is usually referred to as a Lyapunov of drift condition (on w). Under Assumption 2.1, using (2.2) we have

$$\sum_{j \in S} q_{ij}(t, \varphi) w(j) \leq -cw(i) + b \quad \forall i \in S, \ t \geq 0 \qquad (2.8)$$

for every $\varphi \in \Phi$.

The monotonicity of w is not strictly necessary (for a weaker formulation of Assumption 2.1 above, see [44, Assumption A] and [52, Assumption 2.2]). The condition (2.7) is similar to the nonexplosivity condition in [111] for homogeneous Markov processes.

By [44, Theorem 3.1] or [52, Theorem 2.3], if Assumption 2.1 holds then for each $\varphi \in \Phi$ there exists a unique Q^φ-process which, in addition, is regular. Therefore, there exists a right-continuous Markov process $\{x^\varphi(t)\}_{t \geq 0}$ on S with transition rates given by $q_{ij}(t, \varphi)$. We call $\{x^\varphi(t)\}_{t \geq 0}$ the *controlled Markov chain* associated with the randomized Markov policy $\varphi \in \Phi$.

Let $S^{[s,\infty)}$, for some $s \geq 0$, be the family of functions from $[s, \infty)$ to S, which are interpreted as the sample paths of a Markov chain with values in S.

Given an initial state $i \in S$, let P_i^φ be the probability measure on $S^{[0,\infty)}$ defined by the Q^φ-process $\{P_{ij}^\varphi(0, t)\}_{j \in S, t \geq 0}$, that is, P_i^φ gives the distribution of the state process $\{x^\varphi(t)\}_{t \geq 0}$ when $x(0) = i$, and let E_i^φ be the corresponding expectation operator.

Given $s \geq 0$ and $i \in S$, let $P_{s,i}^\varphi$ be the probability measure on $S^{[s,\infty)}$ defined by $\{P_{ij}^\varphi(s, t)\}_{j \in S, t \geq s}$, and let $E_{s,i}^\varphi$ be the associated expectation operator. The probability $P_{s,i}^\varphi$ yields the distribution of the state process $\{x^\varphi(t)\}_{t \geq s}$ conditional on $x^\varphi(s) = i$.

If there is no risk of confusion, we will write $\{x(t)\}$ in lieu of $\{x^\varphi(t)\}$. This is because, since P_i^φ and $P_{s,i}^\varphi$ are probability measures on the space of sample paths $S^{[0,\infty)}$ and $S^{[s,\infty)}$, it is redundant to write, for instance, $E_i^\varphi[w(x^\varphi(t))]$.

Additional notation Given the Lyapunov function w in Assumption 2.1, we define $\mathcal{B}_w(S)$ as the family of functions $u : S \to \mathbb{R}$ such that

$$\|u\|_w := \sup_{i \in S}\{|u(i)|/w(i)\}$$

is finite. It is easily seen that the mapping $u \mapsto \|u\|_w$ is a norm on $\mathcal{B}_w(S)$, with respect to which $\mathcal{B}_w(S)$ is a Banach space. We call $\| \cdot \|_w$ the *w-norm* or *w-weighted norm*.

Also, if the function $u : K \to \mathbb{R}$ is such that $a \mapsto u(i, a)$ is measurable on $A(i)$ for every $i \in S$ and, furthermore, the mapping

$$i \mapsto \sup_{a \in A(i)} |u(i, a)| \in \mathcal{B}_w(S), \tag{2.9}$$

we will write $u \in \mathcal{B}_w(K)$.

In the forthcoming, we will also use notations such as

$$\| \cdot \|_{w^2}, \quad \mathcal{B}_{w^2}(S), \quad \mathcal{B}_{w^2}(K), \quad \mathcal{B}_{w+w'}(S), \quad \text{and} \quad \mathcal{B}_{w+w'}(K), \tag{2.10}$$

where $w' : S \to \mathbb{R}$ is a nonnegative function; these are given the obvious definitions.

Finally, given a bounded function $u : S \to \mathbb{R}$, we define its 1-norm as the usual supremum norm, i.e.,

$$\|u\|_1 := \sup_{i \in S} |u(i)|. \tag{2.11}$$

(Note that the term "1-norm" is consistent with the definition of w-norm.) The Banach space of functions on S with finite 1-norm, that is, the space of bounded functions on S, is denoted by $\mathcal{B}_1(S)$.

Stability and expected growth   Given a Markov policy $\varphi \in \Phi$ and an initial state $i \in S$ at time 0, let T be the sojourn time at i, that is,

$$T := \inf_{t \geq 0} \{x(t) \neq i\}.$$

The stability condition on the transition rates (2.4) ensures that T is positive almost surely. In fact, for any given $t \geq 0$,

$$P_i^{\varphi}\{T > t\} = P_i^{\varphi}\{x(s) = i, \ \forall 0 \leq s \leq t\}$$

$$= \exp\left\{ \int_0^t q_{ii}(s, \varphi) ds \right\}$$

$$\geq \exp\{-t q(i)\}, \tag{2.12}$$

where the last inequality follows from (2.4). Therefore, the time spent by the state process at the state i is stochastically bounded below — uniformly on Φ — by an exponential distribution with parameter $q(i)$.

Assumption 2.1 yields the following important fact. Given $\varphi \in \Phi$ and an initial state $i \in S$ at time $s \geq 0$, the growth of the state process is "bounded" in the sense that

$$E_{s,i}^{\varphi}[w(x(t))] \leq e^{-c(t-s)} w(i) + \frac{b}{c}(1 - e^{-c(t-s)}) \quad \forall t \geq s. \tag{2.13}$$

When $c = 0$, the inequality (2.13) becomes

$$E^{\varphi}_{s,i}[w(x(t))] \leq w(i) + b(t - s).$$

For a proof of (2.13) we refer to [45, Lemma 3.2] or [52, Lemma 6.3]. This proof makes use of the construction of the Q^{φ}-process starting from the Q^{φ}-matrix. As we shall see later (see Remark 2.4 below), the inequality (2.13) can also be derived, under additional hypotheses, from the Dynkin formula.

The Dynkin formula   Suppose that Assumption 2.1 holds and consider a Markov policy $\varphi \in \Phi$. The infinitesimal generator of the state process $x^{\varphi}(\cdot)$ is the operator $u \mapsto L^{t,\varphi}u$ defined, for every function $u : S \to \mathbb{R}$, as

$$(L^{t,\varphi}u)(i) := \sum_{j \in S} q_{ij}(t, \varphi)u(j) \quad \forall i \in S, \; t \geq 0. \tag{2.14}$$

If $\varphi \in \Phi_s$ and $f \in \mathbb{F}$ are stationary policies, then we write

$$(L^{\varphi}u)(i) := \sum_{j \in S} q_{ij}(\varphi)u(j) \quad \text{and} \quad (L^f u)(i) := \sum_{j \in S} q_{ij}(f)u(j) \tag{2.15}$$

for all $i \in S$.

In addition to Assumption 2.1, we impose the following condition.

Assumption 2.2. Let w be the Lyapunov function in Assumption 2.1 and suppose that there exists a nonnegative function $w' : S \to \mathbb{R}$ such that $q(i)w(i) \leq w'(i)$ for every $i \in S$ and, furthermore, there exist constants $c' \in \mathbb{R}$ and $b' \geq 0$ such that

$$\sum_{j \in S} q_{ij}(a)w'(j) \leq -c'w'(i) + b' \quad \forall (i, a) \in K.$$

Assumption 2.2 allows us to derive an expression similar to (2.13) but this time for w':

$$E^{\varphi}_{s,i}[w'(x(t))] \leq e^{-c'(t-s)}w'(i) + \frac{b'}{c'}(1 - e^{-c'(t-s)}) \tag{2.16}$$

for $\varphi \in \Phi$, $i \in S$, and $0 \leq s \leq t$. Now, recalling (2.8), observe that for every $u \in \mathcal{B}_w(S)$, $i \in S$, and $t \geq 0$

$$\left| \sum_{j \in S} q_{ij}(t, \varphi)u(j) \right| \leq \|u\|_w \cdot \left(-q_{ii}(t, \varphi)w(i) + \sum_{j \neq i} q_{ij}(t, \varphi)w(j) \right) \tag{2.17}$$

$$\leq \|u\|_w \cdot \left(-2q_{ii}(t, \varphi)w(i) + \sum_{j \in S} q_{ij}(t, \varphi)w(j) \right)$$

$$\leq \|u\|_w \cdot (2q(i)w(i) - cw(i) + b)$$
$$\leq \|u\|_w \cdot (2w'(i) - cw(i) + b). \tag{2.18}$$

In particular, by (2.13), (2.14), and (2.16),

$$E_i^\varphi \left[\int_0^t |(L^{s,\varphi}u)(x(s))|ds \right] < \infty$$

for every $i \in S$ and $t \geq 0$. This yields the so-called Dynkin formula in the following proposition (see, for instance, [32, p. 132], [52, Appendix C.3], and [64, Lemma 2.1(b)]).

Proposition 2.3. *Suppose that Assumptions 2.1 and 2.2 hold, and fix an arbitrary $\varphi \in \Phi$ and $u \in \mathcal{B}_w(S)$. Then, for every initial state $i \in S$ and every $s \geq 0$,*

$$E_{s,i}^\varphi [u(x(t))] - u(i) = E_{s,i}^\varphi \left[\int_s^t (L^{v,\varphi}u)(x(v))dv \right] \quad \forall t \geq s.$$

In particular, for $s = 0$,

$$E_i^\varphi [u(x(t))] - u(i) = E_i^\varphi \left[\int_0^t (L^{v,\varphi}u)(x(v))dv \right] \quad \forall t \geq 0.$$

The Dynkin formula is also valid for time-dependent functions $u: S \times [0, \infty) \to \mathbb{R}$ provided that appropriate integrability conditions hold. The formula is particularly simple, and very useful, for a function of the form $(i,t) \mapsto e^{\alpha t}u(i)$ for $\alpha \in \mathbb{R}$ and $u \in \mathcal{B}_w(S)$. In this case, the Dynkin formula is

$$E_i^\varphi [e^{\alpha t}u(x(t))] - u(i) = E_i^\varphi \left[\int_0^t e^{\alpha v}(\alpha u + L^{v,\varphi}u)(x(v))dv \right] \tag{2.19}$$

for every $t \geq 0$; see [52, Appendix C.3] and [64, Lemma 2.1]. Note that letting $\alpha = 0$ in (2.19) we obtain Dynkin's formula in the simpler form given in Proposition 2.3.

Remark 2.4. Using (2.14) we can rewrite (2.8) as

$$L^{t,\varphi}w \leq -cw + b.$$

From this inequality and taking $\alpha = c$ in (2.19) we obtain (2.13). The hypotheses needed are, however, stronger (application of the Dynkin formula requires Assumption 2.2), while (2.13) is deduced in [45, 52] by using a direct argument.

2.4.   Exponential ergodicity

In this section, we consider a stationary policy $\varphi \in \Phi_s$ or $f \in \mathbb{F}$, so that, by (2.2), $q_{ij}(t, \varphi) \equiv q_{ij}(\varphi)$ is time-invariant, and similarly for f. Hence the corresponding state Markov process is time-homogeneous, that is, the transition function verifies that

$$P_{ij}^{\varphi}(s, t) =: P_{ij}^{\varphi}(t - s) \quad \text{for } i, j \in S \quad \text{and} \quad t \geq s \geq 0.$$

In particular, $P_{ij}^{\varphi}(t) = P_{ij}^{\varphi}(0, t)$.

A probability measure μ on S is said to be an *invariant probability measure* for the transition function $\{P_{ij}^{\varphi}(t)\}$ (or for $\varphi \in \Phi_s$) if

$$\mu(j) = \sum_{i \in S} \mu(i) P_{ij}^{\varphi}(t) \quad \forall j \in S, \ t \geq 0.$$

It is so-named because, for every $t \geq 0$, the distribution of $x^{\varphi}(t)$ is μ provided that the initial state of the process is chosen with distribution μ. More explicitly, if μ is the distribution of $x^{\varphi}(0)$, then μ is also the distribution of $x^{\varphi}(t)$ for every $t > 0$.

Some of our results concern convergence of expected values, such as

$$E_i^{\varphi}[u(x(t))] \to \mu(u) \quad \text{as } t \to \infty, \tag{2.20}$$

for every $i \in S$, where $u : S \to \mathbb{R}$ and

$$\mu(u) := \sum_{j \in S} u(j) \mu(j). \tag{2.21}$$

We will focus on two features of the above convergence: first, the set of functions u for which (2.20) holds and, second, the speed or rate of convergence in (2.20).

Irreducibility   We say that a stationary policy $\varphi \in \Phi_s$ is irreducible if the corresponding state process is irreducible, in the sense that for any two distinct states i and j we have $P_{ij}^{\varphi}(t) > 0$ for some $t > 0$. Equivalently, for every pair of states $i \neq j$ there exist distinct states $i = i_1, \ldots, i_m = j$ such that

$$q_{i_1 i_2}(\varphi) \cdot q_{i_2 i_3}(\varphi) \cdots q_{i_{m-1} i_m}(\varphi) > 0.$$

Therefore, irreducibility means that starting from any state $i \in S$, the state process can reach any other state $j \in S$ in a finite time, with positive probability.

Our next result gives a sufficient condition for the existence of an invariant probability measure.

Theorem 2.5. *Given an irreducible policy $\varphi \in \Phi_s$, suppose that there exist constants $c > 0, b \geq 0$, a finite set $C \subset S$, and a Lyapunov function w on S such that*

$$\sum_{j \in S} q_{ij}(\varphi) w(j) \leq -cw(i) + b \cdot \mathbf{I}_C(i) \quad \forall i \in S.$$

Then:

(i) *There exists a unique invariant probability measure μ_φ for φ, and $\mu_\varphi(w)$ is finite (recall the notation in (2.21)).*

(ii) *The probabilities $\{\mu_\varphi(i)\}_{i \in S}$ are the unique nonnegative solution of*

$$\begin{cases} 0 = \sum_{i \in S} \mu_\varphi(i) q_{ij}(\varphi) \quad \forall j \in S, \\ 1 = \sum_{j \in S} \mu_\varphi(j). \end{cases}$$

Moreover, $\mu_\varphi(i) > 0$ for every $i \in S$.

(iii) *The Markov process $\{x^\varphi(t)\}$ is w-exponentially ergodic, that is, there exist positive constants δ and R, depending on φ, such that for every $u \in \mathcal{B}_w(S)$, $i \in S$, and $t \geq 0$*

$$|E_i^\varphi[u(x(t))] - \mu_\varphi(u)| \leq Re^{-\delta t} \|u\|_w w(i).$$

Proof. Statement (i) follows from the fact that $\{x^\varphi(t)\}$ is positive Harris recurrent; see [111, Theorem 4.2]. Statement (ii) is derived from [6, Proposition 5.4.1], while (iii) is a consequence of [111, Theorem 4.6]. □

Hence, under the hypotheses of Theorem 2.5, the limit in (2.20) holds for every $u \in \mathcal{B}_w(S)$, and convergence is exponential.

Harmonic functions Given a stationary policy $\varphi \in \Phi_s$ and a function $u \in \mathcal{B}_w(S)$, we say that u is *subharmonic* for φ if

$$(L^\varphi u)(i) \leq 0 \quad \forall i \in S, \tag{2.22}$$

where we use the notation introduced in (2.15). Similarly, we say that u in $\mathcal{B}_w(S)$ is *superharmonic* for $\varphi \in \Phi_s$ if $(L^\varphi u)(i) \geq 0$ for each $i \in S$. Finally, we say that u in $\mathcal{B}_w(S)$ is *harmonic* for $\varphi \in \Phi_s$ if $(L^\varphi u)(i) = 0$ for all $i \in S$.

Our next result shows that, under appropriate conditions, subharmonic, superharmonic, and harmonic functions are constant.

Proposition 2.6. *Suppose that the hypotheses of Theorem 2.5 and Assumption 2.2 are satisfied, and let $\varphi \in \Phi_s$ be an irreducible policy. If a* *function in $\mathcal{B}_w(S)$ is either subharmonic, superharmonic, or harmonic for $\varphi \in \Phi_s$, then it is constant.*

Proof. We only prove the case where $u \in \mathcal{B}_w(S)$ is subharmonic for $\varphi \in \Phi_s$. The result for superharmonic and harmonic functions is proved similarly.

Given an arbitrary initial state $i \in S$, by Dynkin's formula in Proposition 2.3, we have

$$E_i^{\varphi}[u(x(t))] = u(i) + E_i^{\varphi}\left[\int_0^t (L^{\varphi}u)(x(v))dv\right] \quad \forall t \geq 0.$$

Since u is subharmonic for φ, we have $L^{\varphi}u \leq 0$, and this yields

$$E_i^{\varphi}[u(x(t))] \leq u(i) \quad \forall t \geq 0.$$

By Theorem 2.5(iii), $E_i^{\varphi}[u(x(t))] \to \mu_{\varphi}(u)$ as $t \to \infty$. Hence,

$$\mu_{\varphi}(u) \leq u(i) \quad \forall i \in S.$$

This shows that the function u is greater than or equal to its μ_{φ}-expectation; hence, it is constant μ_{φ}-a.s. On the other hand, by Theorem 2.5(ii), we have $\mu_{\varphi}(i) > 0$ for all $i \in S$. We conclude that the function u is constant. $\qquad\square$

Obviously, the converse of Proposition 2.6 is true. Indeed, since the system's transition rates are conservative — see (2.3) — a constant function on S is harmonic for every stationary policy.

Sufficient conditions for uniform exponential ergodicity As we shall see below, the exponential ergodicity in Theorem 2.5(iii) does not suffice for our purposes because the constants δ and R may depend on $\varphi \in \Phi_s$. Therefore, we would like to have conditions under which we can obtain *uniform w-exponential ergodicity* of the control model, that is, the existence of positive constants δ and R such that

$$\sup_{\varphi \in \Phi_s} |E_i^{\varphi}[u(x(t))] - \mu_{\varphi}(u)| \leq Re^{-\delta t}\|u\|_w w(i) \qquad (2.23)$$

for every $i \in S$, $u \in \mathcal{B}_w(S)$, and $t \geq 0$. We note that sometimes the condition (2.23) will be needed only for deterministic stationary policies $f \in \mathbb{F} \subset \Phi_s$.

Now we propose two sets of hypotheses leading to uniform w-exponential ergodicity of CMCs. The first one is based on monotonicity conditions.

Assumption 2.7.

(a) There exists a Lyapunov function w on S, and constants $c > 0$ and $b \geq 0$ such that

$$\sum_{j \in S} q_{ij}(a) w(j) \leq -cw(i) + b \cdot \mathbf{I}_{\{0\}}(i) \quad \forall (i, a) \in K.$$

(b) Each policy $f \in \mathbb{F}$ is irreducible.
(c) For each $f \in \mathbb{F}$, the state process $\{x^f(t)\}$ is stochastically ordered in its initial value, i.e.,

$$\sum_{j=k}^{\infty} q_{ij}(f) \leq \sum_{j=k}^{\infty} q_{i+1,j}(f)$$

for every $i, k \in S$ with $k \neq i + 1$.
(d) For each $f \in \mathbb{F}$ and states $0 < i < j$, the process $\{x^f(t)\}$ can travel with positive probability from i to $\{j, j+1, \ldots\}$ without passing through the state zero. Equivalently, there exist nonzero distinct states $i = i_1, i_2, \ldots, i_m \geq j$ such that

$$q_{i_1 i_2}(f) \cdots q_{i_{m-1} i_m}(f) > 0.$$

Observe that, under Assumption 2.7, the hypotheses of Theorem 2.5 hold and, therefore, $\{x^f(t)\}$ is w-exponentially ergodic for each $f \in \mathbb{F}$. Our next result, which uses the notation in (2.21), gives *uniform* exponential ergodicity. It is a direct consequence of [106, Theorem 2.2], and so we omit its proof.

Theorem 2.8. *If Assumption 2.7 holds, then*

$$\sup_{f \in \mathbb{F}} |E_i^f[u(x(t))] - \mu_f(u)| \leq 2(1 + b/c)e^{-ct}\|u\|_w w(i)$$

for every $i \in S, u \in \mathcal{B}_w(S)$, and $t \geq 0$, where μ_f is the unique invariant probability measure for $f \in \mathbb{F}$. Thus the control model is w-exponentially ergodic uniformly on $\mathbb{F}$.

Remark 2.9. If Assumption 2.7 holds for every stationary $\varphi \in \Phi_s$ in lieu of $f \in \mathbb{F}$, then we obtain w-exponential ergodicity uniformly on Φ_s:

$$\sup_{\varphi \in \Phi_s} |E_i^\varphi[u(x(t))] - \mu_\varphi(u)| \leq 2(1 + b/c)e^{-ct}\|u\|_w w(i)$$

for every $i \in S$, $u \in \mathcal{B}_w(S)$, and $t \geq 0$, where μ_φ is the unique invariant probability measure for $\varphi \in \Phi_s$.

Note that, in Theorem 2.8, we have obtained an explicit expression for the exponential rate of convergence, which is precisely the same constant c as in Assumption 2.7(a), and also for the multiplicative constant R in (2.23), which equals $2(1 + b/c)$.

Our next sufficient condition for uniform exponential ergodicity imposes some compactness and continuity requirements, together with suitable Lyapunov conditions.

Assumption 2.10.

(a) There exist constants $c > 0$ and $b \geq 0$, a finite set $C \subset S$, and a Lyapunov function w on S such that
$$\sum_{j \in S} q_{ij}(a)w(j) \leq -cw(i) + b \cdot \mathbf{I}_C(i) \quad \forall\, (i,a) \in K.$$

(b) Each $f \in \mathbb{F}$ is irreducible.

(c) For each $i \in S$, the set $A(i)$ is compact and $a \mapsto q_{ij}(a)$ is continuous on $A(i)$ for every $i, j \in S$.

(d) There exists a Lyapunov function w' on S such that w'/w is monotone nondecreasing and $\lim_{j \to \infty} w'(j)/w(j) = \infty$. Moreover, for some constants $\tilde{c} \in \mathbb{R}$ and $\tilde{b} \geq 0$,
$$\sum_{j \in S} q_{ij}(a)w'(j) \leq -\tilde{c}w'(i) + \tilde{b} \quad \forall\, (i,a) \in K.$$

Obviously, Assumption 2.10(a) is weaker than Assumption 2.7(a) because the set C in the former assumption is allowed to be any finite subset of S. Also note that Assumption 2.10(d) gives (cf. (2.13) and (2.16))
$$E_i^{\varphi}[w'(x(t))] \leq e^{-\tilde{c}t}w'(i) + \frac{\tilde{b}}{\tilde{c}}(1 - e^{-\tilde{c}t}) \tag{2.24}$$
for every $\varphi \in \Phi$, $i \in S$, and $t \geq 0$.

Finally, observe that possible choices for w' in Assumption 2.10(d) are $w' = w^{1+\varepsilon}$, for some ε, and $w' = 1 + w \log w$. In general, the condition in Assumption 2.10(d) will be "easy to verify" for functions w' which grow slowly. For instance, this condition will be satisfied "more easily" for $w' = 1 + w \log w$ than for $w' = w^{1+\varepsilon}$.

The proof of our next result is postponed to Sec. 2.5.

Theorem 2.11. *If Assumption 2.10 is satisfied then the control model is w-exponentially ergodic uniformly on $\mathbb{F}$, that is, there exist constants $\delta > 0$ and $R > 0$ such that*
$$\sup_{f \in \mathbb{F}} |E_i^f[u(x(t))] - \mu_f(u)| \leq Re^{-\delta t}\|u\|_w w(i)$$

for every $i \in S$, $u \in \mathcal{B}_w(S)$, *and* $t \geq 0$, *where* μ_f *is the unique invariant probability measure for* $f \in \mathbb{F}$.

Remark 2.12. If the class $\mathbb{F}$ is replaced with Φ_s in Assumption 2.10(b), then uniform w-exponential ergodicity in Theorem 2.11 holds in the class of stationary policies Φ_s.

To conclude this section, we state a useful result that relates a Lyapunov condition on w as the one in Assumption 2.10(a) to a similar condition on w^γ — which is also a Lyapunov function — for $0 < \gamma < 1$.

Theorem 2.13. *Suppose that* w *is a Lyapunov function on* S *for which there exist constants* $c > 0$ *and* $b \geq 0$, *and a finite set* $C \subset S$ *such that*

$$\sum_{j \in S} q_{ij}(a) w(j) \leq -cw(i) + b \cdot \mathbf{I}_C(i) \quad \forall (i, a) \in K, \qquad (2.25)$$

where the transition rates $q_{ij}(a)$ *are assumed to be conservative and stable (see Sec. 2.2). Then, for every* $0 < \gamma < 1$, *there exist constants* $\hat{c} = \hat{c}(\gamma) > 0$ *and* $\hat{b} = \hat{b}(\gamma) \geq 0$ *such that*

$$\sum_{j \in S} q_{ij}(a) w^\gamma(j) \leq -\hat{c} w^\gamma(i) + \hat{b} \cdot \mathbf{I}_C(i) \quad \forall (i, a) \in K. \qquad (2.26)$$

Proof. Fix an arbitrary state $i \in S$, and choose a constant $R > c$. Let $x := R + q(i)$. The inequality (2.25) can be rewritten as

$$\frac{1}{x} \left(\sum_{j \neq i} q_{ij}(a) w(j) + (x + q_{ii}(a)) w(i) \right) \leq \frac{x - c}{x} w(i) + \frac{b}{x} \cdot \mathbf{I}_C(i).$$

By Jensen's inequality, it follows that

$$\sum_{j \in S} q_{ij}(a) w^\gamma(j) + x w^\gamma(i) \leq x((x - c)x^{-1} w(i) + bx^{-1} \mathbf{I}_C(i))^\gamma.$$

Using the inequality $(y + z)^\gamma \leq y^\gamma + z^\gamma$, we obtain

$$\sum_{j \in S} q_{ij}(a) w^\gamma(j) \leq [(x - c)^\gamma x^{1-\gamma} - x] w^\gamma(i) + b^\gamma x^{1-\gamma} \mathbf{I}_C(i).$$

The concavity of the function $y \mapsto y^\gamma$ implies that $y^\gamma - z^\gamma \leq \gamma(y - z)z^{\gamma-1}$ whenever $0 \leq y < z$. Hence,

$$\sum_{j \in S} q_{ij}(a) w^\gamma(j) \leq -\gamma c w^\gamma(i) + b^\gamma x^{1-\gamma} \mathbf{I}_C(i) \quad \forall (i, a) \in K.$$

Hence, letting

$$\hat{c}(\gamma) := \gamma c \quad \text{and} \quad \hat{b}(\gamma) := b^\gamma \max_{i \in C}\{(R + q(i))^{1-\gamma}\},$$

it follows that (2.26) holds. This completes the proof. $\qquad \square$

2.5. Proof of Theorem 2.11

Before proceeding with the proof of Theorem 2.11 itself, we establish the following useful fact. The result of Lemma 2.14 below can be deduced by applying the Dynkin formula (2.19) to the function $(i,t) \mapsto e^{ct}w(i)$. This, however, requires imposing Assumption 2.2. So, our arguments here provide a direct proof of Lemma 2.14 that does not use the Dynkin formula.

Lemma 2.14. *Consider the control model $\mathcal{M}$ in Sec. 2.2 and suppose that there exists a Lyapunov function w on S, constants $c \in \mathbb{R}$ and $b \geq 0$, and a subset $C \subseteq S$ such that*

$$\sum_{j \in S} q_{ij}(a)w(j) \leq -cw(i) + b \cdot \mathbf{I}_C(i) \quad \forall\, (i,a) \in K. \tag{2.27}$$

Then, for every control policy $\varphi \in \Phi$, and all $0 \leq s \leq t$ and $i \in S$,

$$E_{s,i}^{\varphi}[w(x(t))] \leq e^{-c(t-s)}w(i) + b \int_s^t e^{-c(t-u)} P_{iC}^{\varphi}(s,u)du,$$

where $P_{iC}^{\varphi}(s,u) := \sum_{k \in C} P_{ik}^{\varphi}(s,u)$ is the probability that $x^{\varphi}(u)$ is in C conditional on $x^{\varphi}(s) = i$.

Proof.   As a consequence of (2.27), there exists a sequence of subsets $\{C_n\}_{n \geq 1} \subseteq S$ such that $C_n \downarrow C$ and, in addition, $S - C_n$ is finite. (To see this, note that $S - C$ is countable, and so we can write $S - C = \{x_1, x_2, \ldots\}$. Then, choose $C_n = C \cup \{x_n, x_{n+1}, \ldots\}$.) Obviously, for every $n \geq 1$ we have

$$\sum_{j \in S} q_{ij}(a)w(j) \leq -cw(i) + b \cdot \mathbf{I}_{C_n}(i) \quad \forall\, (i,a) \in K. \tag{2.28}$$

In what follows, let $\varphi \in \Phi$ be an arbitrary control policy. To simplify the notation, since the policy φ remains fixed, the transition rates $q_{ij}(s,\varphi)$ will be simply denoted by $q_{ij}(s)$.

We know (see [45, Lemma 3.1] or [52, Lemma 6.2]) that

$$(s,i,t) \mapsto \sum_{j \in S} P_{ij}^{\varphi}(s,t)w(j), \quad \text{for } 0 \leq s \leq t \quad \text{and} \quad i \in S,$$

is the minimal nonnegative solution of the inequality

$$h(s,i,t) \geq \sum_{k \neq i} \int_s^t e^{\int_s^u q_{ii}(v)dv} q_{ik}(u)h(u,k,t)du + e^{\int_s^t q_{ii}(v)dv}w(i). \tag{2.29}$$

For a given $n \geq 1$, define the function

$$h_n(s,i,t) := e^{-c(t-s)}w(i) + b \int_s^t e^{-c(t-u)} P_{iC_n}^{\varphi}(s,u)du$$

for $0 \le s \le t$ and $i \in S$. The next step of this proof consists in showing that the function h_n satisfies the inequality (2.29). For notational convenience, we let $h_n := h_n^1 + h_n^2$, where

$$h_n^1(s, i, t) := e^{-c(t-s)} w(i) \quad \text{and} \quad h_n^2(s, i, t) := b \int_s^t e^{-c(t-u)} P_{iC_n}^{\varphi}(s, u) du.$$

We note that

$$\sum_{k \ne i} \int_s^t e^{\int_s^u q_{ii}(v) dv} q_{ik}(u) h_n^1(u, k, t) du$$

equals

$$e^{-c(t-s)} \int_s^t e^{\int_s^u [c + q_{ii}(v)] dv} \left(\sum_{k \ne i} q_{ik}(u) w(k) \right) du,$$

which, by (2.28), is less than or equal to

$$e^{-c(t-s)} \int_s^t e^{\int_s^u [c + q_{ii}(v)] dv} (-(c + q_{ii}(u)) w(i) + b \cdot \mathbf{I}_{C_n}(i)) du$$

$$= e^{-c(t-s)} w(i) - e^{\int_s^t q_{ii}(v) dv} w(i)$$

$$+ b e^{-ct} \cdot \mathbf{I}_{C_n}(i) \int_s^t e^{cu + \int_s^u q_{ii}(v) dv} du. \tag{2.30}$$

On the other hand, $\sum_{k \ne i} \int_s^t e^{\int_s^u q_{ii}(v) dv} q_{ik}(u) h_n^2(u, k, t) du$ equals

$$b \sum_{k \ne i} \int_s^t e^{\int_s^u q_{ii}(v) dv} q_{ik}(u) \int_u^t e^{-c(t-z)} P_{kC_n}^{\varphi}(u, z) dz \; du$$

$$= b e^{-ct} \int_s^t e^{cz} \int_s^z e^{\int_s^u q_{ii}(v) dv} \sum_{k \ne i} q_{ik}(u) P_{kC_n}^{\varphi}(u, z) du \; dz. \tag{2.31}$$

The Kolmogorov backward equations [52, Proposition C.4] give, for every $i, j \in S$, $s \le z$, and almost every u in $[s, z]$,

$$\frac{\partial P_{ij}^{\varphi}(u, z)}{\partial u} = -\sum_{k \in S} q_{ik}(u) P_{kj}^{\varphi}(u, z). \tag{2.32}$$

By the conservative property of the transition rates and since $S - C_n$ is finite, we derive that for almost all u in $[s, z]$

$$\sum_{k \ne i} q_{ik}(u) P_{kC_n}^{\varphi}(u, z) = -q_{ii}(u) P_{iC_n}^{\varphi}(u, z) - \frac{\partial P_{iC_n}^{\varphi}(u, z)}{\partial u}.$$

It follows that (2.31) equals

$$-be^{-ct}\int_s^t e^{cz}\int_s^z e^{\int_s^u q_{ii}(v)dv}\left(q_{ii}(u)P_{iC_n}^\varphi(u,z)+\frac{\partial P_{iC_n}^\varphi(u,z)}{\partial u}\right)du\ dz$$

$$=-be^{-ct}\int_s^t e^{cz}(e^{\int_s^z q_{ii}(v)dv}\cdot \mathbf{I}_{C_n}(i)-P_{iC_n}^\varphi(s,z))dz,$$

which in turn equals

$$b\int_s^t e^{-c(t-z)}P_{iC_n}^\varphi(s,z)dz-be^{-ct}\cdot \mathbf{I}_{C_n}(i)\int_s^t e^{cz+\int_s^z q_{ii}(v)dv}dz. \qquad (2.33)$$

Combining (2.30) and (2.33) yields that the function h_n verifies the inequality (2.29). Therefore, by Lemma 6.2 in [52], for every $0\le s\le t$ and $i\in S$,

$$\sum_{j\in S}P_{ij}^\varphi(s,t)w(j)\le e^{-c(t-s)}w(i)+b\int_s^t e^{-c(t-u)}P_{iC_n}^\varphi(s,u)du.$$

Since n is arbitrary, by dominated convergence we can take the limit as $n\to\infty$ so that

$$\sum_{j\in S}P_{ij}^\varphi(s,t)w(j)\le e^{-c(t-s)}w(i)+b\int_s^t e^{-c(t-u)}P_{iC}^\varphi(s,u)du,$$

which completes the proof of the lemma. $\qquad\square$

Some comments on the proof of Lemma 2.14 are in order.

Remark 2.15. By letting $C=S$ in (2.27), Lemma 2.14 yields the inequality (2.13).

The fact that $S-C_n$ is finite is used to avoid uniform convergence issues in the interchange of the sum over $j\in S-C_n$ and the derivative $\frac{\partial}{\partial u}$ in (2.32).

To prove Theorem 2.11 we use a uniform geometric ergodicity result for *discrete-time* CMCs taken from [25]. The idea is to approximate our continuous-time control model by a discrete-time "skeleton" that satisfies the conditions in [25] for uniform geometric ergodicity. This will lead to our result in Theorem 2.11.

In the remainder of this section we suppose that Assumption 2.10 is satisfied. First, we introduce some notation. Let $C(1):=\{z\in \mathcal{B}_1(S):\|z\|_1\le 1\}$, where the 1-norm is defined in (2.11).

Lemma 2.16. *Suppose that the transition rates of the control model are uniformly bounded, that is, $q:=\sup_{i\in S}q(i)<\infty$ (recall the notation introduced in (2.1)). Then the following results hold.*

(i) *For each $f \in \mathbb{F}$, L^f defined in (2.15) is a linear operator on $B_w(S)$ that satisfies*

$$\|L^f z\|_w \leq (2q+b)\|z\|_w,$$

with b as in Assumption 2.10(a).

(ii) *If $f_n \to f$, then $\lim_{n\to\infty} \sup_{z\in C(1)} \|L^{f_n} z - L^f z\|_w = 0$.*

(iii) *If a sequence $\{z_n\}$ in $C(1)$ converges pointwise to $z \in C(1)$ (i.e., $z_n(k) \to z(k)$ for every $k \in S$), then $\lim_{n\to\infty} \|L^f z_n - L^f z\|_w = 0$ for every $f \in \mathbb{F}$.*

Proof. *Proof of (i).* Given $f \in \mathbb{F}$, $z \in \mathcal{B}_w(S)$, and $i \in S$, we deduce from (2.17) and Assumption 2.10(a) that

$$|(L^f z)(i)| \leq \|z\|_w (2qw(i) - cw(i) + b).$$

Hence, since $c > 0$, (i) follows.

Proof of (ii). Fix $\varepsilon > 0$ and a sequence $\{f_n\}$ in $\mathbb{F}$ converging to $f \in \mathbb{F}$. Since $w(i) \to \infty$ as $i \to \infty$, there exists $i_0 \in S$ such that

$$q/w(i) < \varepsilon/4 \quad \forall i > i_0.$$

Also, there exists $j_0 > i_0$ such that

$$\sup_{a\in A(i)} \sum_{j=j_0}^{\infty} q_{ij}(a) < \varepsilon/4 \quad \text{for every } 0 \leq i \leq i_0.$$

Indeed, $\lim_{j\to\infty} \sum_{k=j}^{\infty} q_{ik}(a) = 0$ because the transition rates are conservative. The uniform convergence follows from the continuity of the $q_{ij}(a)$ and Dini's theorem. Finally, there exists n_0 such that $n > n_0$ implies

$$|q_{ij}(f_n) - q_{ij}(f)| < \frac{\varepsilon}{2j_0} \quad \text{for } 0 \leq i \leq i_0 \quad \text{and} \quad 0 \leq j < j_0.$$

Therefore, given $z \in C(1)$ and $i \in S$, from (2.15) we obtain

$$|(L^{f_n} z - L^f z)(i)| \leq \sum_{j\in S} |q_{ij}(f_n) - q_{ij}(f)| \leq 4q < \varepsilon w(i)$$

if $i > i_0$. Now, if $0 \leq i \leq i_0$ and $n > n_0$, then

$$|(L^{f_n} z - L^f z)(i)| \leq \sum_{0\leq j < j_0} |q_{ij}(f_n) - q_{ij}(f)| + \sum_{j\geq j_0} (q_{ij}(f_n) + q_{ij}(f)) < \varepsilon.$$

Hence, given $\varepsilon > 0$, there exists n_0 such that, for $n > n_0$ and for every $z \in C(1)$,

$$\|L^{f_n} z - L^f z\|_w < \varepsilon,$$

which proves the stated result.

Proof of (iii). The proof of this statement uses arguments similar to those of part (ii) and, therefore, we omit it. $\qquad\square$

Fix $f \in \mathbb{F}$ and $j \in S$, and define $z^f(t) \in C(1)$ as

$$(z^f(t))(i) := P_{ij}^f(t) \quad \forall i \in S, \ t \geq 0,$$

where $P_{ij}^f(t)$ is the transition function of $\{x^f(t)\}$. Note that we do not make j explicit in the notation. The following result requires again the hypothesis of Lemma 2.16.

Lemma 2.17. *Suppose that the transition rates of the control model are uniformly bounded. Then the following statements hold.*

(i) *For every $f \in \mathbb{F}$ and $j \in S$, $t \mapsto z^f(t)$ is $C^1[0, \infty)$ and it is a solution of the differential equation*

$$\frac{d}{dt} z^f(t) = L^f z^f(t) \quad \forall t \geq 0, \tag{2.34}$$

where continuity and differentiability are in the w-norm, not in a componentwise sense.

(ii) *For each $t \geq 0$ and $i, j \in S$, the mapping $f \mapsto P_{ij}^f(t)$ is continuous on $\mathbb{F}$.*

Proof. *Proof of* (i). Note that the differential equation (2.34), which corresponds to the Kolmogorov backward equation, holds componentwise.

From the componentwise continuity of $z^f(t)$ in t and Lemma 2.16(iii), it follows that $t \mapsto Q^f z^f(t)$ is continuous on $\mathcal{B}_w(S)$. Now, choose a sequence of numbers $h_n \to 0$ and apply the mean value theorem (componentwise) to $(z^f(t + h_n) - z^f(t))/h_n$. By an argument mimicking the proof of Lemma 2.16(ii), we obtain that the derivative (2.34) holds as a limit in the w-norm. This completes the proof of statement (i).

Proof of (ii). Given a $C^1[0, \infty)$ function y taking values in a Banach space, as a consequence of the mean value theorem on Banach spaces (see, e.g., [28, Theorem 8.5.4]) we have

$$\|y(t) - y(0)\| \leq \int_0^t \|y'(s)\| ds \quad \forall t \geq 0. \tag{2.35}$$

Now we fix an arbitrary $f \in \mathbb{F}$ and a sequence $\{f_n\} \subseteq \mathbb{F}$ converging to f. By (2.35) and statement (i), and noting that $z^{f_n}(0) = z^f(0)$, we have

$$\|z^{f_n}(t) - z^f(t)\|_w \leq \int_0^t \|L^{f_n} z^{f_n}(s) - L^f z^f(s)\|_w ds \quad \forall t \geq 0. \tag{2.36}$$

Observe now that $\|L^{f_n} z^{f_n}(s) - L^f z^f(s)\|_w$ is bounded by

$$\|L^{f_n} z^{f_n}(s) - L^{f_n} z^f(s)\|_w + \|L^{f_n} z^f(s) - L^f z^f(s)\|_w,$$

which, by Lemma 2.16(i), is in turn bounded by

$$(2q + b)\|z^{f_n}(s) - z^f(s)\|_w + \|L^{f_n} z^f(s) - L^f z^f(s)\|_w.$$

Given $\varepsilon > 0$, Lemma 2.16(ii) ensures that there exists n_0 such that

$$\|L^{f_n} z^f(s) - L^f z^f(s)\|_w \leq \varepsilon \quad \forall n \geq n_0, \ s \geq 0.$$

Combining these bounds with (2.36), we obtain

$$\|z^{f_n}(t) - z^f(t)\|_w \leq (2q + b) \int_0^t \|z^{f_n}(s) - z^f(s)\|_w ds + \varepsilon t$$

for all $n \geq n_0$ and $t \geq 0$. Therefore, by Gronwall's inequality,

$$\|z^{f_n}(t) - z^f(t)\|_w \leq \varepsilon T e^{(2q+b)T} \quad \forall n \geq n_0, \ 0 \leq t \leq T,$$

for arbitrary $T > 0$. In particular, for $n \geq n_0$,

$$|P_{ij}^{f_n}(t) - P_{ij}^f(t)| \leq w(i)\varepsilon T e^{(2q+b)T},$$

which establishes that $f \mapsto P_{ij}^f(t)$ is continuous on $\mathbb{F}$. $\qquad\square$

In the following results we drop the boundedness hypothesis on the transition rates made in Lemmas 2.16 and 2.17.

Given $m \in S$ and $f \in \mathbb{F}$, let

$$T_m := \min\{s \geq 0 : x^f(s) \geq m\},$$

and define the stopped process $\{x_m^f(t)\}$ as

$$x_m^f(t) := x^f(t \wedge T_m),$$

where "$\wedge$" stands for *minimum*. Hence, $\{x_m^f(t)\}$ is a homogeneous Markov process, and its transition probabilities are denoted by $P_{ij}^{f,m}(t)$, for $t \geq 0$ and $i, j \in S$.

Lemma 2.18. *Given $i, j \in S$ and $t \geq 0$,*

$$\lim_{m \to \infty} \sup_{f \in \mathbb{F}} |P_{ij}^f(t) - P_{ij}^{f,m}(t)| = 0.$$

Proof. Fix $m \in S$ and $f \in \mathbb{F}$. Observe that

$$P_{ij}^f(t) - P_{ij}^{f,m}(t) = E_i^f[\mathbf{I}_{\{x^f(t)=j\}} - \mathbf{I}_{\{x_m^f(t)=j\}}]$$

$$= E_i^f[\mathbf{I}_{\{t \geq T_m\}}(\mathbf{I}_{\{x^f(t)=j\}} - \mathbf{I}_{\{x_m^f(t)=j\}})].$$

Therefore,

$$|P_{ij}^f(t) - P_{ij}^{f,m}(t)| \leq 2P_i^f\{t \geq T_m\}. \tag{2.37}$$

Also

$$\{t \geq T_m\} = \left\{ \sup_{0 \leq s \leq t} \{x^f(s)\} \geq m \right\} \subseteq \left\{ \sup_{0 \leq s \leq t} \{w(x^f(s))\} \geq w(m) \right\},$$

with $c > 0$ as in Assumption 2.10(a). Thus, from the nonexplosion condition CD0 and Theorem 2.1 in [111], we have

$$P_i^f \{t \geq T_m\} \leq w(i)e^{(b+c)t}/w(m).$$

This fact and (2.37) yield

$$|P_{ij}^f(t) - P_{ij}^{f,m}(t)| \leq 2w(i)e^{(b+c)t}/w(m) \quad \forall f \in \mathbb{F},$$

and the stated result follows. $\qquad\square$

Proposition 2.19. *For every $t \geq 0$ and $i, j \in S$, the function $f \mapsto P_{ij}^f(t)$ is continuous on $\mathbb{F}$.*

Proof. For every $m \in S$, the transition rates of $\{x_m^f(t)\}$ are bounded by $\max_{0 \leq i < m} q(i)$ and, therefore, by Lemma 2.17(ii), $f \mapsto P_{ij}^{f,m}(t)$ is continuous on $\mathbb{F}$. The continuity of $f \mapsto P_{ij}^f(t)$ is now a straightforward consequence of Lemma 2.18. $\qquad\square$

Proposition 2.20. *For every fixed $t \geq 0$ and $i \in S$, the function $f \mapsto \sum_{j \in S} P_{ij}^f(t)w(j)$ is continuous on $\mathbb{F}$.*

Proof. In what follows, the Lyapunov function w' is taken from Assumption 2.10(d). Given $j \in S$ and $f \in \mathbb{F}$, by the monotonicity of w'/w we have

$$\sum_{k=j}^{\infty} P_{ik}^f(t)w(k) = \sum_{k=j}^{\infty} P_{ik}^f(t)w'(k) \cdot \frac{w(k)}{w'(k)}$$

$$\leq \frac{w(j)}{w'(j)} \cdot \sum_{k=j}^{\infty} P_{ik}^f(t)w'(k)$$

$$\leq \frac{w(j)}{w'(j)} \cdot E_i^f[w'(x(t))]$$

$$\leq \frac{w(j)}{w'(j)} \cdot (e^{-\tilde{c}t}w'(i) + (\tilde{b}/\tilde{c})(1 - e^{-\tilde{c}t})), \qquad (2.38)$$

where (2.38) follows from (2.24). Therefore, since $w(j)/w'(j) \to 0$,

$$\lim_{j \to \infty} \sup_{f \in \mathbb{F}} \sum_{k=j}^{\infty} P_{ik}^f(t)w(k) = 0. \qquad (2.39)$$

Equivalently, the series $\sum_{j \in S} P_{ij}^f(t)w(j)$ of continuous functions on $\mathbb{F}$ (recall Proposition 2.19) converges uniformly on $\mathbb{F}$ and, consequently, it is itself continuous on $\mathbb{F}$. $\qquad\square$

In Assumption 2.10(a) we can suppose, without loss of generality, that the finite set C is of the form $C = \{0, 1, \ldots, j_0\}$ for some $j_0 \geq 0$. We use this notation in our next result.

Proposition 2.21. *There exists $\Delta > 0$ such that*

$$\sum_{j > j_0} P_{ij}^f(t) w(j) \leq e^{-ct/2} w(i)$$

for every $i \in S$, $0 \leq t \leq \Delta$, and $f \in \mathbb{F}$, with c as in Assumption 2.10(a).

Proof. From Assumption 2.10(a) and Lemma 2.14, we obtain

$$\sum_{j=0}^{\infty} P_{ij}^f(t) w(j) \leq e^{-ct} w(i) + b e^{-ct} \sum_{j=0}^{j_0} \int_0^t P_{ij}^f(s) e^{cs} ds,$$

for every $f \in \mathbb{F}$, $i \in S$, and $t \geq 0$. This inequality may be rewritten as

$$\sum_{j > j_0} P_{ij}^f(t) w(j) \leq e^{-ct/2} w(i) R(t, i, f),$$

where

$$R(t, i, f) := e^{-ct/2} \left(1 + \frac{b}{w(i)} \sum_{j=0}^{j_0} \int_0^t P_{ij}^f(s) e^{cs} ds - \frac{e^{ct}}{w(i)} \sum_{j=0}^{j_0} P_{ij}^f(t) w(j) \right).$$

Thus, to prove the proposition it suffices to show that there exists $\Delta > 0$ such that $R(t, i, f) \leq 1$ for every $0 \leq t \leq \Delta$, $i \in S$, and $f \in \mathbb{F}$; or, equivalently,

$$\frac{b}{w(i)} \sum_{j=0}^{j_0} \int_0^t P_{ij}^f(s) e^{cs} ds - \frac{e^{ct}}{w(i)} \sum_{j=0}^{j_0} P_{ij}^f(t) w(j) \leq e^{ct/2} - 1. \tag{2.40}$$

We will consider two cases, depending on whether $0 \leq i \leq j_0$ or $i > j_0$.

Case $0 \leq i \leq j_0$. To establish (2.40), it suffices to show that

$$b \cdot \sum_{j=0}^{j_0} \int_0^t P_{ij}^f(s) e^{cs} ds \leq e^{ct} \sum_{j=0}^{j_0} P_{ij}^f(t) w(j). \tag{2.41}$$

Now, on the one hand we have

$$b \cdot \sum_{j=0}^{j_0} \int_0^t P_{ij}^f(s) e^{cs} ds \leq b \int_0^t e^{cs} ds \leq b t e^{ct}. \tag{2.42}$$

On the other hand, since $0 \leq i \leq j_0$,

$$e^{ct} \sum_{j=0}^{j_0} P_{ij}^f(t) w(j) \geq e^{ct} P_{ii}^f(t) w(i) \geq e^{ct} e^{-q(i)t} w(i). \tag{2.43}$$

where we have used (2.12). By (2.42) and (2.43) we conclude that, for (2.41) to hold, it suffices that

$$bt \leq w(i)e^{-q(i)t},$$

which is true provided that t is small enough. In conclusion, there exists $\Delta_1 > 0$ such that $R(t, i, f) \leq 1$ for every $0 \leq t \leq \Delta_1$, $0 \leq i \leq j_0$, and $f \in \mathbb{F}$.

Case $i > j_0$. To prove (2.40) we will establish that

$$\frac{b}{w(i)} \sum_{j=0}^{j_0} \int_0^t P_{ij}^f(s)e^{cs}ds \leq e^{ct/2} - 1,$$

and noting that $e^{ct/2} - 1 \geq ct/2$, it suffices to show that

$$\frac{b}{w(i)} \sum_{j=0}^{j_0} \int_0^t P_{ij}^f(s)e^{cs}ds \leq ct/2. \tag{2.44}$$

To this end, first observe that

$$\frac{b}{w(i)} \sum_{j=0}^{j_0} \int_0^t P_{ij}^f(s)e^{cs}ds \leq \frac{bte^{ct}}{w(i)},$$

so that there exists I_0 such that, for $0 \leq t \leq \Delta_1$ and $i > I_0$, (2.44) indeed holds.

Suppose now that $j_0 < i \leq I_0$. Since $i > j_0$,

$$\sum_{j=0}^{j_0} P_{ij}^f(s) \leq 1 - e^{q_{ii}(f)s} \leq -q_{ii}(f)s \leq q(i)s.$$

Consequently, the left-hand side of (2.44) is bounded above by

$$be^{ct}t^2q(i)/w(i).$$

It is now clear that by choosing t small enough, (2.44) holds for $j_0 < i \leq I_0$. Thus we have shown that there exists $\Delta_2 > 0$ such that $R(t, i, f) \leq 1$ for $0 \leq t \leq \Delta_2$, $f \in \mathbb{F}$, and $i > j_0$.

To conclude, $R(t, i, f) \leq 1$ for every $0 \leq t \leq \min\{\Delta_1, \Delta_2\}$, $i \in S$, and $f \in \mathbb{F}$, as we wanted to prove. $\square$

We now consider a "skeleton" or time-discretization of our continuous-time Markov control model. We suppose that the system is observed at times $0, \Delta, 2\Delta, \ldots$, where $\Delta > 0$ is taken from Proposition 2.21. Then we consider a discrete-time control model with the following elements:

- The state space is S.

- The action space is the compact metric space $\mathbb{F}$ (with the topology of componentwise convergence), which is the same for every $i \in S$.
- The transition probabilities are $P_{ij}^f(\Delta)$, for $i, j \in S$, and $f \in \mathbb{F}$.

The class of deterministic stationary policies $f \in \mathbb{F}$ in the continuous-time model corresponds to the family of deterministic stationary *constant* policies (that is, the chosen action $f \in \mathbb{F}$ is the same at each state $i \in S$) in the discrete-time model. Therefore, the class of deterministic stationary policies for the skeleton control model is larger than $\mathbb{F}$.

Uniform ergodicity for such discrete-time CMCs has been proved in Key Theorem II in [25], from which our next result is a direct consequence. We recall that μ_f denotes the invariant probability measure of the state process $\{x^f(t)\}$.

Proposition 2.22. *Suppose that Assumption* 2.10 *holds, and let* $\Delta > 0$ *be as in Proposition* 2.21. *Then there exist positive constants* $\beta < 1$ *and* H *such that*

$$\sum_{j \in S} |P_{ij}^f(k\Delta) - \mu_f(j)| w(j) \le H\beta^k w(i)$$

for every $i \in S$, $k \ge 0$ *integer, and* $f \in \mathbb{F}$.

Proof. First of all we observe that, as a consequence of Assumption 2.10(b), the Markov chain $\{x^f(k\Delta)\}_{k \ge 0}$ is irreducible and aperiodic for each $f \in \mathbb{F}$.

Now, by Propositions 2.19, 2.20, and 2.21, the discretized control model verifies the following properties:

(i) For each $i, j \in S$, the function $f \mapsto P_{ij}^f(\Delta)$ is continuous on $\mathbb{F}$.
(ii) For each $i \in S$, the function $f \mapsto \sum_{j \in S} P_{ij}^f(\Delta) w(j)$ is continuous on $\mathbb{F}$.
(iii) There exists a constant $\gamma < 1$ and $j_0 \in S$ such that

$$\sum_{j > j_0} P_{ij}^f(\Delta) w(j) \le \gamma w(i)$$

for every $f \in \mathbb{F}$ and $i \in S$ (in [25], $C = \{0, 1, \ldots, j_0\}$ is called a taboo set).

Hence, the stated result follows from Key Theorem II in [25, p. 544]. $\square$

Now we are ready to prove our main result.

Proof of Theorem 2.11. Let $i \in S$, $u \in \mathcal{B}_w(S)$, and $f \in \mathbb{F}$. Given $t \ge 0$, let $0 \le s < \Delta$ and $k \ge 0$ integer be such that $t = k\Delta + s$. The following holds.

$$|E_i^f u(x(t)) - \mu_f(u)| = \left| \sum_{j \in S} (P_{ij}^f(t) - \mu_f(j)) u(j) \right|$$

$$\leq \|u\|_w \sum_{j \in S} w(j) \cdot |P_{ij}^f(t) - \mu_f(j)|$$

$$= \|u\|_w \sum_{j \in S} w(j) \cdot \left| \sum_{i' \in S} P_{ii'}^f(s) (P_{i'j}^f(k\Delta) - \mu_f(j)) \right|,$$

by the Chapman–Kolmogorov equation (2.6). Therefore,

$$|E_i^f u(x(t)) - \mu_f(u)| \leq \|u\|_w \sum_{i' \in S} P_{ii'}^f(s) \sum_{j \in S} w(j) |P_{i'j}^f(k\Delta) - \mu_f(j)|$$

$$\leq H\beta^k \|u\|_w \sum_{i' \in S} P_{ii'}^f(s) w(i') \quad [\text{Proposition 2.22}]$$

$$\leq H\beta^{-1} (\beta^{1/\Delta})^t \|u\|_w \sum_{i' \in S} P_{ii'}^f(s) w(i').$$

Finally, by (2.13),

$$|E_i^f u(x(t)) - \mu_f(u)| \leq H\beta^{-1} (\beta^{1/\Delta})^t \|u\|_w (1 + b/c) w(i)$$

for every $i \in S$, $u \in \mathcal{B}_w(S)$, and $t \geq 0$, which proves the statement of Theorem 2.11 with $R := H\beta^{-1}(1 + b/c)$ and $\delta := -(\log \beta)/\Delta$. $\quad\square$

It is worth noting that Theorem 2.11 also holds if instead of Assumption 2.10(d) we *suppose* that the uniform integrability condition (2.39) holds for every fixed $i \in S$ and $t \geq 0$ (see [129, Theorem 2.5]). This uniform integrability condition, however, seems difficult to verify in practice, and so we have preferred to use the more easily verifiable Assumption 2.10(d).

We conclude this section with a comment about Remark 2.12. If we replace $\mathbb{F}$ with Φ_s in Assumption 2.10(b), and recalling that Φ_s is compact with the topology of componentwise weak convergence (2.5), the proof of the uniform ergodicity on Φ_s is a verbatim copy of that of the uniform ergodicity on $\mathbb{F}$.

2.6. Conclusions

The construction of a Q-process, starting from a Q-matrix, has been studied by Anderson [6], Guo and Hernández-Lerma in [44, 45], and also discussed in Appendices B and C in [52]. In particular, in the latter references, the Lyapunov or drift condition in Assumption 2.1 is proposed and its relation to the expected growth (2.13) of the controlled Markov process is proved.

Also, regarding the existence of the controlled Markov process, in [44, 45] the class of admissible policies is restricted to the policies $\varphi \in \Phi$ such that, for all fixed states $i, j \in S$, the function

$$t \mapsto q_{ij}(t, \varphi),$$

defined in (2.2), is *continuous*. This requirement is from the paper by Feller [35]. In particular, this means that some "simple" polices such as, for instance, step polices are not admissible. In a recent paper, Ye, Guo, and Hernández-Lerma [167] show that the aforementioned continuity of the transition rates can be relaxed, and it suffices that some mild measurability and integrability conditions are satisfied. Here, we have followed this approach.

The Lyapunov or drift conditions in Assumption 2.1 and in Theorem 2.5 are related to the nonexplosivity, recurrence, and positive recurrence conditions explored by Meyn and Tweedie [111]. In particular, as noted in Theorem 2.5, they yield w-exponential ergodicity of stationary policies. As already mentioned, however, exponential ergodicity for each *single* stationary policy does not suffice to study the average reward optimality criterion; see Secs 3.4 and 3.5 below. Indeed, we need w-exponential ergodicity *uniformly* in the class of stationary policies.

To this end, we propose two different sufficient conditions yielding w-exponential uniform ergodicity; see Theorems 2.8 and 2.11. The advantage of the monotonicity and irreducibility conditions in Theorem 2.8 is that the rate δ of convergence to zero, as well as the multiplicative constant R in (2.23), have *known* values, while these values are unknown under the hypotheses of Theorem 2.11. On the other hand, the indicator function in the drift condition in Assumption 2.7(a) is necessarily $\mathbf{I}_{\{0\}}$, whereas, in Assumption 2.10(a), it can be the indicator function of any finite subset C of S. An important feature of both uniform ergodicity assumptions is that they are imposed on the primitive data of the control problem, that is, they are imposed directly on the transition rates of the system. Therefore, in principle, it is easy to verify (or discard) them.

Chapter 3

Basic Optimality Criteria

3.1. Introduction

This chapter is devoted to the study of the so-called basic optimality criteria for continuous-time CMCs. More precisely, we deal with the total expected reward optimality criterion for finite horizon CMCs, while, for infinite horizon CMCs, we analyze the total expected discounted reward and the long-run expected average reward criteria. In later chapters we consider optimality criteria that are modifications or combinations of these three.

In this book, we focus on the *dynamic programming* approach, which was introduced by Bellman [11]. This means that the optimal value v^* of the control problem (under the various optimality criteria) is characterized by means of an optimality equation (also referred to as a dynamic programming or Bellman equation). Typically, this equation is of the form

$$v^*(i) = \max_{a \in A(i)} \mathbf{H}(i, a) \quad \forall\, i \in S,$$

where $i \in S$ stands for the initial state of the system, and $\mathbf{H} : K \to \mathbb{R}$ is a suitably defined operator, which depends on the optimality criterion. An optimal policy $f^* \in \mathbb{F}$ for the corresponding optimality criterion is usually a policy attaining the maximum in the dynamic programming equation, that is,

$$v^*(i) = \mathbf{H}(i, f^*(i)) \quad \forall\, i \in S.$$

In addition to dynamic programming and its variants (such as, e.g., viscosity solutions [37]), there exist other approaches to study control problems. For instance, a common approach is to transform the control problem into a linear programming problem on suitably defined (actually, infinite-dimensional) spaces; see, for instance, Puterman [138], and

Hernández-Lerma and Lasserre [68, 70]. In Chapter 8, below, we will use this approach to deal with a constrained CMC.

Our main aim being the study of infinite horizon optimality criteria, we do not make an in-depth analysis of the finite horizon control problem. The reader interested in the details is referred to [64] and [130].

The total expected discounted reward optimality criterion, introduced by Howard [77], assumes that the rewards earned by the controller are depreciated at a constant rate $\alpha > 0$. That is, if the controller receives a payoff, say, $r(t)$ at time t, then this payoff is brought to its present value $e^{-\alpha t}r(t)$. The goal of the decision-maker is to obtain a policy such that the present value of his/her total expected reward, that is, $\int_0^\infty e^{-\alpha t}r(t)dt$, is maximal. This optimality criterion is among the most popular because of its straightforward economical interpretation.

Under the long-run expected average reward optimality criterion, analyzed as well by Howard [77], the decision-maker wants to maximize his/her long-run average reward

$$\lim_{T\to\infty} \frac{1}{T} \int_0^T r(t)dt.$$

Thus, this optimality criterion is only concerned with the asymptotic rewards earned by the controller, and it is widely used in, e.g., queueing systems and telecommunication networks.

The usual tool used to analyze the average reward optimality criterion is the *vanishing discount* approach, which consists in letting the discount rate $\alpha \downarrow 0$ in the α-discounted dynamic programming equation. Here, we use this approach which, in addition, allows us to establish an interesting Abelian theorem for CMCs showing the relation existing between discounted and average control problems.

Finally, let us mention that the analysis of the average reward optimality criterion relies heavily on the chain structure of the controlled Markov process. Here, we deal with an ergodic control model for which the controlled process $\{x^\varphi(t)\}$ for stationary $\varphi \in \Phi_s$ is irreducible. Dealing with, e.g., transient or multichain control models is beyond the scope of this book. Such models have been analyzed by Puterman [138], and also by Guo and Hernández-Lerma [52].

Observe that, in the optimality criteria that we consider here, the rewards are earned by the controller *continuously in time*. There exist other approaches in which the rewards are earned only at certain (discrete) times; the interested reader is referred to the works by Lippman [104], Puterman [138], or Sennott [147].

The contents of this chapter are the following. In Sec. 3.2 we briefly analyze CMCs with a finite horizon. Section 3.3 is devoted to the discounted optimality criterion and, in particular, we establish the *discounted reward optimality equation* and the existence of discount optimal policies. We follow [44, 53]. The average reward optimality criterion is studied in Sec. 3.4, after [45]. In this section, we give important definitions such as the *bias* of a stationary policy and the *Poisson equation*. The vanishing discount approach in Sec. 3.5 allows us to establish the *average reward optimality equation*, as well as the existence of average optimal policies. In Sec. 3.6, we study *pathwise* average optimality. Section 3.7 makes an interesting link between average optimality and finite horizon control problems. Our conclusions are stated in Sec. 3.8.

Finally, let us make an assumption that will be needed throughout this chapter. Note that Assumptions 3.1(a) and 3.1(b) below are the same as Assumptions 2.1 and 2.2, respectively, but we restate them here for ease of reference.

Assumption 3.1. There exists a Lyapunov function w on S such that:

(a) For some constants $c \in \mathbb{R}$ and $b \geq 0$,

$$\sum_{j \in S} q_{ij}(a)w(j) \leq -cw(i) + b \quad \forall\, (i, a) \in K.$$

(b) There exists a nonnegative function $w' : S \to \mathbb{R}$ such that $q(i)w(i) \leq w'(i)$ for each $i \in S$ and, in addition, there exist constants $c' \in \mathbb{R}$ and $b' \geq 0$ such that

$$\sum_{j \in S} q_{ij}(a)w'(j) \leq -c'w'(i) + b' \quad \forall\, (i, a) \in K.$$

(c) There exists a constant $M > 0$ such that $|r(i, a)| \leq Mw(i)$ for every $(i, a) \in K$. (In particular, we can write $r \in \mathcal{B}_w(K)$; recall (2.9).)

By the results in Sec. 2.3, the above conditions guarantee the existence of the controlled Markov chain $\{x^\varphi(t)\}$ for $\varphi \in \Phi$, and the application of the Dynkin formula (Proposition 2.3) for functions in $\mathcal{B}_w(S)$. We will also need the notation (cf. (2.2))

$$r(t, i, \varphi) := \int_{A(i)} r(i, a)\varphi_t(da|i) \quad \forall\, i \in S,\ \varphi \in \Phi,\ t \geq 0, \qquad (3.1)$$

which, for stationary policies $f \in \mathbb{F}$ or $\varphi \in \Phi_s$, will be simply written as

$$r(i, f) := r(i, f(i)) \quad \text{or} \quad r(i, \varphi) := \int_{A(i)} r(i, a)\varphi(da|i)$$

for $i \in S$, respectively.

Throughout this chapter, we suppose that Assumption 3.1 is satisfied.

3.2.   The finite horizon case

As already mentioned, in this book our interest is mainly focused on infinite horizon control problems. For the sake of completeness, however, in this section we briefly analyze finite horizon control problems.

We will consider a control problem with finite horizon $[0, T]$ for some fixed $T > 0$. The controller wishes to maximize the *total expected reward* over the time horizon $[0, T]$. More precisely, let $h \in \mathcal{B}_w(S)$ be a terminal reward function, and consider an initial state $i \in S$ and a Markov policy $\varphi \in \Phi$. The corresponding total expected reward with terminal payoff h is defined as:

$$J_T(i, \varphi, h) := E_i^\varphi \left[\int_0^T r(t, x(t), \varphi)dt + h(x(T)) \right].$$

When there is no terminal payoff function, we write

$$J_T(i, \varphi) := E_i^\varphi \left[\int_0^T r(t, x(t), \varphi)dt \right] \quad \forall \, i \in S, \; \varphi \in \Phi. \qquad (3.2)$$

Obviously, when dealing with a finite horizon control problem, the requirements in the definition of a control policy $\varphi \in \Phi$ can be restricted to the time interval $[0, T]$.

By (3.1) and Assumption 3.1(c), $|r(t, i, \varphi)| \leq Mw(i)$ for every $t \geq 0$ and $i \in S$. Hence, since h is in $\mathcal{B}_w(S)$,

$$|J_T(i, \varphi, h)| \leq M \int_0^T E_i^\varphi[w(x(t))]dt + \|h\|_w E_i^\varphi[w(x(T))].$$

This inequality and (2.13) yield that, for every $i \in S$, the optimal reward function

$$J_T^*(i, h) := \sup_{\varphi \in \Phi} J_T(i, \varphi, h)$$

is finite. We say that a policy $\varphi^* \in \Phi$ is *optimal for the finite horizon control problem on* $[0, T]$ *with terminal payoff* $h \in \mathcal{B}_w(S)$ if

$$J_T(i, \varphi^*, h) = J_T^*(i, h) \quad \forall \, i \in S.$$

The weaker notion of ε-optimality is often considered, although we will focus on optimality (or 0-optimality).

It is sometimes useful to consider the control problem over the time interval $[s, T]$, for $0 \le s \le T$. More precisely, for $0 \le s \le T$, $i \in S$, $h \in \mathcal{B}_w(S)$, and $\varphi \in \Phi$, let

$$J_T(s, i, \varphi, h) := E_{s,i}^{\varphi} \left[\int_s^T r(t, x(t), \varphi)dt + h(x(T)) \right],$$

and let

$$J_T^*(s, i, h) := \sup_{\varphi \in \Phi} J_T(s, i, \varphi, h) \tag{3.3}$$

be the corresponding optimal total expected reward function.

Define $C_w^1([0, T] \times S)$ as the set of functions $v : [0, T] \times S \to \mathbb{R}$ such that $s \mapsto v(s, i)$ is continuously differentiable on $[0, T]$ for each $i \in S$ and, in addition,

$$\max_{s \in [0,T]} |v(s, \cdot)| \in \mathcal{B}_w(S).$$

We say that a function $v \in C_w^1([0, T] \times S)$ is a solution of the *finite horizon optimality equation* if it satisfies, for all $i \in S$ and $0 \le s \le T$,

$$0 = \sup_{a \in A(i)} \left\{ r(i, a) + \frac{\partial v}{\partial s}(s, i) + \sum_{j \in S} q_{ij}(a)v(s, j) \right\}, \tag{3.4}$$

and, moreover,

$$v(T, i) = h(i) \quad \forall \, i \in S. \tag{3.5}$$

In Theorem 3.4 below we prove that the function $J_T^*(s, i, h)$ in (3.3) is the solution of the finite horizon optimality equation, and that this equation also allows us to derive optimal policies for the finite horizon control problem.

Remark 3.2. An optimality equation such as (3.4) and (3.5) is also known as a *dynamic programming equation* or a *Bellman equation*.

To state our next result, we impose the following compactness-continuity hypotheses.

Assumption 3.3.

(a) For each $i \in S$, the set $A(i)$ is compact.

(b) For every $i, j \in S$, the functions $a \mapsto q_{ij}(a)$, $a \mapsto r(i,a)$, and $a \mapsto \sum_{j \in S} q_{ij}(a)w(j)$ are continuous on $A(i)$, where w is the Lyapunov function in Assumption 3.1.

Now we are ready to state our main result in this section.

Theorem 3.4. *Suppose that Assumptions 3.1 and 3.3 hold, and consider a control problem with finite horizon $[0, T]$ and final payoff function $h \in \mathcal{B}_w(S)$. Suppose, in addition, that the function $J_T^*(\cdot, \cdot, h)$ in (3.3) is in $C_w^1([0, T] \times S)$. Then the following statements hold:*

(i) *The function $(s, i) \mapsto J_T^*(s, i, h)$ is the unique solution in $C_w^1([0, T] \times S)$ of the finite horizon optimality equation (3.4) and (3.5).*

(ii) *There exists a nonstationary deterministic policy $\varphi \in \Phi$ such that*

$$J_T(s, i, \varphi, h) = J_T^*(s, i, h) \quad \forall\, i \in S,\ 0 \le s \le T.$$

That is, φ is an optimal policy for the finite horizon problem on $[s, T]$ (for every $0 \le s \le T$) with terminal payoff h.

The proof of Theorem 3.4 is quite long and technical. For details, refer to [130, Theorem 3.1]. The following remark, however, is in order. The function

$$(s, i, a) \mapsto r(i, a) + \sum_{j \in S} q_{ij}(a) J_T^*(s, j, h)$$

is continuous on $[0, T] \times K$ (this follows, in particular, from Assumption 3.3(b) and Dini's theorem). Therefore, by a result for measurable selectors (see Sec. XI.3, Theorems 1 and 2, in [12]), there exists a nonstationary deterministic policy $\varphi \in \Phi$ (that is, a measurable function $\varphi : [0, T] \times S \to A$ with $\varphi(s, i) \in A(i)$ for every $i \in S$ and $0 \le s \le T$) such that

$$\max_{a \in A(i)} \left\{ r(i, a) + \sum_{j \in S} q_{ij}(a) J_T^*(s, j, h) \right\} = r(s, i, \varphi) + \sum_{j \in S} q_{ij}(s, \varphi) J_T^*(s, j, h)$$

for every $0 \le s \le T$ and $i \in S$. This policy φ is the optimal policy referred to in part (ii) of Theorem 3.4.

In Theorem 3.28 below we prove that, for a suitable choice of the final payoff function $h \in \mathcal{B}_w(S)$, the optimal reward $J_T^*(s, i, h)$ is an affine function of s and T, and, furthermore, there exist optimal deterministic *stationary* policies.

Theorem 3.4 states the so-called *Bellman's principle* of optimality, according to which if a policy is optimal for the time horizon $[0, T]$, then it is also optimal on $[s, T]$ for *any* time $0 \leq s \leq T$. Let us also mention that discrete-time finite horizon control problems are efficiently solved by using the *backward induction* principle, also known as the *dynamic programming algorithm*. For continuous-time models, however, there is no such algorithm.

3.3. The infinite horizon discounted reward

Recall that we are supposing that Assumption 3.1 is satisfied.

For a control problem with infinite time horizon, in this section we suppose that the rewards earned by the controller are depreciated at a constant discount rate $\alpha > 0$. Therefore, the controller has to determine a policy with the maximal *total expected discounted reward*. This leads to the following definitions.

3.3.1. *Definitions*

Given a Markov policy $\varphi \in \Phi$ and an initial state $i \in S$, the total expected discounted reward (or discounted reward, for short) is defined as

$$V_\alpha(i, \varphi) := E_i^\varphi \left[\int_0^\infty e^{-\alpha t} r(t, x(t), \varphi) dt \right]. \tag{3.6}$$

By Assumption 3.1(c),

$$|V_\alpha(i, \varphi)| \leq M \int_0^\infty e^{-\alpha t} E_i^\varphi [w(t)] dt.$$

Therefore, as a consequence of (2.13), the above defined discounted reward is finite if $\alpha + c > 0$, where c is the constant in Assumption 3.1. More precisely,

$$|V_\alpha(i, \varphi)| \leq \frac{M w(i)}{\alpha + c} + \frac{bM}{\alpha(\alpha + c)} \quad \forall\, i \in S, \; \varphi \in \Phi. \tag{3.7}$$

Hence, $V_\alpha(\cdot, \varphi)$ is in $\mathcal{B}_w(S)$ and, moreover, the optimal discounted reward

$$V_\alpha^*(i) := \sup_{\varphi \in \Phi} V_\alpha(i, \varphi) \quad \forall\, i \in S \qquad (3.8)$$

verifies

$$|V_\alpha^*(i)| \leq \frac{Mw(i)}{\alpha + c} + \frac{bM}{\alpha(\alpha + c)} \quad \forall\, i \in S. \qquad (3.9)$$

Thus V_α^* is also in $\mathcal{B}_w(S)$.

We say that a policy $\varphi^* \in \Phi$ is *discount optimal* for the discount rate α (or α-discount optimal) if

$$V_\alpha(i, \varphi^*) = V_\alpha^*(i) \quad \forall\, i \in S.$$

It is sometimes useful to define the total expected discounted reward as in (3.6) but for a reward rate $u \neq r$. Let $u : K \to \mathbb{R}$ be such that $a \mapsto u(i, a)$ is measurable on $A(i)$ for each $i \in S$, and such that, for some constant $M' > 0$, $|u(i, a)| \leq M'w(i)$ for each $(i, a) \in K$, where w is the Lyapunov function in Assumption 3.1 (equivalently, using the notation in (2.9), we can write $u \in \mathcal{B}_w(K)$). Given $\varphi \in \Phi$ and $i \in S$, let

$$V_\alpha(i, \varphi, u) := E_i^\varphi \left[\int_0^\infty e^{-\alpha t} u(t, x(t), \varphi) dt \right], \qquad (3.10)$$

where $u(t, i, \varphi)$ is defined as in (3.1), that is,

$$u(t, i, \varphi) := \int_{A(i)} u(i, a) \varphi_t(da|i). \qquad (3.11)$$

3.3.2.   The discounted reward optimality equation

The next lemma uses the notation $L^{t,\varphi} v$ introduced in (2.14).

Lemma 3.5. *Suppose that Assumption 3.1 holds and let the discount rate $\alpha > 0$ be such that $\alpha + c > 0$.*

(i) *If $v \in \mathcal{B}_w(S)$ and $\varphi \in \Phi$ verify*

$$\alpha v(i) \geq r(t, i, \varphi) + (L^{t,\varphi} v)(i) \quad \forall\, i \in S,\ t \geq 0, \qquad (3.12)$$

then $v(i) \geq V_\alpha(i, \varphi)$ for every $i \in S$. If the inequality $\geq$ in (3.12) is replaced with $\leq$, then $v(i) \leq V_\alpha(i, \varphi)$ for every $i \in S$.

(ii) *If $v \in \mathcal{B}_w(S)$ and $\varphi \in \Phi_s$ verify*

$$\alpha v(i) \geq r(i, \varphi) + (L^\varphi v)(i) \quad \forall\, i \in S, \qquad (3.13)$$

with strict inequality at some $i_0 \in S$, then $v(i) \geq V_\alpha(i, \varphi)$ for every $i \in S$, and $v(i_0) > V_\alpha(i_0, \varphi)$. If the inequality $\geq$ in (3.13) is replaced with

$\le$, and, in addition, it is strict for some $i_0 \in S$, then $v(i) \le V_\alpha(i, \varphi)$ for every $i \in S$, and $v(i_0) < V_\alpha(i_0, \varphi)$.

Proof. *Proof of* (*i*). By the Dynkin formula (2.19), for every $i \in S$ and $t \ge 0$,

$$E_i^\varphi[e^{-\alpha t}v(x(t))] - v(i) = E_i^\varphi\left[\int_0^t e^{-\alpha s}(L^{s,\varphi}v - \alpha v)(x(s))ds\right]$$

$$\le -E_i^\varphi\left[\int_0^t e^{-\alpha s}r(s, x(s), \varphi)ds\right].$$

As a consequence of (2.13) and the fact that $\alpha + c > 0$, taking the limit as $t \to \infty$ yields

$$E_i^\varphi[e^{-\alpha t}v(x(t))] \to 0$$

and

$$E_i^\varphi\left[\int_0^t e^{-\alpha s}r(s, x(s), \varphi)ds\right] \to V_\alpha(i, \varphi).$$

The stated result now follows. For the reverse inequality, the proof is similar.

Proof of (*ii*). Observe that for some $\varepsilon > 0$

$$\alpha v(i) \ge r(i, \varphi) + (L^\varphi v)(i) + \varepsilon \mathbf{I}_{\{i_0\}}(i) \quad \forall\, i \in S.$$

By the Dynkin formula (2.19) again, given $t \ge 0$ and $i \in S$,

$$E_i^\varphi[e^{-\alpha t}v(x(t))] - v(i) = E_i^\varphi\left[\int_0^t e^{-\alpha s}(L^\varphi v - \alpha v)(x(s))ds\right]$$

$$\le -E_i^\varphi\left[\int_0^t e^{-\alpha s}(r(x(s), \varphi) + \varepsilon \mathbf{I}_{\{i_0\}}(x(s)))ds\right].$$

As in the proof of (i), we obtain

$$E_i^\varphi[e^{-\alpha t}v(x(t))] \to 0 \quad \text{and} \quad E_i^\varphi\left[\int_0^t e^{-\alpha s}r(x(s), \varphi)ds\right] \to V_\alpha(i, \varphi)$$

as $t \to \infty$. Consequently,

$$V_\alpha(i, \varphi) \le v(i) - \varepsilon \cdot E_i^\varphi\left[\int_0^\infty e^{-\alpha s}\mathbf{I}_{\{i_0\}}(x(s))ds\right]$$

$$= v(i) - \varepsilon \cdot \int_0^\infty e^{-\alpha s}P_{ii_0}^\varphi(s)ds. \tag{3.14}$$

As a consequence of (3.14), we have — as we already know from part (i) — that $V_\alpha(i, \varphi) \leq v(i)$ for all $i \in S$. Suppose now that $i_0 \in S$ is the initial state of the system. If T denotes the first jump of the process $\{x^\varphi(t)\}_{t \geq 0}$ then, by (2.12), for all $s \geq 0$,

$$P^\varphi_{i_0 i_0}(s) \geq P^\varphi_{i_0}\{T > s\} \geq e^{-q(i_0)s}.$$

Therefore, by (3.14),

$$V_\alpha(i_0, \varphi) \leq v(i_0) - \frac{\varepsilon}{\alpha + q(i_0)},$$

and $V_\alpha(i_0, \varphi) < v(i_0)$ follows. (In fact, we have that $V_\alpha(i, \varphi) < v(i)$ for all $i \in S$ such that i_0 can be reached from i with positive probability.)   □

As a direct consequence of Lemma 3.5, if for some $v \in \mathcal{B}_w(S)$ and $\varphi \in \Phi_s$,

$$\alpha v(i) = r(i, \varphi) + \sum_{j \in S} q_{ij}(\varphi) v(j) \quad \forall \, i \in S,$$

then $v(i) = V_\alpha(i, \varphi)$ for every $i \in S$. Our next result shows that, in fact, the solution in $\mathcal{B}_w(S)$ to the above equation indeed exists.

Lemma 3.6. *Suppose that Assumption 3.1 holds and that the discount rate $\alpha > 0$ satisfies $\alpha + c > 0$. Given an arbitrary stationary policy φ in Φ_s, $V_\alpha(\cdot, \varphi)$ is the unique solution in $\mathcal{B}_w(S)$ of the equations*

$$\alpha v(i) = r(i, \varphi) + \sum_{j \in S} q_{ij}(\varphi) v(j) \quad \forall \, i \in S. \tag{3.15}$$

Proof.   As already mentioned, and as a consequence of Lemma 3.5, if $v \in \mathcal{B}_w(S)$ is a solution of (3.15) then, necessarily, $v = V_\alpha(\cdot, \varphi)$.

Conversely, for each $i \in S$ we have that

$$\sum_{j \in S} q_{ij}(\varphi) V_\alpha(j, \varphi) = \sum_{j \in S} q_{ij}(\varphi) \int_0^\infty e^{-\alpha t} \sum_{k \in S} P^\varphi_{jk}(t) r(k, \varphi) dt$$

$$= \sum_{k \in S} r(k, \varphi) \int_0^\infty e^{-\alpha t} \sum_{j \in S} q_{ij}(\varphi) P^\varphi_{jk}(t) dt \tag{3.16}$$

$$= \sum_{k \in S} r(k, \varphi) \int_0^\infty e^{-\alpha t} \frac{dP^\varphi_{ik}(t)}{dt} dt \tag{3.17}$$

$$= -r(i, \varphi) + \alpha \sum_{k \in S} r(k, \varphi) \int_0^\infty e^{-\alpha t} P^\varphi_{ik}(t) dt \tag{3.18}$$

$$= -r(i, \varphi) + \alpha V_\alpha(i, \varphi), \tag{3.19}$$

where the interchange of sums and integrals in (3.16) follows from (2.18), while (3.17) is a consequence of Kolmogorov's backward equations [6, 52], and (3.18) is obtained after integration by parts. Finally, (3.19) yields that $V_\alpha(\cdot, \varphi)$ is a solution of (3.15). This completes the proof of the lemma. $\square$

Our next result, which is taken from [44, Theorem 3.2] and Theorems 3.1 and 3.2 in [53], shows that V_α^* is the solution of a suitably defined optimality equation, and it also proves the existence of stationary deterministic discount optimal policies.

Theorem 3.7. *Suppose that Assumption 3.1 is satisfied, and consider a discount rate $\alpha > 0$ such that $\alpha + c > 0$. Then the following statements hold:*

(i) *There exist solutions $u \in \mathcal{B}_w(S)$ of the* <u>*discounted reward optimality equation*</u> *(DROE)*

$$\alpha u(i) = \sup_{a \in A(i)} \left\{ r(i,a) + \sum_{j \in S} q_{ij}(a) u(j) \right\} \quad \forall\, i \in S.$$

Suppose, in addition, that Assumption 3.3 is satisfied. Then:

(ii) *The optimal discounted reward V_α^* is the unique solution in $\mathcal{B}_w(S)$ of the DROE*

$$\alpha V_\alpha^*(i) = \max_{a \in A(i)} \left\{ r(i,a) + \sum_{j \in S} q_{ij}(a) V_\alpha^*(j) \right\} \quad \forall\, i \in S. \qquad (3.20)$$

(iii) *A deterministic stationary policy $f \in \mathbb{F}$ is α-discount optimal if and only if, for every $i \in S$, $f(i) \in A(i)$ attains the maximum in the right-hand side of (3.20), i.e.,*

$$\alpha V_\alpha^*(i) = r(i,f) + \sum_{j \in S} q_{ij}(f) V_\alpha^*(j) \quad \forall\, i \in S$$

(and such a policy indeed exists).

Proof. Proof of (i). Given $i \in S$, choose $m(i) > 0$ such that $m(i) > q(i)$, and define

$$p_{ij}(a) := \frac{q_{ij}(a)}{m(i)} + \delta_{ij} \quad \text{for } a \in A(i) \text{ and } j \in S$$

(recall that δ_{ij} denotes the Kronecker delta). By the conservative property of the transition rates $q_{ij}(a)$ (see the definition of the control model $\mathcal{M}$

in Sec. 2.2), we have that $\{p_{ij}(a)\}_{j \in S}$ is a probability measure for every $(i, a) \in K$.

We define the following operator: for $u \in \mathcal{B}_w(S)$,

$$(Tu)(i) := \sup_{a \in A(i)} \left\{ \frac{r(i, a)}{\alpha + m(i)} + \frac{m(i)}{\alpha + m(i)} \sum_{j \in S} p_{ij}(a)u(j) \right\} \quad \forall \, i \in S.$$

Let

$$u_0(i) := -\frac{bM}{\alpha(\alpha + c)} - \frac{M}{\alpha + c}w(i) \quad \forall \, i \in S$$

(here, the constants b, c, and M are as in Assumption 3.1), and define

$$u_{n+1} := Tu_n \quad \forall \, n \geq 0.$$

Next, we give the main steps of the proof (for details, we refer to [44, Theorem 3.2], [52, Theorem 6.6], or [53, Theorem 3.1]). First of all, note that the operator T is monotone; that is, given u and v in $\mathcal{B}_w(S)$ such that $u \leq v$ (componentwise), then $Tu \leq Tv$. Secondly, a direct calculation shows that $u_1 \geq u_0$. It then follows that $\{u_n\}$ is a monotone sequence. Moreover, it can be proved that the sequence $\{u_n\}$ is bounded in the w-norm. So, let u^* be the pointwise limit of $\{u_n\}$, which, besides, is in $\mathcal{B}_w(S)$.

Using an induction argument, we can prove that $Tu^* \geq u_n$ for every $n \geq 0$, and thus $Tu^* \geq u^*$. To prove the reverse inequality, first note that, by the definition of u_{n+1}, for each $(i, a) \in K$ we have

$$u_{n+1}(i) \geq \frac{r(i, a)}{\alpha + m(i)} + \frac{m(i)}{\alpha + m(i)} \sum_{j \in S} p_{ij}(a)u_n(j).$$

Using the extension of the Fatou lemma in [70, Lemma 8.3.7], we can take the limit as $n \to \infty$ in this inequality. This yields

$$u^*(i) \geq \frac{r(i, a)}{\alpha + m(i)} + \frac{m(i)}{\alpha + m(i)} \sum_{j \in S} p_{ij}(a)u^*(j) \quad \forall \, (i, a) \in K,$$

and $u^* \geq Tu^*$ follows.

Thus we have shown that $u^* = Tu^*$. Finally, a simple calculation shows that $u^* = Tu^*$ is indeed equivalent to the DROE in part (i).

Proof of (ii). If $u \in \mathcal{B}_w(S)$ is a solution of the DROE, then for every $(i, a) \in K$

$$\alpha u(i) \geq r(i, a) + \sum_{j \in S} q_{ij}(a)u(j).$$

Given $\varphi \in \Phi$ and $t \geq 0$, and integrating the previous inequality with respect to $\varphi_t(da|i)$, we obtain

$$\alpha u(i) \geq r(t, i, \varphi) + \sum_{j \in S} q_{ij}(t, \varphi) u(j) \quad \forall\, i \in S.$$

By Lemma 3.5(i), $u(i) \geq V_\alpha(i, \varphi)$ for each $i \in S$. Since $\varphi \in \Phi$ is arbitrary, it follows that $u(i) \geq V_\alpha^*(i)$ for all $i \in S$.

By the continuity and compactness hypotheses in Assumption 3.3, the DROE can indeed be written as

$$\alpha u(i) = \max_{a \in A(i)} \left\{ r(i, a) + \sum_{j \in S} q_{ij}(a) u(j) \right\} \quad \forall\, i \in S,$$

that is, with "max" rather than "sup". In particular, there exists $f \in \mathbb{F}$ such that

$$\alpha u(i) = r(i, f) + \sum_{j \in S} q_{ij}(f) u(j) \quad \forall\, i \in S$$

and so, Lemma 3.6 implies that $u(i) = V_\alpha(i, f)$ for all $i \in S$. This shows that $u = V_\alpha^*$ and, besides, that f is a discount optimal policy.

Proof of (iii). We have already shown that any $f \in \mathbb{F}$ attaining the maximum in the DROE is α-discount optimal.

Let us now prove the converse result. We proceed by contradiction. Hence, suppose that $f \in \mathbb{F}$ is a discount optimal policy that does not attain the maximum in the DROE, that is, we have

$$\alpha V_\alpha^*(i) \geq r(i, f) + \sum_{j \in S} q_{ij}(f) V_\alpha^*(j) \quad \forall\, i \in S$$

with strict inequality for some $i_0 \in S$. Then, by Lemma 3.5(ii), we have $V_\alpha^*(i_0) > V_\alpha(i_0, f)$, which contradicts the α-discount optimality of f. The proof of Theorem 3.7 is now complete. $\qquad\square$

The usual approach for establishing the DROE for discrete-time models is by means of the *value iteration* procedure. This approach shows, under suitable hypotheses, that the optimal discounted reward is the unique fixed point of a contraction operator and it is, therefore, obtained as the limit of the successive compositions of this operator. For such discrete-time models, the convergence of the value iteration procedure is shown to be geometric. For the continuous-time case, it is proved in [130] that this convergence is exponential. We note that the proof of Theorem 3.7 also uses a fixed point approach, though the operator T is not a contraction.

3.3.3.   *The uniformization technique*

For continuous-time control models with both finite state and action spaces, the DROE was established by Howard [77], Rykov [144], Miller [112], and Veinott [161] under different sets of hypotheses. These results were generalized by Kakumanu [86] to a control model with denumerable state and action spaces with bounded transition and reward rates.

In this latter case (when the transition and the reward rates are both bounded) the *uniformization technique* used by many authors, e.g., Howard [77], Veinott [161], Serfozo [149], Puterman [138], or Bertsekas [13], allows us to transform the continuous-time control model into a discrete-time one. This procedure is as follows. Let

$$\mathbf{q} := \sup_{i \in S} \{q(i)\} \quad \text{and} \quad \mathbf{r} := \sup_{(i,a) \in K} \{|r(i,a)|\}$$

(recall that the notation $q(i)$ was introduced in (2.1)), and define

$$\tilde{p}_{ij}(a) := \frac{q_{ij}(a)}{\mathbf{q}} + \delta_{ij}, \quad \tilde{r}(i,a) := \frac{r(i,a)}{\alpha + \mathbf{q}}, \quad \text{and} \quad \beta := \frac{\mathbf{q}}{\alpha + \mathbf{q}},$$

for $(i,a) \in K$. Then, by analogy with $\mathcal{M}$ in Sec. 2.2, we can consider a discrete-time model

$$\mathcal{M}_d := \{S, A, (A(i)), (\tilde{p}_{ij}(a)), (\tilde{r}(i,a))\},$$

with transition probabilities $\tilde{p}_{ij}(a)$, reward function $\tilde{r}(i,a)$, and discount factor β. It follows from [68, Theorem 4.2.3] that the DROE for this discrete-time model, that is,

$$u(i) = \sup_{a \in A(i)} \left\{ \tilde{r}(i,a) + \beta \sum_{j \in S} \tilde{p}_{ij}(a) u(j) \right\} \quad \forall\, i \in S,$$

is *equivalent* to the continuous-time DROE (3.20). (Note that our definition of $p_{ij}(a)$ in the proof of Theorem 3.7 is somehow similar to that of $\tilde{p}_{ij}(a)$ above.)

Therefore, the most interesting cases arise when the reward or the transition rates are *unbounded*. This issue has been addressed by, e.g., Wu and Zhang [165], Hernández-Lerma and Govindan [67], Guo and Zhu [57],

Guo and Hernández-Lerma [44, 48], or Prieto-Rumeau and Hernández-Lerma [130].

3.3.4. *A continuity theorem for discounted rewards*

We conclude this section by proving that the total expected discounted reward $V_\alpha(i, \varphi)$ is a continuous function of $\varphi \in \Phi_s$. First of all, we need a preliminary lemma.

Lemma 3.8. *Let Assumptions* 3.1 *and* 3.3 *be satisfied. Consider a sequence of stationary policies* $\{\varphi_n\} \subseteq \Phi_s$, *and a bounded sequence* $\{h_n\}$ *in* $\mathcal{B}_w(S)$ *(that is,* $\sup_n \|h_n\|_w < \infty$*) that converges pointwise to* $h \in \mathcal{B}_w(S)$.

(i) *If* $\{\varphi_n\}$ *converges to some* $\varphi \in \Phi_s$ *in the topology of componentwise weak convergence (recall* (2.5)*), then for every* $i \in S$

$$\lim_{n \to \infty} \sum_{j \in S} q_{ij}(\varphi_n) h_n(j) = \sum_{j \in S} q_{ij}(\varphi) h(j).$$

Equivalently (using the notation in (2.15)*),*

$$(L^{\varphi_n} h_n)(i) \longrightarrow (L^\varphi h)(i) \quad \forall\, i \in S.$$

(ii) *If* $h \equiv 0$ *(and regardless* $\{\varphi_n\}$ *converges) then, for every* $i \in S$,

$$\lim_{n \to \infty} \sum_{j \in S} q_{ij}(\varphi_n) h_n(j) = 0.$$

Proof.   We prove part (i). Fix a state $i \in S$. By Assumption 3.3(b), the sequence of continuous functions $\sum_{j \le k} q_{ij}(a) w(j)$ converges monotonically (for $k \ge i$) to the continuous function $\sum_{j \in S} q_{ij}(a) w(j)$. Hence, by Dini's theorem, the convergence is uniform on the compact set $A(i)$. Alternatively, $\sum_{j > k} q_{ij}(a) w(j)$ decreases uniformly to 0 on $A(i)$.

Consequently, given $\varepsilon > 0$ we can choose $k_0 \ge i$ with

$$\sup_{a \in A(i)} \left\{ \sum_{j > k_0} q_{ij}(a) w(j) \right\} \le \varepsilon / 3C,$$

where C is such that $\sup_n \|h_n\|_w \le C$. Therefore,

$$\left| \sum_{j > k_0} q_{ij}(\varphi_n) h_n(j) \right| \le C \cdot \sup_{a \in A(i)} \left\{ \sum_{j > k_0} q_{ij}(a) w(j) \right\} \le \varepsilon / 3,$$

and

$$\left| \sum_{j > k_0} q_{ij}(\varphi) h(j) \right| \leq C \cdot \sup_{a \in A(i)} \left\{ \sum_{j > k_0} q_{ij}(a) w(j) \right\} \leq \varepsilon/3.$$

Therefore, we have

$$\left| (L^{\varphi_n} h_n)(i) - (L^{\varphi} h)(i) \right| \leq \left| \sum_{j \leq k_0} [q_{ij}(\varphi_n) h_n(j) - q_{ij}(\varphi) h(j)] \right| + 2\varepsilon/3. \quad (3.21)$$

By Assumption 3.3(b) and the convergences

$$\varphi_n(\cdot|i) \xrightarrow{w} \varphi(\cdot|i), \quad \text{and} \quad h_n(j) \to h(j) \quad \text{for } j \leq k_0,$$

it follows that we can choose n large enough such that

$$\left| (L^{\varphi_n} h_n)(i) - (L^{\varphi} h)(i) \right| \leq \varepsilon,$$

and statement (i) follows.

Starting from (3.21), part (ii) is proved similarly. $\qquad\square$

We next state our continuity theorem.

Theorem 3.9. *We suppose that Assumptions* 3.1 *and* 3.3 *are satisfied, and we consider a discount rate* $\alpha > 0$ *with* $\alpha + c > 0$. *If the sequence of stationary policies* $\{\varphi_n\}_{n \geq 1}$ *converges to* $\varphi \in \Phi_{\mathrm{s}}$, *then*

$$\lim_{n \to \infty} V_\alpha(i, \varphi_n) = V_\alpha(i, \varphi) \quad \forall\, i \in S.$$

Proof. We know from Lemma 3.6 that $h_n := V_\alpha(\cdot, \varphi_n)$ is the unique solution in $\mathcal{B}_w(S)$ of the equation

$$\alpha h_n(i) = r(i, \varphi_n) + \sum_{j \in S} q_{ij}(\varphi_n) h_n(j) \quad \forall\, i \in S. \quad (3.22)$$

The sequence h_n being bounded in $\mathcal{B}_w(S)$ (see (3.7)), choose an arbitrary subsequence n' such that $\{h_{n'}\}$ converges pointwise to some $h \in \mathcal{B}_w(S)$.

By Lemma 3.8(i), taking the limit in (3.22) through the subsequence n', we obtain

$$\alpha h(i) = r(i, \varphi) + \sum_{j \in S} q_{ij}(\varphi) h(j) \quad \forall\, i \in S.$$

From Lemma 3.6 again we conclude that $h(i) = V_\alpha(i, \varphi)$ for all $i \in S$.

Therefore, $\lim_{n'} V_\alpha(i, \varphi_{n'}) = V_\alpha(i, \varphi)$, and it is now easily shown that $\lim_n V_\alpha(i, \varphi_n) = V_\alpha(i, \varphi)$ for all $i \in S$. □

Similar continuity theorems for long-run expected average rewards are obtained in Sec. 3.4 below.

3.4. The long-run expected average reward

Suppose that Assumption 3.1 is satisfied, and let $J_T(i, \varphi)$ be as in (3.2). The *long-run expected average reward* (or average reward in short) of a Markov policy $\varphi \in \Phi$ is defined as

$$J(i, \varphi) := \liminf_{T \to \infty} \frac{1}{T} J_T(i, \varphi). \tag{3.23}$$

Concerning this definition, some remarks are in order. In general, the limit of $J_T(i, \varphi)/T$ as $T \to \infty$ might not exist. Then, it is a standard convention in control problems to define the average reward of a policy φ as a "lim inf", as in (3.23), which can be interpreted as the "worst" long-run average return that is to be maximized (as in a minimax setting). Similarly, when dealing with *cost rates*, instead of *reward rates*, the long-run expected average cost of a policy is defined as a "lim sup"; see, e.g., (8.8) below.

The average reward of φ is also referred to as the *gain* of φ. If the constant c in Assumption 3.1(a) is positive (see Assumption 3.11(a) below), then it follows from (2.13) and Assumption 3.1(c) that

$$|J(i, \varphi)| \le bM/c \quad \forall\, i \in S,\ \varphi \in \Phi. \tag{3.24}$$

In particular, the optimal average reward

$$J^*(i) := \sup_{\varphi \in \Phi} J(i, \varphi)$$

is finite for every $i \in S$. A Markov policy $\varphi^* \in \Phi$ is said to be *average (reward) optimal* or *gain optimal* if

$$J^*(i) = J(i, \varphi^*) \quad \forall\, i \in S.$$

Our next result is the average reward counterpart of Lemma 3.5(i).

Lemma 3.10. *Suppose that Assumption 3.1 holds, where the constant c in Assumption 3.1(a) is positive. If there exist $g \in \mathbb{R}$, $h \in \mathcal{B}_w(S)$, and $\varphi \in \Phi$*

such that

$$g \geq r(t, i, \varphi) + \sum_{j \in S} q_{ij}(t, \varphi) h(j) \quad \forall\, i \in S,\ t \geq 0,$$

then $g \geq J(i, \varphi)$ for all $i \in S$.

Proof. The Dynkin formula in Proposition 2.3 yields, for $T \geq 0$,

$$E_i^{\varphi}[h(x(T))] - h(i) = E_i^{\varphi}\left[\int_0^T (L^{t,\varphi} h)(x(t)) dt\right]$$

$$\leq E_i^{\varphi}\left[\int_0^T [g - r(t, x(t), \varphi)] dt\right]$$

$$= gT - J_T(i, \varphi).$$

Therefore, for every $T > 0$, we have

$$\frac{J_T(i, \varphi)}{T} \leq g - \frac{E_i^{\varphi}[h(x(T))] - h(i)}{T}.$$

Taking the lim inf as $T \to \infty$ in this inequality gives, by (2.13) and definition (3.23), that $J(i, \varphi) \leq g$ for each $i \in S$ (here, we use the fact that the constant c in Assumption 3.1(a) is positive). $\qquad \square$

When dealing with average reward optimality, we will impose the next conditions, which are supposed to hold in the remainder of this chapter.

Assumption 3.11. Part (a) of Assumption 3.1 is replaced with the following.

(a) There exist constants $c > 0$ and $b \geq 0$, and a finite set $C \subset S$ such that

$$\sum_{j \in S} q_{ij}(a) w(j) \leq -cw(i) + b \cdot \mathbf{I}_C(i) \quad \forall\, (i, a) \in K.$$

The next conditions also hold.

(b) Each $f \in \mathbb{F}$ is irreducible (in particular, together with (a), this ensures the existence of the invariant probability measure μ_f; see Theorem 2.5).

(c) The control model is w-exponentially ergodic uniformly on $\mathbb{F}$, that is, there exist constants $\delta > 0$ and $R > 0$ such that

$$\sup_{f \in \mathbb{F}} |E_i^f[u(x(t))] - \mu_f(u)| \leq Re^{-\delta t} \|u\|_w w(i).$$

for every $i \in S$, $u \in \mathcal{B}_w(S)$, and $t \geq 0$ (recall the notation introduced in (2.21)).

The condition in (a) is taken from Assumption 2.10(a). Sufficient conditions for the w-exponential ergodicity in (c) are given in Theorems 2.8 and 2.11. Under Assumptions 3.11(a)–(b) and as a consequence of Theorem 2.5(iii), for each $f \in \mathbb{F}$ the average reward $J(i, f)$ becomes

$$J(i, f) = \lim_{T \to \infty} \frac{1}{T} J_T(i, f) = \sum_{j \in S} r(j, f)\mu_f(j) =: g(f), \qquad (3.25)$$

which does not depend on the initial state $i \in S$.

The following important remark concerns an extension of (3.25) to a randomized stationary policy $\varphi \in \Phi_s$.

Remark 3.12. As a consequence of Theorem 2.5 we have the following. If $\varphi \in \Phi_s$ is an irreducible stationary policy, then its average reward is constant and it verifies

$$J(i, \varphi) = \lim_{T \to \infty} \frac{1}{T} J_T(i, \varphi) = \sum_{j \in S} r(j, \varphi)\mu_\varphi(j) =: g(\varphi) \quad \forall \, i \in S.$$

The bias Given a deterministic stationary policy $f \in \mathbb{F}$, we define the *bias* of f as

$$h_f(i) := \int_0^\infty [E_i^f[r(x(t), f)] - g(f)]dt \qquad (3.26)$$

for each $i \in S$. By the uniform ergodicity Assumption 3.11(c) for the function $i \mapsto r(i, f)$, together with Assumption 3.1(c), we obtain

$$|E_i^f[r(x(t), f)] - g(f)| \leq RMe^{-\delta t}w(i),$$

and so, for each $i \in S$, $|h_f(i)| \leq RMw(i)/\delta$. This shows that, for each $f \in \mathbb{F}$, the bias h_f is in $\mathcal{B}_w(S)$ and, further,

$$\sup_{f \in \mathbb{F}} \|h_f\|_w \leq RM/\delta. \qquad (3.27)$$

The exponential ergodicity assumption and the fact that $\mu_f(w) < \infty$ ensure that, when taking the μ_f-expectation in (3.26), interchange of the integrals is allowed and, therefore,

$$\mu_f(h_f) = \int_0^\infty \sum_{i \in S} \mu_f(i)[E_i^f[r(x(t), f)] - g(f)]dt = 0, \qquad (3.28)$$

where the latter equality is due to the invariance of μ_f.

We note that uniform ergodicity in Assumption 3.11(c) is not strictly necessary to define the bias of a policy $f \in \mathbb{F}$. Indeed, by Assumptions 3.11(a)–(b), each single policy $f \in \mathbb{F}$ is w-exponentially ergodic (see Theorem 2.5(iii)), and so we can as well define the bias of f as in (3.26). If we drop Assumption 3.11(c), however, the bound (3.27), which is uniform on $\mathbb{F}$, might not hold.

Remark 3.13. We can derive an alternative expression of the bias h_f in terms of the total expected reward of the policy f on $[0, T]$. Indeed, we have:

$$h_f(i) = \int_0^T [E_i^f[r(x(t), f)] - g(f)]dt + \int_T^\infty [E_i^f[r(x(t), f)] - g(f)]dt$$

$$= J_T(i, f) - g(f)T + \int_T^\infty \sum_{j \in S} P_{ij}^f(T)[E_j^f[r(x(t-T), f)] - g(f)]dt$$

[by the Chapman–Kolmogorov equation]

$$= J_T(i, f) - g(f)T + \sum_{j \in S} P_{ij}^f(T) \int_T^\infty [E_j^f[r(x(t-T), f)] - g(f)]dt$$

$$= J_T(i, f) - g(f)T + \sum_{j \in S} P_{ij}^f(T) \int_0^\infty [E_j^f[r(x(t), f)] - g(f)]dt$$

$$= J_T(i, f) - g(f)T + E_i^f[h_f(x(T))].$$

Hence, we have shown that

$$J_T(i, f) = g(f)T + h_f(i) - E_i^f[h_f(x(T))] \tag{3.29}$$

for all $f \in \mathbb{F}$, $i \in S$, and $T \geq 0$. The expression (3.29) will be useful below, and also in Sec. 3.7 and in later chapters.

The Poisson equation   Next, we study the characterization of the bias by means of a system of linear equations. Given $f \in \mathbb{F}$, we say that the pair $(g, h) \in \mathbb{R} \times \mathcal{B}_w(S)$ is a solution of the *Poisson equation for f* if

$$g = r(i, f) + \sum_{j \in S} q_{ij}(f)h(j) \quad \forall\, i \in S. \tag{3.30}$$

The proof of the following result uses the fact that [44, Lemma 6.1] for every $u \in \mathcal{B}_w(S)$, $f \in \mathbb{F}$, and $i \in S$

$$\lim_{T \downarrow 0} \frac{E_i^f[u(x(T))] - u(i)}{T} = \sum_{j \in S} q_{ij}(f)u(j). \tag{3.31}$$

This is actually a consequence of (2.15), according to which $Q^f = [q_{ij}(f)]$ can be identified with the infinitesimal generator L^f of $\{x^f(t)\}_{t \geq 0}$.

Proposition 3.14. *Suppose that Assumptions* 3.1 *and* 3.11(a) *hold, and let $f \in \mathbb{F}$ be an irreducible policy. The solutions of the Poisson equation for $f \in \mathbb{F}$ are of the form $(g(f), h_f + z\mathbf{1})$ for any constant $z \in \mathbb{R}$. In particular, $(g(f), h_f)$ is the unique solution of the Poisson equation* (3.30) *for f for which $\mu_f(h) = 0$.*

Proof. First of all, let us prove that $(g(f), h_f)$ is a solution of the Poisson equation (3.30). Given $i \in S$, by (3.29), for every $T > 0$ we have

$$\frac{E_i^f[h_f(x(T))] - h_f(i)}{T} = g(f) - \frac{J_T(i, f)}{T} = g(f) - \frac{E_i^f[\int_0^T r(x(t), f)dt]}{T}.$$

Letting $T \downarrow 0$ in the above equation yields, by (3.31),

$$\sum_{j \in S} q_{ij}(f) h_f(j) = g(f) - r(i, f),$$

which proves that $(g(f), h_f)$ is indeed a solution of the Poisson equation (3.30) for f.

Now, let (g, h) and (g', h') be two solutions in $\mathbb{R} \times \mathcal{B}_w(S)$ of the Poisson equation (3.30). Subtracting both Poisson equations, we obtain

$$g - g' = \sum_{j \in S} q_{ij}(f)(h(j) - h'(j)) = (L^f(h - h'))(i) \quad \forall \, i \in S,$$

where we use the notation in (2.15). This shows that $u := h - h'$ is either subharmonic or superharmonic for f, depending on the sign of $g - g'$; see (2.22). Then, by Proposition 2.6, u is constant (thus showing that h and h' differ by an additive constant) and, therefore, $g = g'$. $\qquad\qquad\square$

Remark 3.15. Suppose, in addition to Assumptions 3.1 and 3.11(a), that $\varphi \in \Phi_s$ is an irreducible randomized stationary policy. We can define the bias $h_\varphi \in \mathcal{B}_w(S)$ of φ as in (3.26), and then the Poisson equation for φ, namely,

$$g(\varphi) = r(i, \varphi) + \sum_{j \in S} q_{ij}(\varphi) h_\varphi(j) \quad \forall \, i \in S$$

as well as Proposition 3.14 still hold.

For discrete-time models, the bias of a policy is given a similar definition (with the integral in (3.26) replaced with a sum). The properties of the discrete-time Poisson equation concerning the existence and uniqueness of its solution are similar to those of the continuous-time Poisson equation studied here. The reader interested in the discrete-time setting can consult [16, 70, 138] and references therein.

Our next result is related to Lemma 3.10 (also, cf. Lemma 3.5(ii)).

Corollary 3.16. *Suppose that Assumptions* 3.1 *and* 3.11(a) *hold, and let* $\varphi \in \Phi_{\mathrm{s}}$ *be an irreducible policy. If there exists a constant* $g \in \mathbb{R}$ *and a function* $h \in \mathcal{B}_w(S)$ *such that*

$$g \geq r(i, \varphi) + \sum_{j \in S} q_{ij}(\varphi)h(j) \quad \forall\, i \in S, \tag{3.32}$$

with strict inequality for some $i_0 \in S$, *then* $g > g(\varphi)$. *Similarly, if*

$$g \leq r(i, \varphi) + \sum_{j \in S} q_{ij}(\varphi)h(j) \quad \forall\, i \in S \tag{3.33}$$

with strict inequality for some $i_0 \in S$, *then* $g < g(\varphi)$.

Proof.   Suppose that (3.32) holds with strict inequality for some $i_0 \in S$. We deduce from Lemma 3.10 and Remark 3.12 that $g \geq g(\varphi)$.

To prove that $g > g(\varphi)$ we proceed by contradiction, and so we suppose that $g = g(\varphi)$. From the Poisson equation for φ (Remark 3.15)

$$g(\varphi) = r(i, \varphi) + \sum_{j \in S} q_{ij}(\varphi)h_\varphi(j) \quad \forall\, i \in S$$

and (3.32) we derive that $h - h_\varphi$ is subharmonic for φ and, therefore, constant; see Proposition 2.6. Consequently, we would have that (3.32) holds with equality for all $i \in S$, which yields a contradiction. This establishes that $g > g(\varphi)$.

The proof for the case (3.33) is similar. $\square$

A continuity theorem   In Theorem 3.9 we proved the continuity on Φ_{s} of the discounted reward function. Now we will prove an analogous result for the long-run expected average reward (or gain) function. We will first show that $f \mapsto g(f)$ is continuous on $\mathbb{F}$ in the topology of componentwise convergence introduced in Sec. 2.2.

Theorem 3.17. *Suppose that the Assumptions 3.1, 3.3, and 3.11 hold, and that, moreover, the sequence $\{f_n\}_{n \geq 1}$ in $\mathbb{F}$ converges to $f \in \mathbb{F}$. Then $g(f_n) \to g(f)$.*

Proof. For every $n \geq 1$, the Poisson equation for f_n is

$$g(f_n) = r(i, f_n) + \sum_{j \in S} q_{ij}(f_n) h_n(j) \quad \forall\, i \in S, \tag{3.34}$$

where $h_n \in \mathcal{B}_w(S)$ is the bias of f_n. Observe now that the sequence $\{g(f_n)\}$ is bounded (see (3.24)), and let n' be any convergent subsequence: $g(f_{n'}) \to \bar{g}$ for some $\bar{g} \in \mathbb{R}$. Since the sequence $\{h_{n'}\}$ is uniformly bounded in $\mathcal{B}_w(S)$ (recall (3.27)), by a diagonal argument there exists $\bar{h} \in \mathcal{B}_w(S)$ and a further subsequence n'' such that

$$h_{n''}(i) \to \bar{h}(i) \quad \forall\, i \in S.$$

By Lemma 3.8(i) and the continuity of $a \mapsto r(i, a)$ in Assumption 3.3(b), we take the limit as $n'' \to \infty$ in (3.34) and we obtain

$$\bar{g} = r(i, f) + \sum_{j \in S} q_{ij}(f) \bar{h}(j) \quad \forall\, i \in S.$$

Therefore, $(\bar{g}, \bar{h})$ is a solution of the Poisson equation for f and, necessarily, $g(f) = \bar{g}$.

This proves that the limit of any convergent subsequence of the bounded sequence $\{g(f_n)\}$ is $g(f)$. Hence, the whole sequence $\{g(f_n)\}$ converges to $g(f)$. $\quad\square$

For randomized stationary policies, we obtain the next result.

Corollary 3.18. *Suppose that Assumptions 3.1, 3.3, and 3.11 hold, where irreducibility in Assumption 3.11(b) is supposed for every $\varphi \in \Phi_s$, and where w-exponential ergodicity in Assumption 3.11(c) is supposed to be uniform on Φ_s. Then (using the notation introduced in Remark 3.12) the mapping*

$$\varphi \mapsto g(\varphi)$$

is continuous on Φ_s with respect to the topology of componentwise weak convergence defined in (2.5).

Proof. The proof makes use of Proposition 3.14 for stationary policies (see Remark 3.15) and it is similar to that of Theorem 3.17. $\quad\square$

The above continuity theorem will be useful when dealing with the vanishing discount approach, which we introduce next.

3.5. The vanishing discount approach to average optimality

The so-called *vanishing discount approach* is a standard technique for dealing with an expected average reward criterion (say, as in (3.23)) by means of the limit, as the discount rate vanishes (that is, $\alpha \downarrow 0$), of the α-discounted rewards. To describe this approach we begin with some general facts.

Under Assumptions 3.1, 3.3, and 3.11, Theorem 3.7 holds for every discount rate $\alpha > 0$ because the constant c in Assumption 3.11(a) is positive.

Fix an arbitrary state $i_0 \in S$ and, for $\alpha > 0$, define the function

$$u_\alpha(i) := V_\alpha^*(i) - V_\alpha^*(i_0) \quad \forall \, i \in S. \tag{3.35}$$

Obviously, by (3.9), u_α is in $\mathcal{B}_w(S)$ for every (fixed) $\alpha > 0$. The following lemma states a stronger fact.

Lemma 3.19. *Under Assumptions* 3.1, 3.3, *and* 3.11,

$$\sup_{\alpha > 0} \|u_\alpha\|_w < \infty.$$

Proof.   For each $\alpha > 0$, let $f_\alpha \in \mathbb{F}$ be an α-discount optimal policy (see Theorem 3.7(iii)). Then, we can express (3.35), for any given $i \in S$, as

$$
\begin{aligned}
u_\alpha(i) &= V_\alpha(i, f_\alpha) - V_\alpha(i_0, f_\alpha) \\
&= \int_0^\infty e^{-\alpha t} E_i^{f_\alpha}[r(x(t), f_\alpha)]dt - \int_0^\infty e^{-\alpha t} E_{i_0}^{f_\alpha}[r(x(t), f_\alpha)]dt \\
&= \int_0^\infty e^{-\alpha t}[E_i^{f_\alpha}[r(x(t), f_\alpha)] - g(f_\alpha)]dt \\
&\quad - \int_0^\infty e^{-\alpha t}[E_{i_0}^{f_\alpha}[r(x(t), f_\alpha)] - g(f_\alpha)]dt.
\end{aligned}
$$

By the uniform exponential ergodicity in Assumption 3.11(c),

$$|E_i^{f_\alpha}[r(x(t), f_\alpha)] - g(f_\alpha)| \le RMe^{-\delta t}w(i)$$

and, therefore,

$$|u_\alpha(i)| \le \frac{RM(w(i) + w(i_0))}{\alpha + \delta} \le \frac{RM(w(i) + w(i_0))}{\delta} \quad \forall \, \alpha > 0, \; i \in S.$$

The stated result easily follows. $\square$

The next result, which can be found in [45, Theorem 4.1], [52, Theorem 7.8], or [53, Theorem 4.1], characterizes the optimal average reward $J^*(i)$ by means of equation (3.36), which also gives necessary and sufficient conditions for the existence of an average reward optimal policy.

Theorem 3.20. *Suppose that Assumptions* 3.1, 3.3, *and* 3.11 *are verified. Then the following statements hold.*

(i) *There exists a solution* $(g^*, h) \in \mathbb{R} \times \mathcal{B}_w(S)$ *to the* <u>*average reward optimality equation*</u> *(AROE)*

$$g^* = \max_{a \in A(i)} \left\{ r(i,a) + \sum_{j \in S} q_{ij}(a) h(j) \right\} \quad \forall\, i \in S. \tag{3.36}$$

(ii) *For every* $i \in S$, $J^*(i) = g^*$, *so* g^* *is the optimal average reward or optimal gain. Hence,* g^* *in* (3.36) *is unique, but* h *is unique up to additive constants.*

(iii) *A deterministic stationary policy* $f \in \mathbb{F}$ *is average reward optimal if and only if* f *attains the maximum in the AROE* (3.36), *that is,*

$$g^* = r(i,f) + \sum_{j \in S} q_{ij}(f) h(j) \quad \forall\, i \in S. \tag{3.37}$$

(iv) *For every* $i \in S$, *let* $A^*(i)$ *be the set of maxima of* (3.36), *that is,*

$$A^*(i) := \left\{ a \in A(i) : g^* = r(i,a) + \sum_{j \in S} q_{ij}(a) h(j) \right\}. \tag{3.38}$$

(By Assumption 3.3, *the sets* $A^*(i)$ *are nonempty and compact for all* $i \in S$.) *An irreducible stationary policy* $\varphi \in \Phi_s$ *is average reward optimal if and only if, for each* $i \in S$, $\varphi(\cdot|i)$ *is supported on* $A^*(i)$, *which means that* $\varphi(A^*(i)|i) = 1$ *for all* $i \in S$.

Proof. Proof of (i). First, we prove the existence of solutions to the AROE (3.36).

Pick an arbitrary state $i_0 \in S$, and let u_α be as in (3.35). It follows from (3.9) that $\alpha V_\alpha^*(i_0)$ is bounded for $\alpha > 0$. So, we choose a sequence $\alpha_n \downarrow 0$ of discount rates such that, for some constant $g^* \in \mathbb{R}$,

$$\lim_{n \to \infty} \alpha_n V_{\alpha_n}^*(i_0) = g^*.$$

Moreover, by Lemma 3.19 and using a diagonal argument, there exists a subsequence $\{\alpha_{n'}\}$ of $\{\alpha_n\}$ such that

$$\lim_{n'} u_{\alpha_{n'}}(i) = h(i) \quad \forall\, i \in S$$

for some function h in $\mathcal{B}_w(S)$. For ease of notation, the subsequence $\{\alpha_{n'}\}$ will be simply denoted by $\{\alpha_n\}$.

The DROE (3.20) can be equivalently written as

$$\alpha V_\alpha^*(i_0) + \alpha u_\alpha(i) = \max_{a \in A(i)} \left\{ r(i,a) + \sum_{j \in S} q_{ij}(a) u_\alpha(j) \right\} \quad \forall\, i \in S.$$

In particular,

$$\alpha_n V_{\alpha_n}^*(i_0) + \alpha_n u_{\alpha_n}(i) \geq r(i,a) + \sum_{j \in S} q_{ij}(a) u_{\alpha_n}(j) \quad \forall\, (i,a) \in K.$$

By Lemmas 3.8(i) and 3.19, we can take the limit as $n \to \infty$ in the above inequality to obtain $g^* \geq r(i,a) + \sum_{j \in S} q_{ij}(a) h(j)$ for every $(i,a) \in K$, which in turn gives

$$g^* \geq \max_{a \in A(i)} \left\{ r(i,a) + \sum_{j \in S} q_{ij}(a) h(j) \right\} \quad \forall\, i \in S. \tag{3.39}$$

Let $f_\alpha \in \mathbb{F}$ attain the maximum in the DROE (3.20) for the discount rate α (see Theorem 3.7(iii)), that is,

$$\alpha V_\alpha^*(i_0) + \alpha u_\alpha(i) = r(i, f_\alpha) + \sum_{j \in S} q_{ij}(f_\alpha) u_\alpha(j) \quad \forall\, i \in S. \tag{3.40}$$

By compactness of $\mathbb{F}$ (which is derived from Assumption 3.3(a)), we deduce that there exists $f \in \mathbb{F}$ and a subsequence $\{\alpha_{n'}\}$ of $\{\alpha_n\}$ such that

$$\lim_{n'} f_{\alpha_{n'}}(i) = f(i) \quad \forall\, i \in S.$$

Taking the limit as $n' \to \infty$ in (3.40) yields, by Lemma 3.8(i),

$$g^* = r(i, f) + \sum_{j \in S} q_{ij}(f) h(j) \quad \forall\, i \in S. \tag{3.41}$$

In particular, $(g^*, h) \in \mathbb{R} \times \mathcal{B}_w(S)$ is a solution of the Poisson equation for f and, by Proposition 3.14, we have

$$g^* = g(f). \tag{3.42}$$

Hence, by (3.39) and (3.41), we conclude that $(g^*, h) \in \mathbb{R} \times \mathcal{B}_w(S)$ is a solution of the AROE.

Proof of (ii). We next show that $J^*(i) = g^*$ for every $i \in S$. By the AROE (3.36),

$$g^* \geq r(i, a) + \sum_{j \in S} q_{ij}(a)h(j) \quad \forall \, (i, a) \in K.$$

Given $\varphi \in \Phi$, $t \geq 0$, and $i \in S$, by integration of the above inequality with respect to $\varphi_t(da|i)$, we have

$$g^* \geq r(t, i, \varphi) + \sum_{j \in S} q_{ij}(t, \varphi)h(j).$$

By Lemma 3.10, $J(i, \varphi) \leq g^*$ for each $\varphi \in \Phi$ and $i \in S$, which combined with (3.42) shows that $J^*(i) = g^*$ for all $i \in S$.

To complete the proof of (ii), let us now show that if $(g^*, h_1) \in \mathbb{R} \times \mathcal{B}_w(S)$ and $(g^*, h_2) \in \mathbb{R} \times \mathcal{B}_w(S)$ are two solutions of the AROE, then h_1 and h_2 are equal up to an additive constant.

Let $f \in \mathbb{F}$ attain the maximum in the AROE for (g^*, h_1), that is,

$$g^* = r(i, f) + \sum_{j \in S} q_{ij}(f)h_1(j) \quad \forall \, i \in S.$$

We then have, by the AROE for (g^*, h_2),

$$g^* \geq r(i, f) + \sum_{j \in S} q_{ij}(f)h_2(j) \quad \forall \, i \in S.$$

Subtracting both expressions and defining $u := h_1 - h_2$, we obtain

$$(L^f u)(i) = \sum_{j \in S} q_{ij}(f)u(j) \geq 0 \quad \forall \, i \in S.$$

This shows that the function $u \in \mathcal{B}_w(S)$ is superharmonic for $f \in \mathbb{F}$, and so, u is constant (by Proposition 2.6). This completes the proof of part (ii).

Proof of (iii). It is clear from Proposition 3.14 that a policy $f \in \mathbb{F}$ that attains the maximum in the AROE, that is, a policy such that (3.37) holds, is average reward optimal.

Conversely, let $f \in \mathbb{F}$ be an average reward optimal policy, that is, such that $g(f) = g^*$. From the AROE (3.36) we deduce that

$$g^* \geq r(i, f) + \sum_{j \in S} q_{ij}(f)h(j) \quad \forall \, i \in S.$$

If the above inequality is strict for some $i_0 \in S$ then, as a consequence of Corollary 3.16, we have $g(f) < g^*$, which is not possible. Hence, f attains the maximum in the AROE.

Proof of (iv). If the irreducible stationary policy $\varphi \in \Phi_s$ is such that $\varphi(\cdot|i)$ is supported on $A^*(i)$ for each $i \in S$, then

$$g^* = r(i, \varphi) + \sum_{j \in S} q_{ij}(\varphi)h(j) \quad \forall\, i \in S,$$

where $(g^*, h) \in \mathbb{R} \times \mathcal{B}_w(S)$ is a solution of the AROE. From Remark 3.15, it follows that $g^* = g(\varphi)$. Consequently, φ is an average reward optimal policy.

Conversely, suppose that the irreducible policy $\varphi \in \Phi_s$ is average reward optimal but, for some $i_0 \in S$, we have that $\varphi(\cdot|i_0)$ is not supported on $A^*(i_0)$. This yields

$$g^* \geq r(i, \varphi) + \sum_{j \in S} q_{ij}(\varphi)h(j) \quad \forall\, i \in S,$$

with strict inequality at i_0. Hence, by Corollary 3.16, we obtain $g(\varphi) < g^*$, which gives a contradiction. Hence, for each $i \in S$, $\varphi(\cdot|i)$ is necessarily supported on the set $A^*(i)$. □

Remark 3.21. Observe that for a deterministic policy $f \in \mathbb{F} \subset \Phi_s$, part (iv) in Theorem 3.20 reduces to part (iii). Also note that, since the solution h of the AROE is unique up to additive constants and $L^f u \equiv 0$ if u is constant, the sets $A^*(i)$ do not depend on the particular solution h of the AROE.

In our next remark we introduce some notation and terminology.

Remark 3.22.

(a) The family of average reward optimal deterministic stationary policies is denoted by $\mathbb{F}_{ao}$. We say that $f \in \mathbb{F}$ is *canonical*, which will be denoted by $f \in \mathbb{F}_{ca}$, if f attains the maximum in the AROE (3.36), or, equivalently, $f(i) \in A^*(i)$ for every $i \in S$.

(b) Theorem 3.20(iii) shows that, in fact, $\mathbb{F}_{ao} = \mathbb{F}_{ca}$. In general, this is *not* the case, so one has $\mathbb{F}_{ca} \subseteq \mathbb{F}_{ao}$ and the inclusion can be strict (see [54] for an example). Our irreducibility hypothesis, however, ensures that $\mathbb{F}_{ao} = \mathbb{F}_{ca}$.

(c) Finally, we say that $(g, h, f) \in \mathbb{R} \times \mathcal{B}_w(S) \times \mathbb{F}$ is a *canonical triplet* if it satisfies (3.36) and (3.37); hence, in particular, $g = g^*$, the function h is unique up to an additive constant, and f is canonical.

At the end of this section (Remark 3.24) we make some bibliographic remarks about the results in Theorem 3.20 above.

As a direct consequence of Proposition 3.14 and Theorem 3.20, we obtain the following result, which we state without proof.

Corollary 3.23. *Let* $(g^*, h) \in \mathbb{R} \times \mathcal{B}_w(S)$ *be a solution of the AROE* (3.36). *If* $f \in \mathbb{F}$ *is canonical, then its bias* h_f *and* h *differ by a constant.*

Comments on the vanishing discount approach As was already noted, the vanishing discount approach is a standard procedure to study average reward optimality. The idea is to analyze α-discounted control problems and then let $\alpha \downarrow 0$. This is in fact the approach we used to derive the AROE in the proof of Theorem 3.20(i).

This approach hinges on a classical result stating that the limit of discount optimal policies (as the discount rate vanishes) is average reward optimal. Under our standing assumptions, this result indeed holds. Its proof is postponed to Theorem 6.10, however, because after introducing a suitable background we will be able to give a simple proof of the result.

The vanishing discount approach is related to the so-called *Abelian theorems*. We say that a sequence $\{b_n\}$ verifies an Abelian theorem if

$$\lim_{z \uparrow 1} (1 - z) \sum_{n=0}^{\infty} b_n z^n = \lim_{n \to \infty} \frac{b_1 + \ldots + b_n}{n}.$$

The continuous-time counterpart is, for a function r satisfying suitable integrability assumptions,

$$\lim_{\alpha \downarrow 0} \alpha \int_0^{\infty} e^{-\alpha t} r(t) dt = \lim_{T \to \infty} \frac{1}{T} \int_0^T r(t) dt. \tag{3.43}$$

In our context, this suggests the following result. Given a policy $f \in \mathbb{F}$,

$$\lim_{\alpha \downarrow 0} \alpha V_\alpha(i, f) = g(f) \quad \forall \, i \in S.$$

This result will be proved in Lemma 6.8 below. In fact, the latter limit also holds when taking the supremum over the family of all policies:

$$\lim_{\alpha \downarrow 0} \alpha V_\alpha^*(i) = g^* \quad \forall \, i \in S,$$

and this will be established in Theorem 6.10. Alternatively, (3.43) follows from standard facts about the asymptotic behavior of semigroups. See, for instance, [22, Theorem 5.1] or [64, pp. 24–27].

We conclude this section with some references on average reward CMCs.

Remark 3.24. The average reward control problem for finite state and action spaces was analyzed by Howard [77] and Miller [112], and later by Lembersky [101]. For countable (nonfinite) state spaces, the case of bounded transition and reward rates has been investigated by Kakumanu [87]; see also [88]. For unbounded reward rates but bounded transition rates, we can cite Haviv and Puterman [62], Lewis and Puterman [102], and Sennott [147]. The case of unbounded reward and transition rates is treated by Guo and Zhu [58], Guo and Cao [43], and Guo and Hernández-Lerma [45]. General results for average reward Markov control processes have been obtained by Doshi [29] and Hernández-Lerma [64]. Finally, let us mention that the uniformization technique can be used as well for average reward CMCs with bounded reward and transition rates.

In Theorem 3.20 we have used the AROE to derive the existence of average optimal policies. This is indeed the usual approach. There exist other possibilities such as the average reward optimality inequality studied by Guo and Liu [54], which can be used when the AROE does not hold, or the convex analytic approach used by Piunovskiy [122, 123], which is also useful for multicriteria and constrained control problems.

3.6. Pathwise average optimality

The issue of sample path (or pathwise) optimality for average reward control problems has been extensively studied. For discrete-time models we may cite, for instance, Hernández-Lerma and Lasserre [70], Hernández-Lerma, Vega-Amaya, and Carrasco [73], Lasserre [96], Mendoza-Pérez and Hernández-Lerma [108], and Zhu and Guo [173]. The continuous-time case has been studied by Guo and Cao [43], while, for instance, Haviv [61] and Ross and Varadarajan [141, 142] have analyzed the issue of pathwise constraints (see Chapter 8). Here, we follow [131].

In Theorem 3.20 above we have shown the existence of optimal policies for the long-run *expected* average reward criterion. Now, under additional hypotheses, we show the existence of optimal policies for the long-run *pathwise* average reward criterion. In what follows, we suppose that Assumptions 3.1, 3.3, and 3.11 are satisfied.

We begin with some definitions. Given a control policy $\varphi \in \Phi$, an initial state $i \in S$, and $T \geq 0$, we define the total pathwise reward on $[0, T]$ as the random variable

$$J_T^0(i, \varphi) := \int_0^T r(t, x(t), \varphi)dt. \tag{3.44}$$

Observe that $E_i^\varphi[J_T^0(i, \varphi)] = J_T(i, \varphi)$ (recall (3.2)). Then we define the long-run pathwise average reward (or pathwise average reward, for short) as

$$J^0(i, \varphi) := \liminf_{T \to \infty} \frac{1}{T} J_T^0(i, \varphi). \tag{3.45}$$

Our next lemma shows that the pathwise average reward of an irreducible stationary policy equals its gain $g(\varphi)$ in Remark 3.12 with probability one. We omit its proof, which is a direct consequence of the *strong law of large numbers for Markov processes*; see [15].

Lemma 3.25. *Suppose that Assumptions* 3.1(c) *and* 3.11(a) *hold, and suppose that* $\varphi \in \Phi_s$ *is irreducible. Then, for each* $i \in S$,

$$J^0(i, \varphi) = \lim_{T \to \infty} \frac{1}{T} \int_0^T r(x(t), \varphi)dt = g(\varphi) \quad P_i^\varphi\text{-a.s.}$$

Under the conditions of Lemma 3.25, we say that a policy $\varphi^* \in \Phi_s$ is *pathwise average optimal* if, for each $\varphi \in \Phi$ and $i \in S$,

$$J^0(i, \varphi) \leq g(\varphi^*) \quad P_i^\varphi\text{-a.s.}$$

In our next result we use the notation $L^{t,\varphi}$ defined in (2.14).

Lemma 3.26. *Suppose that Assumptions* 3.1(a)–(b) *hold, where the constants* c *and* c' *are positive. Given a control policy* $\varphi \in \Phi$, *an initial state* $i \in S$, *and an arbitrary* $h \in B_w(S)$,

$$\frac{1}{n} \int_0^n (L^{t,\varphi}h)(x(t))dt \xrightarrow{p} 0 \quad as \ n \to \infty$$

with respect to P_i^φ (convergence in probability was defined in Sec. 1.4).

Proof. Under our hypotheses,

$$Z_t^\varphi := h(x(t)) - h(x(0)) - \int_0^t (L^{s,\varphi}h)(x(s))ds,$$

for $t \geq 0$, is a P_i^φ-martingale. Now consider the discrete-time martingale difference

$$Z_{n+1}^\varphi - Z_n^\varphi = h(x(n+1)) - h(x(n)) - \int_n^{n+1} (L^{s,\varphi} h)(x(s))ds$$

for $n \geq 0$. Proceeding as in (2.18), we obtain

$$|(L^{s,\varphi} h)(x(s))| \leq R\|h\|_w (w + w')(x(s)) \quad \forall \, s \geq 0,$$

for some constant R. Hence, by (2.13) and (2.16), we deduce that

$$\sup_{n \geq 0} E_i^\varphi |Z_{n+1}^\varphi - Z_n^\varphi| < \infty.$$

By [59, Theorem 2.18], this shows that Z_n^φ/n converges to 0 in probability. Similarly, we can prove that $h(x(n))/n$ also converges in probability to 0. These facts yield the stated result. $\qquad\square$

Our main result in this section is the following.

Theorem 3.27. *Suppose that Assumptions* 3.1, 3.3, *and* 3.11 *hold, where the constant c' in Assumption* 3.1(b) *is positive. Then a stationary policy $f \in \mathbb{F}$ is pathwise average optimal if and only if it is gain optimal.*

Proof. As a consequence of Lemma 3.25, if a policy $f \in \mathbb{F}$ is pathwise average optimal then it is necessarily gain optimal.

Conversely, let us prove that a gain optimal policy $f \in \mathbb{F}$ is pathwise average optimal. Fix $\varphi \in \Phi$ and $i \in S$. Let $(g^*, h) \in \mathbb{R} \times \mathcal{B}_w(S)$ be a solution of the AROE (3.36). We have, for every $s \geq 0$,

$$r(s, x(s), \varphi) + (L^{s,\varphi} h)(x(s)) \leq g^*,$$

where $g^* = g(f)$ is the optimal gain, and thus

$$\frac{1}{T} \int_0^T r(s, x(s), \varphi)ds + \frac{1}{T} \int_0^T (L^{s,\varphi} h)(x(s))ds \leq g^* \quad \forall \, T > 0. \quad (3.46)$$

By the convergence in probability in Lemma 3.26, there exists a sequence $\{T_n\}$ such that $T_n \to \infty$ and

$$\lim_{n \to \infty} \frac{1}{T_n} \int_0^{T_n} (L^{s,\varphi} h)(x(s))ds = 0 \quad P_i^\varphi\text{-a.s.} \quad (3.47)$$

Therefore,

$$J^0(i, \varphi) = \liminf_{T \to \infty} \frac{1}{T} \int_0^T r(s, x(s), \varphi) ds$$

$$\leq \liminf_{n \to \infty} \frac{1}{T_n} \int_0^{T_n} r(s, x(s), \varphi) ds$$

$$\leq g^* \quad P_i^\varphi\text{-a.s.} \quad [\text{by } (3.46) \text{ and } (3.47)].$$

This completes the proof. ☐

If the conditions in Assumption 3.11 hold for Φ_s, then Theorem 3.27 remains valid if $\mathbb{F}$ is replaced with Φ_s. Indeed, it is evident that a gain optimal policy in Φ_s is also pathwise average optimal, whereas the converse is trivial, by Lemma 3.25.

3.7. Canonical triplets and finite horizon control problems

We recall from Remark 3.22(c) that $(g, h, f) \in \mathbb{R} \times \mathcal{B}_w(S) \times \mathbb{F}$ is said to be a canonical triplet if, for every $i \in S$,

$$g = \max_{a \in A(i)} \left\{ r(i, a) + \sum_{j \in S} q_{ij}(a) h(j) \right\}$$

$$= r(i, f) + \sum_{j \in S} q_{ij}(f) h(j).$$

In this case, Theorem 3.20 gives that g equals the optimal gain g^*, that the policy f is canonical, and that h is unique up to additive constants.

We know from Theorem 3.4 that, in general, an optimal policy for a finite horizon control problem on $[0, T]$ is deterministic and nonstationary. In our next result we show that, if the final payoff function is taken from a canonical triplet, then *stationary* deterministic optimal policies exist and, besides, the optimal total expected reward is an affine function of T.

In the following theorem we use the notation in Sec. 3.2. In particular, see (3.2) and (3.3).

Theorem 3.28. *Suppose that Assumptions 3.1, 3.3, and 3.11 hold. The triplet $(g, h, f) \in \mathbb{R} \times \mathcal{B}_w(S) \times \mathbb{F}$ is canonical if and only if for every $T \geq 0$, $0 \leq s \leq T$, and $i \in S$*

$$J_T^*(s, i, h) = (T - s)g + h(i) = J_T(s, i, f, h). \tag{3.48}$$

In particular, f is optimal for each finite horizon control problem with final payoff h.

Proof.   *Proof of sufficiency.* Pick an arbitrary $f' \in \mathbb{F}$. Letting $s = 0$ in (3.48) we obtain, for every $i \in S$,

$$J_T(i, f') + E_i^{f'}[h(x(T))] = J_T(i, f', h) \leq Tg + h(i)$$
$$= J_T(i, f, h) = J_T(i, f) + E_i^f[h(x(T))].$$

Dividing the inequality

$$J_T(i, f') + E_i^{f'}[h(x(T))] \leq J_T(i, f) + E_i^f[h(x(T))]$$

by T and letting $T \to \infty$ yields

$$g(f') \leq g = g(f),$$

where we have used the bound (2.13) for $h \in \mathcal{B}_w(S)$, and (3.25). Since f' is arbitrary, it follows that $g = g^*$ and that f is average reward optimal (and, hence, canonical; see Theorem 3.20(iii)). Also, from (3.29) and the equation

$$Tg + h(i) = J_T(i, f) + E_i^f[h(x(T))],$$

we deduce that

$$h(i) - E_i^f[h(x(T))] = h_f(i) - E_i^f[h_f(x(T))] \quad \forall \, i \in S,$$

where $h_f \in \mathcal{B}_w(S)$ is the bias of f. Letting $T \to \infty$, by ergodicity we obtain that

$$h - \mu_f(h) = h_f.$$

Hence, h and h_f differ by a constant and, by Corollary 3.23, it follows that (g, h) is a solution of the AROE (3.36), thus showing that (g, h, f) is a canonical triplet.

   Proof of necessity. If (g, h, f) is a canonical triplet, then the AROE gives, for every $(i, a) \in K$,

$$\sum_{j \in S} q_{ij}(a)h(j) \leq g - r(i, a).$$

Fix an arbitrary $\varphi \in \Phi$ and $t \geq 0$. Integrating the above inequality with respect to $\varphi_t(da|i)$ yields (using the notations of (2.14) and (3.1))

$$(L^{t,\varphi}h)(i) \leq g - r(t,i,\varphi) \quad \forall\, i \in S, \ t \geq 0. \tag{3.49}$$

Hence, for every $i \in S$, $0 \leq s \leq T$, and $\varphi \in \Phi$,

$$J_T(s,i,\varphi,h) = J_T(s,i,\varphi) + E^{\varphi}_{s,i}[h(x(T))]$$

$$= J_T(s,i,\varphi) + h(i) + E^{\varphi}_{s,i}\left[\int_s^T (L^{t,\varphi}h)(x(t))dt\right]$$

[by Dynkin's formula (Proposition 2.3)]

$$\leq J_T(s,i,\varphi) + h(i) + g(T-s)$$

$$- E^{\varphi}_{s,i}\left[\int_s^T r(t,x(t),\varphi)dt\right] \tag{3.50}$$

$$= g(T-s) + h(i),$$

where (3.50) is derived from (3.49).

Taking $\varphi = f$, the inequalities (3.49) and (3.50) become equalities because f is canonical. Thus

$$J_T(s,i,\varphi,h) \leq g(T-s) + h(i) = J_T(s,i,f,h)$$

for every $0 \leq s \leq T$, $i \in S$, and $\varphi \in \Phi$. The desired result now follows. $\square$

It is worth noting that, for fixed T, the function $v(s,i) := g(T-s)+h(i)$ (see (3.48)) is a solution of the finite horizon optimality equation (3.4) and (3.5). In this case, (3.4) is precisely the AROE.

3.8. Conclusions

In this chapter, we have presented a thorough analysis of the discounted reward and the average reward optimality criteria. We have based our approach on the dynamic programming equations: the DROE and the AROE. We have also studied pathwise average optimality, and we have given interesting characterizations and interpretations of canonical triplets, which will be useful when dealing with bias optimality in Chapter 5. We have mainly focused on the unbounded reward and unbounded transition rates case, so that the uniformization technique cannot be used in our context.

With respect to the computation of the optimal rewards and policies for the optimality criteria studied in this chapter, let us mention that for *discrete-time* CMCs there exist two general approximation algorithms, *value iteration* and *policy iteration*. For *continuous-time* CMCs, however, there is no computationally efficient value iteration algorithm. In contrast, the policy iteration algorithm and its various modifications turn out to be a powerful tool to solve numerically discounted and average reward CMCs. The policy iteration algorithm is studied in Chapter 4, while applications of practical interest of this algorithm are shown in Chapter 9.

Let us now give a motivation for the so-called refined or advanced optimality criteria that will be studied from Chapter 5 onwards. For a control policy $\varphi \in \Phi$ and an initial state $i \in S$, the total expected α-discounted reward and the long-run expected average reward are defined as

$$V_\alpha(i, \varphi) = E_i^\varphi \left[\int_0^\infty e^{-\alpha t} r(t, x(t), \varphi) dt \right]$$

and

$$J(i, \varphi) = \liminf_{T \to \infty} \frac{1}{T} E_i^\varphi \left[\int_0^T r(t, x(t), \varphi) dt \right].$$

It is clear that the "early returns" are the most relevant when using the discounted optimality criterion. In contrast, the long-run average reward depends only on the asymptotic behavior of the process $\{x^\varphi(t)\}$, and it is not affected by the returns earned on any finite time interval. Therefore, it is natural to question whether there exists an optimality criterion for which the "early returns" and the "asymptotic returns" are both relevant. Obviously, the total expected reward optimality criterion

$$J_\infty(i, \varphi) := E_i^\varphi \left[\int_0^\infty r(t, x(t), \varphi) dt \right]$$

satisfies this requirement, although $J_\infty(i, \varphi)$ will be infinite except for some particular CMCs (those, for instance, which are absorbed by a recurrent class of states with zero reward rate). Note, however, that, informally, we have

$$J_\infty(i, \varphi) = \lim_{T \to \infty} E_i^\varphi \left[\int_0^T r(t, x(t), \varphi) dt \right]$$

$$= \lim_{\alpha \downarrow 0} E_i^\varphi \left[\int_0^\infty e^{-\alpha t} r(t, x(t), \varphi) dt \right].$$

Hence, intuitively, there are two ways of defining an intermediate optimality criterion:

(i) either by maximizing the growth of the finite horizon total expected rewards $J_T(i, \varphi)$ as $T \to \infty$ (this corresponds precisely to the overtaking and bias optimality approach; see Chapter 5);

(ii) or by finding policies that are α-discount optimal as the discount rate α vanishes, i.e., $\alpha \downarrow 0$ (this leads to the sensitive discount optimality criteria in Chapter 6, and the Blackwell optimality criterion in Chapter 7).

Chapter 4

Policy Iteration and Approximation Theorems

4.1. Introduction

The *policy iteration algorithm* is used to find an optimal policy in a CMC model. Loosely speaking, this algorithm starts from an arbitrary initial policy, say $f_0 \in \mathbb{F}$, and then determines a sequence of policies $\{f_n\}_{n \geq 0}$ in $\mathbb{F}$ with increasing expected discounted or average reward. For instance, when dealing with the average reward optimality criterion, the policy iteration algorithm yields a sequence of policies with increasing gains:

$$g(f_0) < g(f_1) < g(f_2) \dots.$$

If the state and action spaces are *finite* then, under mild conditions, this algorithm finds an optimal policy in a finite number of steps, that is,

$$g(f_0) < g(f_1) < \cdots < g(f_{N-1}) < g(f_N) = g^*$$

for some $N \geq 0$. Otherwise, in the case of *nonfinite* state and action spaces, under suitable assumptions this algorithm converges to the optimal gain, that is, $g(f_n) \uparrow g^*$ (a precise statement is given in Sec. 4.2 below).

The policy iteration algorithm was introduced by Howard [77] and Bellman [11], and it has been extensively studied for *discrete-time* CMCs. For instance, in case that the state and action spaces are finite, this algorithm is used to find discounted reward, average reward, n-discount (see Chapter 6), and Blackwell (see Chapter 7) optimal policies. These results can be found in Puterman's book [138]. For general discrete-time control models, the convergence of the policy iteration algorithm has been investigated by Hernández-Lerma and Lasserre [68–70]. It is also worth mentioning that policy iteration is related to Newton's method for finding the zeroes of a function [138], and also to the simplex algorithm in the linear programming formulation of a control problem [163].

On the other hand, the policy iteration algorithm has not been thoroughly analyzed for *continuous-time* models. Prieto-Rumeau and Hernández-Lerma studied the policy iteration algorithm for *denumerable state* CMCs under the discounted reward optimality criterion [134], while Guo and Hernández-Lerma studied this algorithm for average reward CMCs; see [45]. The state space being denumerable, however, finite convergence of the algorithm is not guaranteed: the optimal discounted reward and gain, and the optimal policies are obtained "in the limit." Let us also mention that Fleming [36] proposed a policy iteration algorithm for *controlled diffusions*.

In Sec. 4.2 of this chapter, we deal with the policy iteration algorithm for discounted and average reward denumerable state CMCs. Our analysis of the algorithm is borrowed from [45, 53, 134].

The second topic analyzed in this chapter (see Secs 4.3 and 4.4) is the approximation of continuous-time CMCs under the discounted and average reward optimality criteria. This subject is concerned with a so-called original CMC model $\mathcal{M}$ whose optimal (discounted or average) reward and optimal policies we want to approximate. To this end, we have at hand a sequence of control models $\{\mathcal{M}_n\}_{n \geq 1}$ that are approximations, in a suitable defined sense, of $\mathcal{M}$. The approximation theorems give conditions under which the optimal rewards and policies of $\{\mathcal{M}_n\}_{n \geq 1}$ converge to those of $\mathcal{M}$. We base our approach on the results by Prieto-Rumeau and Hernández-Lerma [135] for the discounted reward optimality criterion, and by Prieto-Rumeau and Lorenzo [136] for average reward problems. Similar approximation issues have been analyzed in, e.g., [63, Chapter 6] for discrete-time discounted control problems, and in [3, 5, 158] for discrete-time control problems with constraints.

A case of particular interest (from a computational and practical point of view) is when the state space of the original control model $\mathcal{M}$ is denumerable and the approximating control models $\mathcal{M}_n$ have finite state spaces (so that they can be explicitly solved by using the policy iteration algorithm). Based on [135, 136], finite state approximations are studied also in Secs 4.3 and 4.4. We note that in Chapter 9 we apply these approximation results to controlled population models, and we provide some numerical computations.

Now we make some bibliographic remarks about approximation theorems. They are particularly relevant because, as can be seen in the books by Bertsekas [13], Puterman [138], or Sennott [147], there are several approaches to show the existence of optimal policies, but it is not clear at all how to *compute* or *approximate* them (and their expected rewards). More precisely, in [13, Chapter 5] and [147], continuous-time CMCs with bounded reward and transition rates are analyzed by using the uniformization technique, which is not applicable in our case because we deal with unbounded reward and transition rates.

We conclude this chapter with some final comments in Sec. 4.5.

4.2. The policy iteration algorithm

4.2.1. *Discounted reward problems*

We consider a control model $\mathcal{M}$ as in Sec. 2.2, and we suppose that Assumptions 3.1 and 3.3 are verified. In addition, we consider a discount rate $\alpha > 0$ such that $\alpha + c > 0$, where the constant c is taken from Assumption 3.1(a).

Following [134], we define the policy iteration algorithm.

Step I. Choose an arbitrary policy $f_0 \in \mathbb{F}$ and let $n = 0$.

Step II. Determine the total expected discounted reward $V_\alpha(\cdot, f_n)$ of the policy f_n as the unique solution in $\mathcal{B}_w(S)$ of

$$\alpha V_\alpha(i, f_n) = r(i, f_n) + \sum_{j \in S} q_{ij}(f_n) V_\alpha(j, f_n) \quad \forall \, i \in S;$$

recall Lemma 3.6.

Step III. Find $f_{n+1} \in \mathbb{F}$ such that, for each $i \in S$,

$$r(i, f_{n+1}) + \sum_{j \in S} q_{ij}(f_{n+1}) V_\alpha(j, f_n)$$

$$= \max_{a \in A(i)} \left\{ r(i, a) + \sum_{j \in S} q_{ij}(a) V_\alpha(j, f_n) \right\},$$

letting $f_{n+1}(i) = f_n(i)$ if possible.

If $f_{n+1} = f_n$, then stop the algorithm because f_n is an α-discount optimal policy. Otherwise, replace n with $n + 1$ and return to Step II.

In our next result we prove the convergence of the policy iteration algorithm. At this point we recall that, by Assumption 3.3(a), the set $\mathbb{F}$ of deterministic stationary policies is compact with the topology of componentwise convergence introduced in Sec. 2.2. In particular, the sequence

$\{f_n\}_{n\geq 0}$ has limit policies, meaning that there exist a subsequence $\{n'\}$ and a policy $f \in \mathbb{F}$ such that $f_{n'} \to f$.

Theorem 4.1. *Under Assumptions* 3.1 *and* 3.3, *and provided that the discount rate* $\alpha > 0$ *verifies that* $\alpha + c > 0$, *with* c *as in Assumption* 3.1(a), *one of the following conditions* (i) *or* (ii) *holds.*

(i) *For some* $n \geq 0$ *we have* $f_{n+1} = f_n$. *In this case,* f_n *is a discount optimal policy and* $V_\alpha(\cdot, f_n) = V_\alpha^*$.

(ii) *For every* $n \geq 0$, *we have* $f_{n+1} \neq f_n$. *Then,*
$$\lim_{n \to \infty} V_\alpha(i, f_n) = V_\alpha^*(i) \quad \forall\, i \in S.$$

In addition, every limit policy of $\{f_n\}_{n\geq 0}$ *is discount optimal (and such a limit policy indeed exists because* $\mathbb{F}$ *is compact).*

Proof.   Suppose first that for some $n \geq 0$ we have $f_{n+1} = f_n$. Then, by Step III of the algorithm and as a consequence of Lemma 3.6,

$$\max_{a \in A(i)} \left\{ r(i,a) + \sum_{j \in S} q_{ij}(a) V_\alpha(j, f_n) \right\} = r(i, f_n) + \sum_{j \in S} q_{ij}(f_n) V_\alpha(j, f_n)$$
$$= \alpha V_\alpha(i, f_n),$$

for all $i \in S$. It follows that $V_\alpha(\cdot, f_n)$ is a solution of the DROE. Theorem 3.7 gives that $V_\alpha(\cdot, f_n)$ equals V_α^*, the optimal discounted reward, and, moreover, that f_n is a discount optimal policy.

So, in what follows, we will suppose that $f_{n+1} \neq f_n$ for all $n \geq 0$. First of all, we will show that the sequence $\{V_\alpha(\cdot, f_n)\}_{n\geq 0}$ is monotone. To this end, note that for all $i \in S$

$$r(i, f_{n+1}) + \sum_{j \in S} q_{ij}(f_{n+1}) V_\alpha(j, f_n) = \max_{a \in A(i)} \left\{ r(i,a) + \sum_{j \in S} q_{ij}(a) V_\alpha(j, f_n) \right\}$$
$$\geq r(i, f_n) + \sum_{j \in S} q_{ij}(f_n) V_\alpha(j, f_n) \quad (4.1)$$
$$= \alpha V_\alpha(i, f_n).$$

We deduce from Lemma 3.5(i) that $V_\alpha(\cdot, f_{n+1}) \geq V_\alpha(\cdot, f_n)$. Therefore, by the bound (3.7), we conclude that the sequence $\{V_\alpha(\cdot, f_n)\}_{n\geq 0}$ converges monotonically to some $u \in \mathcal{B}_w(S)$ with $u \leq V_\alpha^*$.

By Step III of the policy iteration algorithm, for all $n \geq 0$ and $(i,a) \in K$ we have

$$r(i, f_{n+1}) + \sum_{j \in S} q_{ij}(f_{n+1}) V_\alpha(j, f_n) \geq r(i,a) + \sum_{j \in S} q_{ij}(a) V_\alpha(j, f_n). \quad (4.2)$$

On the other hand, $r(i, f_{n+1}) + \sum_{j \in S} q_{ij}(f_{n+1}) V_\alpha(j, f_n)$ equals

$$\alpha V_\alpha(i, f_{n+1}) + \sum_{j \in S} q_{ij}(f_{n+1})(V_\alpha(j, f_n) - V_\alpha(j, f_{n+1})).$$

Therefore, (4.2) can be rewritten as

$$\alpha V_\alpha(i, f_{n+1}) + \sum_{j \in S} q_{ij}(f_{n+1})(V_\alpha(j, f_n) - V_\alpha(j, f_{n+1}))$$

$$\geq r(i, a) + \sum_{j \in S} q_{ij}(a) V_\alpha(j, f_n) \tag{4.3}$$

for all $n \geq 0$ and $(i, a) \in K$. By (3.7), note that the sequence $\{V_\alpha(\cdot, f_n) - V_\alpha(\cdot, f_{n+1})\}_{n \geq 0}$ is bounded in $\mathcal{B}_w(S)$, and, moreover, it converges pointwise to 0. Hence, by parts (ii) and (i) of Lemma 3.8, respectively, taking the limit as $n \to \infty$, we have

$$\sum_{j \in S} q_{ij}(f_{n+1})(V_\alpha(j, f_n) - V_\alpha(j, f_{n+1})) \to 0$$

and

$$\sum_{j \in S} q_{ij}(a) V_\alpha(j, f_n) \to \sum_{j \in S} q_{ij}(a) u(j).$$

It then follows from (4.3) that

$$\alpha u(i) \geq r(i, a) + \sum_{j \in S} q_{ij}(a) u(j) \quad \forall \, (i, a) \in K. \tag{4.4}$$

By Lemma 3.5(i) (recall also the proof of Theorem 3.7(ii)) we have, thus, shown that $u \geq V_\alpha^*$. Hence, we conclude that $u = V_\alpha^*$.

Suppose now that, for some subsequence $\{n'\}$, the policies $f_{n'}$ converge pointwise to some $f \in \mathbb{F}$. By the continuity result in Theorem 3.9 it follows that, for every $i \in S$,

$$V_\alpha(i, f) = \lim_{n'} V_\alpha(i, f_{n'}) = V_\alpha^*(i),$$

which shows that f is a discount optimal policy. $\qquad \square$

Remark 4.2. In the proof of Theorem 4.1, and as a consequence of Lemma 3.6, we have

$$\alpha V_\alpha^*(i) = r(i, f) + \sum_{j \in S} q_{ij}(f) V_\alpha^*(j) \quad \forall \, i \in S,$$

which, combined with (4.4) for $u = V_\alpha^*$, yields that V_α^* is indeed a solution of the DROE. Hence, this is an alternative proof of the existence of a solution to the DROE, based on the policy iteration algorithm, while these results have been proved in Theorem 3.7 by using the value iteration algorithm (and under weaker hypotheses: note, in particular, that Assumption 3.3 is not needed to establish the existence of a solution of the DROE).

Remark 4.3. In the policy iteration algorithm, we observe that if $f_{n+1}(i_0) \neq f_n(i_0)$ for some $i_0 \in S$, then $V_\alpha(i_0, f_{n+1}) > V_\alpha(i_0, f_n)$. Indeed, the inequality in (4.1) is strict for $i_0 \in S$, and so

$$r(i, f_{n+1}) + \sum_{j \in S} q_{ij}(f_{n+1})V_\alpha(j, f_n) \geq \alpha V_\alpha(i, f_n) \quad \forall\, i \in S$$

with strict inequality at i_0. Hence, as a consequence of Lemma 3.5(ii), we have $V_\alpha(i_0, f_{n+1}) > V_\alpha(i_0, f_n)$. So, in this case, there is a *strict* improvement of the discounted reward function.

We note that, to use the policy iteration algorithm, one has to solve at each step n an infinite system of linear equations to determine the discounted reward of the policy f_n (recall Step II of the algorithm), and then compute a denumerable number of maxima (see Step III). This means that, in practice, this algorithm cannot be used to solve a given control problem.

Nevertheless, by making an appropriate truncation of the control model $\mathcal{M}$, the results in Sec. 4.3 below show that we can use the policy iteration algorithm to approximate numerically the optimal discounted reward and policies of $\mathcal{M}$; see also the examples in Chapter 9.

4.2.2. *Average reward problems*

Now we suppose that the control model $\mathcal{M}$ satisfies Assumptions 3.1, 3.3, and 3.11. Next, we define the policy iteration algorithm to solve an average reward control problem. We follow [45] and [53]. The algorithm consists of the following steps.

Step I. Choose an arbitrary policy $f_0 \in \mathbb{F}$. Set $n = 0$.

Step II. Determine the gain $g_n \in \mathbb{R}$ and the bias $h_n \in \mathcal{B}_w(S)$ of the policy f_n by solving the *Poisson equation* for f_n, that is,

$$g_n = r(i, f_n) + \sum_{j \in S} q_{ij}(f_n)h_n(j) \quad \forall\, i \in S, \qquad (4.5)$$

subject to $\mu_{f_n}(h_n) = 0$. (Recall Proposition 3.14.)

Step III. Find $f_{n+1} \in \mathbb{F}$ such that, for each $i \in S$,

$$r(i, f_{n+1}) + \sum_{j \in S} q_{ij}(f_{n+1})h_n(j)$$

$$= \max_{a \in A(i)} \left\{ r(i, a) + \sum_{j \in S} q_{ij}(a)h_n(j) \right\}$$

letting $f_{n+1}(i) = f_n(i)$ if possible.

Is $f_{n+1} = f_n$? If "yes," then stop the algorithm because f_n is average reward optimal. If "not," then replace n with $n+1$ and return to Step II.

The convergence of the policy iteration algorithm is proved now. By the way, we note that the proof below is different from that in [45, 53].

Theorem 4.4. *Suppose that Assumptions 3.1, 3.3, and 3.11 hold. Let $\{f_n\} \subseteq \mathbb{F}$ be the sequence of policies obtained by application of the policy iteration algorithm. Then one of the following conditions* (i) *or* (ii) *hold.*

(i) *If $f_{n+1} = f_n$ for some $n \geq 0$, then $g_n = g^*$ and f_n is an average reward optimal policy.*

(ii) *For every $n \geq 0$ we have $f_{n+1} \neq f_n$. In this case, $\lim_n g_n = g^*$ and any limit policy in $\mathbb{F}$ of $\{f_n\}$ is average reward optimal.*

Proof. Suppose that the hypothesis of (i) holds. Then by (4.5),

$$g_n = r(i, f_n) + \sum_{j \in S} q_{ij}(f_n) h_n(j)$$

$$= \max_{a \in A(i)} \left\{ r(i, a) + \sum_{j \in S} q_{ij}(a) h_n(j) \right\} \quad \forall\, i \in S.$$

So, (g_n, h_n) is a solution of the AROE and, from Theorem 3.20, we obtain $g_n = g^*$. That is, f_n is a gain optimal policy.

Now, to prove (ii), let us assume that $f_{n+1} \neq f_n$ for all $n \geq 0$. By Step III of the algorithm we have, for each $i \in S$,

$$r(i, f_{n+1}) + \sum_{j \in S} q_{ij}(f_{n+1}) h_n(j) = \max_{a \in A(i)} \left\{ r(i, a) + \sum_{j \in S} q_{ij}(a) h_n(j) \right\}$$

$$\geq r(i, f_n) + \sum_{j \in S} q_{ij}(f_n) h_n(j) = g_n, \quad (4.6)$$

where the rightmost equality in (4.6) follows from (4.5) and the inequality in (4.6) is strict for at least one $i \in S$. Hence, Corollary 3.16 yields that $g_n < g_{n+1}$.

The increasing sequence $\{g_n\}$ is bounded by the optimal gain g^*. Therefore, it converges to some $\bar{g} \leq g^*$. Now observe that, by the Poisson equation,

$$g_{n+1} = r(i, f_{n+1}) + \sum_{j \in S} q_{ij}(f_{n+1}) h_{n+1}(j) \quad \forall\, i \in S,$$

and so, as in (4.6),

$$g_{n+1} + \sum_{j \in S} q_{ij}(f_{n+1})(h_n - h_{n+1})(j) \geq g_n. \tag{4.7}$$

On the other hand, for every $(i, a) \in K$ and $n \geq 0$,

$$g_{n+1} + \sum_{j \in S} q_{ij}(f_{n+1})(h_n - h_{n+1})(j) \geq r(i, a) + \sum_{j \in S} q_{ij}(a)h_n(j). \tag{4.8}$$

Since the sequence $\{h_n\}$ of bias functions is bounded in the w-norm (see (3.27)), there exists a subsequence $\{n'\}$, functions $h, \bar{h} \in \mathcal{B}_w(S)$, and a policy $\bar{f} \in \mathbb{F}$ such that

$$h_{n'} \to h, \quad h_{n'} - h_{n'+1} \to \bar{h}, \quad \text{and} \quad f_{n'+1} \to \bar{f}$$

as $n' \to \infty$.

By Lemma 3.8(i), we can take the limit as $n' \to \infty$ in (4.7) and we obtain

$$\bar{g} + \sum_{j \in S} q_{ij}(\bar{f})\bar{h}(j) \geq \bar{g} \quad \forall i \in S.$$

Consequently, $\bar{h}$ is superharmonic for $\bar{f}$ and, by Proposition 2.6, $\bar{h}$ is constant. On the other hand, also by Lemma 3.8(i) we can take the limit as $n' \to \infty$ in (4.8) so as to obtain

$$\bar{g} \geq r(i, a) + \sum_{j \in S} q_{ij}(a)h(j) \quad \forall (i, a) \in K.$$

If follows from Lemma 3.10 (see also the proof of Theorem 3.20(ii)) that $\bar{g} \geq g^*$. This establishes that $\lim_n g_n = g^*$.

To complete the proof of the theorem, suppose that $\{n'\}$ is a subsequence such that $f_{n'} \to \bar{f}$ for some policy $\bar{f} \in \mathbb{F}$. By the continuity Theorem 3.17,

$$g(\bar{f}) = \lim_{n'} g_{n'} = g^*,$$

and, consequently, $\bar{f}$ is a gain optimal policy. $\qquad \square$

As noted for discounted reward problems, to use the policy iteration algorithm one has to solve an infinite system of linear equations and compute a denumerable number of maxima. So, the policy iteration algorithm cannot be used in practice. As we shall see, however, by suitably truncating the control model $\mathcal{M}$ we will be able, in Sec. 4.4 and in Chapter 9, to obtain approximations of the optimal gain and policies.

4.3. Approximating discounted reward CMCs

Let

$$\mathcal{M} := \{S, A, (A(i)), (q_{ij}(a)), (r(i, a))\}$$

be a CMC model as in Sec. 2.2 whose optimal discounted reward and optimal policies we want to approximate. Suppose also that $\{\mathcal{M}_n\}_{n \geq 1}$ is a sequence of control models that converges to $\mathcal{M}$ in a suitably defined sense. In this section, we propose conditions ensuring that the optimal control policies and the optimal discounted rewards of $\mathcal{M}_n$ converge, as $n \to \infty$, to the corresponding optimal control policies and the optimal discounted reward of $\mathcal{M}$. This section is based on [135].

The approximating control models   We consider a sequence of control models

$$\mathcal{M}_n := \{S_n, A, (A_n(i)), (q_{ij}^n(a)), (r_n(i, a))\}$$

for $n = 1, 2, \ldots$ The control model $\mathcal{M}$ is interpreted as the *original control model*, that is, the one whose optimal policies and optimal discounted reward we want to approximate; the control models $\mathcal{M}_n$, for $n \geq 1$, are thus the *approximating control models*.

The elements of the control models $\mathcal{M}_n$, for $n \geq 1$, are the following:

- The state space S_n, which is assumed to be a subset of S. (Note that S_n may be finite or infinite.)
- The action space A, which is the same as the action space of the control model $\mathcal{M}$.
- The action sets $A_n(i)$, for $i \in S_n$, which are measurable subsets of $A(i)$. Also, let K_n be the family of state-action pairs for $\mathcal{M}_n$, that is,

$$K_n := \{(i, a) \in S \times A : i \in S_n, a \in A_n(i)\}.$$

- The transition rates $q_{ij}^n(a)$, assumed to be measurable on $A_n(i)$ for each fixed $i, j \in S_n$. The transition rates are conservative and stable, that is, $\sum_{j \in S_n} q_{ij}^n(a) = 0$ for all $(i, a) \in K_n$, and

$$q_n(i) := \sup_{a \in A_n(i)} \{-q_{ii}^n(a)\} < \infty \quad \forall i \in S_n.$$

- The reward rates $r_n(i, a)$, which are measurable functions of $a \in A_n(i)$ for each fixed $i \in S_n$.

Our definitions in Chapter 2, for the original control model $\mathcal{M}$, can be easily modified to account for the control models $\mathcal{M}_n$. For instance, given $u : S_n \to \mathbb{R}$, its w-norm is defined as

$$\|u\|_w := \sup_{i \in S_n} \{|u(i)|/w(i)\}, \tag{4.9}$$

and $\mathcal{B}_w(S_n)$ is the Banach space of functions on S_n with finite w-norm.

The class of Markov policies for the control model $\mathcal{M}_n$ is denoted by Φ_n. Also, we denote by $\mathbb{F}_n$ the set of deterministic stationary policies for the control model $\mathcal{M}_n$; that is, $f : S_n \to A$ is in $\mathbb{F}_n$ if $f(i) \in A_n(i)$ for all $i \in S_n$. Notations such as, e.g., E_i^φ, $\{x^\varphi(t)\}_{t \geq 0}$, or $r_n(t, i, \varphi)$, for $\varphi \in \Phi_n$, $i \in S_n$, and $t \geq 0$, are given the obvious definitions.

Next, we state our hypotheses for the original control model $\mathcal{M}$ and the sequence $\{\mathcal{M}_n\}_{n \geq 1}$ of approximating control models. Roughly speaking, we will suppose that the conditions in Assumptions 3.1 and 3.3 hold "uniformly" in $n \geq 1$.

Assumption 4.5. There exists a Lyapunov function w on S such that:

(a) For some constants $c \in \mathbb{R}$ and $b \geq 0$,

$$\sum_{j \in S} q_{ij}(a)w(j) \leq -cw(i) + b \quad \forall \, (i, a) \in K,$$

and, furthermore, for every $n \geq 1$,

$$\sum_{j \in S_n} q_{ij}^n(a)w(j) \leq -cw(i) + b \quad \forall \, (i, a) \in K_n.$$

(b) For each $i \in S$, $q(i) \leq w(i)$. Also, for every $n \geq 1$ and $i \in S_n$, $q_n(i) \leq w(i)$.

(c) For some constants $c' \in \mathbb{R}$ and $b' \geq 0$,

$$\sum_{j \in S} q_{ij}(a)w^2(j) \leq -c'w^2(i) + b' \quad \forall \, (i, a) \in K,$$

and, in addition, for every $n \geq 1$,

$$\sum_{j \in S_n} q_{ij}^n(a)w^2(j) \leq -c'w^2(i) + b' \quad \forall \, (i, a) \in K_n.$$

(d) There is a constant $M > 0$ such that $|r(i, a)| \leq Mw(i)$ for all $(i, a) \in K$, and such that, for every $n \geq 1$ and $(i, a) \in K_n$, we also have $|r_n(i, a)| \leq Mw(i)$.

(e) For all $i, j \in S$, the functions $r(i, a)$ and $q_{ij}(a)$ are continuous on $A(i)$. Moreover, for every $n \geq 1$ and $i, j \in S_n$, the functions $r_n(i, a)$ and $q_{ij}^n(a)$ are continuous in $a \in A_n(i)$.

(f) The action sets $A(i)$, for $i \in S$, as well as the action sets $A_n(i)$, for $i \in S_n$ and $n \geq 1$, are compact.

Remark 4.6. Let us make some remarks about this assumption.

(i) The condition in Assumption 4.5(b) is not strictly necessary but, as we shall see, it is usually verified in practice. Assumptions 4.5(b) and (c) mean that Assumption 3.1(b) — which is needed to use Dynkin's formula — is satisfied for $w' = w^2$ by the control models $\mathcal{M}$ and $\mathcal{M}_n$.

(ii) We also note that Assumption 4.5 implies that

$$a \mapsto \sum_{j \in S} q_{ij}(a)w(j)$$

is continuous on $A(i)$ (cf. Assumption 3.3(b)). Indeed, given $i \in S$ and $k \geq i$, for all $a \in A(i)$ we have

$$\sum_{j>k} q_{ij}(a)w(j) \leq \frac{1}{w(k)} \sum_{j>k} q_{ij}(a)w^2(j)$$

$$\leq \frac{1}{w(k)} \left(\sum_{j \in S} q_{ij}(a)w^2(j) - q_{ii}(a)w^2(i) \right)$$

$$\leq \frac{1}{w(k)} \left(-c'w^2(i) + b' + q(i)w^2(i) \right).$$

It follows that

$$\lim_{k \to \infty} \sup_{a \in A(i)} \sum_{j>k} q_{ij}(a)w(j) = 0.$$

Consequently, the series $\sum_{j \in S} q_{ij}(a)w(j)$ of continuous functions converges uniformly on $A(i)$ and, hence, it is itself continuous on $A(i)$. The same argument applies for all the control models $\mathcal{M}_n$.

Let $\alpha > 0$ be a discount rate such that $\alpha + c > 0$, with the constant c as in Assumption 4.5(a). The total expected discounted reward of the policy $\varphi \in \Phi_n$ for the control model $\mathcal{M}_n$, for $n \geq 1$, is defined as

$$V_\alpha^n(i, \varphi) := E_i^\varphi \left[\int_0^\infty e^{-\alpha t} r_n(t, x(t), \varphi) dt \right] \quad \forall\, i \in S_n.$$

Obviously, each control model $\mathcal{M}_n$ satisfies Assumptions 3.1 and 3.3, and, therefore, the results on discount optimality in Chapter 3 hold. In particular, by Theorem 3.7, the optimal discounted reward of the control model $\mathcal{M}_n$, that we will denote by V_α^{*n}, is the unique solution in $\mathcal{B}_w(S_n)$ of the corresponding DROE:

$$\alpha u(i) = \max_{a \in A_n(i)} \left\{ r_n(i,a) + \sum_{j \in S_n} q_{ij}^n(a)u(j) \right\} \quad \forall\, i \in S_n.$$

Moreover, a policy $f \in \mathbb{F}_n$ is discount optimal for $\mathcal{M}_n$ if and only if it attains the maximum in the DROE, that is,

$$\alpha V_\alpha^{*n}(i) = r_n(i,f) + \sum_{j \in S_n} q_{ij}^n(f)V_\alpha^{*n}(j) \quad \forall\, i \in S_n.$$

Furthermore, as in (3.7) and (3.9), we obtain

$$\sup_{n \geq 1, \varphi \in \Phi_n} ||V_\alpha^n(\cdot, \varphi)||_w < \infty \quad \text{and} \quad \sup_n ||V_\alpha^{*n}||_w < \infty. \tag{4.10}$$

This is because the constants b, c, and M are the same for every control model $\mathcal{M}_n$.

We omit the proof of our next lemma, which follows by using the same arguments as in Remark 4.6(ii).

Lemma 4.7. *Suppose that the control models $\mathcal{M}$ and $\mathcal{M}_n$, for $n \geq 1$, verify Assumptions 4.5(b)–(c). Given $i \in S$ and $\varepsilon > 0$ there exists $K > i$ such that for every $a \in A(i)$ we have $\sum_{j \geq K} q_{ij}(a)w(j) < \varepsilon$ and, in addition, for every $n \geq 1$ such that $i \in S_n$, and for all $a \in A_n(i)$,*

$$\sum_{j \in S_n, j \geq K} q_{ij}^n(a)w(j) < \varepsilon.$$

Convergence of control models Our next task is to give a suitable definition of convergence of control models.

Definition 4.8. Let $\mathcal{M}$ and $\{\mathcal{M}_n\}_{n \geq 1}$ be the control models defined above. We say that $\{\mathcal{M}_n\}_{n \geq 1}$ converges to $\mathcal{M}$ as $n \to \infty$, which will be denoted by $\mathcal{M}_n \to \mathcal{M}$, if the following conditions are satisfied:

(i) The sequence of state spaces $\{S_n\}_{n \geq 1}$ is monotone nondecreasing and, in addition, $S_n \uparrow S$. As a consequence, if for each $i \in S$ we define $n(i) := \min\{n \geq 1 : i \in S_n\}$, we have $i \in S_n$ if and only if $n \geq n(i)$.

(ii) For each fixed $(i,a) \in K$ and every subsequence $\{n'\}$, there exists a further subsequence $\{n''\}$ and actions $a_{n''} \in A_{n''}(i)$, for all $n'' \geq n(i)$, such that $a_{n''} \to a$ as $n'' \to \infty$.

(iii) For each fixed $i, j \in S$, if the sequence $a_n \in A_n(i)$, for $n \geq n(i) \vee n(j)$ (here, $\vee$ stands for "maximum"), converges to $a \in A(i)$ as $n \to \infty$, then $q_{ij}^n(a_n) \to q_{ij}(a)$.

(iv) For each fixed $i \in S$, if the sequence $a_n \in A_n(i)$, for $n \geq n(i)$, converges to $a \in A(i)$ as $n \to \infty$, then $r_n(i, a_n) \to r(i, a)$.

The condition in Definition 4.8(ii) is equivalent to the following statement. Given arbitrary $(i, a) \in K$ and $\varepsilon > 0$,

$$B(a, \varepsilon) \cap A_n(i) \neq \emptyset \quad \text{for all except finitely many } n \geq 1, \tag{4.11}$$

where $B(a, \varepsilon)$ is the ball (in the metric of the action space A) centered on a with radius ε. We call the condition in Definition 4.8(ii), or its equivalent formulation in (4.11), an "asymptotically dense" requirement. The conditions in (iii) and (iv) state, roughly speaking, that $r_n(i, a)$ and $q_{ij}^n(a)$ converge to $r(i, a)$ and $q_{ij}(a)$, respectively, uniformly in a for all fixed states i and j.

Remark 4.9. Let us make some comments about Definition 4.8. Note that we allow all the elements of the control models $\mathcal{M}_n$ (namely, the state space, the action sets, and the transition and reward rates) to depend on $n \geq 1$.

When dealing with similar definitions of convergence of control models, the state space is usually allowed to depend on n; see [3, 63, 136]. The transition and reward rates may also depend on n. In this case, the uniform convergence property in Definitions 4.8(iii)–(iv) is a usual requirement; see, for instance, condition (2) in [3, Theorem 6.1], and Assumptions 3.1(c) and 3.3(c) in [5].

We note, however, that, in references [3, 5, 63, 136], the action sets of $\mathcal{M}_n$ are the same as the action sets of the original control model $\mathcal{M}$. So, it seems that allowing the action sets $A_n(i)$ to depend on n is a novel feature and, in this sense, the key property is the "asymptotically dense" condition introduced in (4.11). It is also worth noting that we do not require the action sets $A_n(i)$ to increase with n.

Finally, we should mention that condition (4.11) on the action sets $A_n(i)$ is related to the discretization of the state space made in [63, Sec. 6.3] for a discrete-time Markov control process.

Before giving our main result, we state the following preliminary fact.

Lemma 4.10. *Suppose that the control models $\mathcal{M}_n$, for $n \geq 1$, satisfy Assumption 4.5 as well as Definition 4.8(i). Also, let the discount rate $\alpha > 0$*

verify $\alpha + c > 0$, *and let* $V_\alpha^{*n} \in \mathcal{B}_w(S_n)$ *and* $f_n^* \in \mathbb{F}_n$ *be the optimal discounted reward and a discount optimal policy for* $\mathcal{M}_n$, $n \geq 1$, *respectively. Then, the following statements are verified:*

(i) *There exists a subsequence* $\{n'\}$ *and some* $u \in \mathcal{B}_w(S)$ *such that*

$$\lim_{n' \to \infty} V_\alpha^{*n'}(i) = u(i) \quad \forall \, i \in S.$$

(ii) *There exists a subsequence* $\{n'\}$ *and a policy* $f \in \mathbb{F}$ *with*

$$\lim_{n' \to \infty} f_{n'}^*(i) = f(i) \quad \forall \, i \in S. \tag{4.12}$$

In this case, we say that f *is a limit policy of* $\{f_n^*\}_{n \geq 1}$.

Proof.   The proof of this lemma is based on the standard diagonal argument, and makes use of (4.10). In fact, the only difficulty in this proof relies on the fact that $V_\alpha^{*n}(i)$ and $f_n^*(i)$ are defined only for $n \geq n(i)$, that is, when $i \in S_n$.   □

Theorem 4.11. *Suppose that the control models* $\mathcal{M}$ *and* $\{\mathcal{M}_n\}_{n \geq 1}$ *satisfy Assumption 4.5 and let the discount rate* $\alpha > 0$ *be such that* $\alpha + c > 0$, *with* $c \in \mathbb{R}$ *as in Assumption 4.5(a). If* $\mathcal{M}_n \to \mathcal{M}$ *then the next statements hold:*

(i) *For every* $i \in S$, $\lim_{n \to \infty} V_\alpha^{*n}(i) = V_\alpha^*(i)$.
(ii) *If* $f_n^* \in \mathbb{F}_n$, *for* $n \geq 1$, *is a discount optimal policy for* $\mathcal{M}_n$, *then any limit policy* $f^* \in \mathbb{F}$ *of* $\{f_n^*\}_{n \geq 1}$ *is discount optimal for* $\mathcal{M}$.

Proof.   Suppose that the constant $C > 0$ is such that for all $\varphi \in \Phi$, and all $n \geq 1$ and $\varphi \in \Phi_n$

$$\|V_\alpha(\cdot, \varphi)\|_w \leq C \quad \text{and} \quad \|V_\alpha^n(\cdot, \varphi)\|_w \leq C,$$

respectively (recall (3.7) and (4.10)). We also have $\|V_\alpha^*\|_w \leq C$ and $\|V_\alpha^{*n}\|_w \leq C$ for every $n \geq 1$.

Suppose that $\{n'\}$ is a subsequence such that $\{V_\alpha^{*n'}\}$ converges pointwise to some $u \in \mathcal{B}_w(S)$ (with, necessarily, $\|u\|_w \leq C$), and such that $\{f_{n'}^*\}$ converges to some $f^* \in \mathbb{F}$; recall Lemma 4.10.

Fix an arbitrary state $i \in S$, an action $a \in A(i)$, and $\varepsilon > 0$. By Definition 4.8(ii), there exists a subsequence $n'' \geq n(i)$ of $\{n'\}$ and actions $a_{n''} \in A_{n''}(i)$ such that $a_{n''} \to a$ as $n'' \to \infty$. For simplicity in the notation, and without loss of generality, this subsequence will also be denoted by $\{n'\}$. For such n', from the DROE for the control model $\mathcal{M}_{n'}$ we obtain

$$\alpha V_\alpha^{*n'}(i) \geq r_{n'}(i, a_{n'}) + \sum_{j \in S_{n'}} q_{ij}^{n'}(a_{n'}) V_\alpha^{*n'}(j). \tag{4.13}$$

Now, by Lemma 4.7, there exists $K > i$ (which only depends on i and ε) such that

$$\left| \sum_{j \geq K} q_{ij}(a)u(j) \right| \leq C \sum_{j \geq K} q_{ij}(a)w(j) \leq C\varepsilon$$

and, for all $n' \geq n(i)$

$$\left| \sum_{j \in S_{n'}, j \geq K} q_{ij}^{n'}(a_{n'})V_\alpha^{*n'}(j) \right| \leq C \cdot \sum_{j \in S_{n'}, j \geq K} q_{ij}^{n'}(a_{n'})w(j) \leq C\varepsilon.$$

Hence, from (4.13), we deduce that for all $n' \geq n(i)$

$$\alpha V_\alpha^{*n'}(i) \geq r_{n'}(i, a_{n'}) + \sum_{j \in S_{n'}, j < K} q_{ij}^{n'}(a_{n'})V_\alpha^{*n'}(j)$$

$$+ \sum_{j \geq K} q_{ij}(a)u(j) - 2C\varepsilon. \qquad (4.14)$$

Now observe that, as a consequence of Definitions 4.8(iii) and (iv),

$$r_{n'}(i, a_{n'}) \to r(i, a) \quad \text{and} \quad q_{ij}^{n'}(a_{n'}) \to q_{ij}(a)$$

as $n' \to \infty$. On the other hand, for large n' we have $\{0, 1, \ldots, K-1\} \subseteq S_{n'}$. Finally, for every state $0 \leq j < K$, the limit $V_\alpha^{*n'}(j) \to u(j)$ as $n' \to \infty$ holds. Hence, taking the limit as $n' \to \infty$ in (4.14) yields

$$\alpha u(i) \geq r(i, a) + \sum_{j \in S} q_{ij}(a)u(j) - 2C\varepsilon.$$

Since $\varepsilon > 0$ and $(i, a) \in K$ are arbitrary, it follows that

$$\alpha u(i) \geq r(i, a) + \sum_{j \in S} q_{ij}(a)u(j) \quad \forall (i, a) \in K.$$

Therefore, from Lemma 3.5(i) we conclude that $u \geq V_\alpha^*$.

Let us now prove the reverse inequality. We recall that we are assuming that there is a subsequence $\{n'\}$ and a policy $f^* \in \mathbb{F}$ such that $f_{n'}^*(i) \to f^*(i)$ for all $i \in S$. Fix a state $i \in S$ and suppose that $n \geq n(i)$. The policy $f_n^* \in \mathbb{F}_n$ being discount optimal for the control model $\mathcal{M}_n$, by Theorem 3.7(iii) we have

$$\alpha V_\alpha^{*n}(i) = r_n(i, f_n^*) + \sum_{j \in S_n} q_{ij}^n(f_n^*)V_\alpha^{*n}(j). \qquad (4.15)$$

Given $\varepsilon > 0$, by Lemma 4.7 again, there exists $K > i$ (which depends only on i and ε) such that

$$\left| \sum_{j \geq K} q_{ij}(f^*)u(j) \right| \leq C \sum_{j \geq K} q_{ij}(f^*)w(j) \leq C\varepsilon,$$

where $u \in \mathcal{B}_w(S)$ is the pointwise limit of $\{V_\alpha^{*n'}\}$, and for all $n \geq n(i)$

$$\left| \sum_{j \in S_n, j \geq K} q_{ij}^n(f_n^*) V_\alpha^{*n}(j) \right| \leq C \cdot \sum_{j \in S_n, j \geq K} q_{ij}^n(f_n^*) w(j) \leq C\varepsilon.$$

Thus, as a consequence of (4.15),

$$\alpha V_\alpha^{*n}(i) \leq r_n(i, f_n^*) + \sum_{j \in S_n, j < K} q_{ij}^n(f_n^*) V_\alpha^{*n}(j) + \sum_{j \geq K} q_{ij}(f^*) u(j) + 2C\varepsilon.$$

Letting $n' \to \infty$ in the above inequality, and recalling that (see Definitions 4.8(iii) and (iv))

$$r_{n'}(i, f_{n'}^*) \to r(i, f^*) \quad \text{and} \quad q_{ij}^{n'}(f_{n'}^*) \to q_{ij}(f^*),$$

and also that $V_\alpha^{*n'}(j) \to u(j)$ for all $0 \leq j < K$, it follows that

$$\alpha u(i) \leq r(i, f^*) + \sum_{j \in S} q_{ij}(f^*) u(j) + 2C\varepsilon.$$

The state $i \in S$ and the constant $\varepsilon > 0$ being arbitrary, we conclude that

$$\alpha u(i) \leq r(i, f^*) + \sum_{j \in S} q_{ij}(f^*) u(j) \quad \forall \, i \in S.$$

From Lemma 3.5 we obtain $u \leq V_\alpha(\cdot, f^*) \leq V_\alpha^*$. Combined with the previously established inequality $V_\alpha^* \leq u$, this yields that u equals the optimal discounted reward V_α^*. In addition, we obtain that f^* is discount optimal for the control model $\mathcal{M}$.

Hence, we have shown that the pointwise limit of $\{V_\alpha^{*n'}\}$ through any convergent subsequence is necessarily V_α^*. Therefore, the whole sequence $\{V_\alpha^{*n}\}$ converges pointwise to V_α^*. This proves part (i) of the theorem, while part (ii) is a direct consequence of the arguments above.   □

Application to finite approximations Suppose that $\mathcal{M}$ is a control model whose optimal discounted reward and policies we want to approximate. Typically, one does not have, at hand, a sequence $\{\mathcal{M}_n\}_{n \geq 1}$ of approximating control models. So, a simple way of *explicitly constructing*, starting from $\mathcal{M}$, a sequence $\{\mathcal{M}_n\}_{n \geq 1}$ of control models that converges to $\mathcal{M}$ is by the finite-state and finite-action truncation procedure defined next.

Let $\mathcal{M} = \{S, A, (A(i)), (q_{ij}(a)), (r(i, a))\}$ be a control model assumed to satisfy the conditions in Assumption 4.5. For each $n \geq 1$, define the control model

$$\mathcal{M}_n := \{S_n, A, (A_n(i)), (q_{ij}^n(a)), (r_n(i, a))\}$$

as follows:

- The state space is $S_n := \{0, 1, \ldots, n\}$.
- For each $i \in S_n$, let $A_n(i)$ be a finite subset of $A(i)$. In addition, suppose that the sets $A_n(i)$ verify the condition in Definition 4.8(ii).
- Given states $i \in S_n$ and $0 \leq j < n$, let

$$q_{ij}^n(a) := q_{ij}(a) \quad \text{and} \quad q_{in}^n(a) := \sum_{j \geq n} q_{ij}(a),$$

for $a \in A_n(i)$.
- The reward rate is $r_n(i, a) := r(i, a)$ for $i \in S_n$ and $a \in A_n(i)$.

Proposition 4.12. *Suppose that $\mathcal{M}$ is a control model that satisfies Assumption 4.5. If $\{\mathcal{M}_n\}_{n \geq 1}$ is the sequence of finite state and action control models defined above, then the $\mathcal{M}_n$ satisfy Assumption 4.5 and, moreover, $\mathcal{M}_n \to \mathcal{M}$.*

Proof. Before proceeding with the proof itself, note that the particular definition of $q_{in}^n(a)$ yields

$$\sum_{j \in S_n} q_{ij}^n(a) = 0 \quad \forall\, (i, a) \in K_n,$$

so that the transition rates of $\mathcal{M}_n$ are conservative. Their stability will be proved, below, together with Assumption 4.5(b).

Let us check that Assumption 4.5(a) is satisfied. Recall that the Lyapunov function w, as well as the constants $c \in \mathbb{R}$ and $b \geq 0$, are taken from Assumption 4.5 for the control model $\mathcal{M}$. Given $n \geq 1$ and $(i, a) \in K_n$,

$$\sum_{j \in S_n} q_{ij}^n(a) w(j) = \sum_{0 \leq j \leq n} q_{ij}(a) w(j) + \sum_{j > n} q_{ij}(a) w(n)$$

$$\leq \sum_{0 \leq j \leq n} q_{ij}(a) w(j) + \sum_{j > n} q_{ij}(a) w(j) \qquad (4.16)$$

$$= \sum_{j \in S} q_{ij}(a) w(j) \leq -cw(i) + b, \qquad (4.17)$$

where (4.16) follows from the monotonicity of w and the fact that $q_{ij}(a) \geq 0$ for $j > n \geq i$, while (4.17) is derived from Assumption 4.5(a) for the control model $\mathcal{M}$. Hence, Assumption 4.5(a) is satisfied, and, obviously, Assumption 4.5(c) is derived similarly.

Concerning Assumption 4.5(b), note that given $i \in S_n$ and $a \in A_n(i)$, we have $-q_{ii}^n(a) = -q_{ii}(a) \leq w(i)$ for $0 \leq i < n$, while

$$-q_{nn}^n(a) = -\sum_{j \geq n} q_{nj}(a) \leq -q_{nn}(a) \leq w(n).$$

Therefore, for every $n \geq 1$ and $i \in S_n$, we have $q_n(i) \leq w(i)$.

Assumption 4.5(d) is a straightforward consequence of the definition of r_n. Finally, Assumptions 4.5(e) and (f) hold because the action sets $A_n(i)$, for $n \geq 1$ and $i \in S_n$, are finite.

To conclude the proof of this proposition, let us check that $\mathcal{M}_n \rightarrow \mathcal{M}$ as $n \rightarrow \infty$. It is clear that the conditions in Definitions 4.8(i) and (ii) are satisfied. Regarding (iii) and (iv), suppose that, given $i, j \in S$, the sequence $a_n \in A_n(i)$, for $n \geq i$, converges to $a \in A(i)$. Since for large n we have $q_{ij}^n(a_n) = q_{ij}(a_n)$, the convergence $q_{ij}^n(a_n) \rightarrow q_{ij}(a)$ follows from the continuity of the transition rates $q_{ij}(a)$. In the same way, we can prove that $r_n(i, a_n)$ converges to $r(i, a)$.

Hence, we have shown that $\mathcal{M}_n \rightarrow \mathcal{M}$. $\qquad\qquad\square$

By using the policy iteration algorithm (see Sec. 4.2.1), we can *explicitly* obtain the optimal discounted reward V_α^{*n} of $\mathcal{M}_n$, as well as the corresponding optimal policies $f_n^* \in \mathbb{F}_n$. This is because, the state space S_n and the action sets $A_n(i)$ being finite, the set $\mathbb{F}_n$ of deterministic stationary policies is finite, and so, the policy iteration algorithm converges in a finite number of steps.

Finally, Theorem 4.11 ensures that $V_\alpha^{*n} \rightarrow V_\alpha^*$, the optimal discounted reward of $\mathcal{M}$, and that any limit policy in $\mathbb{F}$ of $\{f_n^*\}_{n \geq 1}$ is discount optimal for $\mathcal{M}$. Numerical applications of the truncation procedure described above will be shown in Chapter 9.

4.4.   Approximating average reward CMCs

In this section we study approximation theorems such as those in Sec. 4.3, but now for CMCs under the average reward optimality criterion. Let

$$\mathcal{M} := \{S, A, (A(i)), (q_{ij}(a)), (r(i, a))\}$$

be the *original control model*, whose optimal gain and optimal policies we want to approximate, and let $\{\mathcal{M}_n\}_{n \geq 1}$ be a sequence of *approximating control models* that converge to $\mathcal{M}$ in a suitably defined sense. In this section, we propose conditions ensuring that the optimal control policies and the optimal gains of $\mathcal{M}_n$ converge, as $n \rightarrow \infty$, to the corresponding optimal control policies and the optimal gain of $\mathcal{M}$.

The approximating control models We consider a sequence of continuous-time CMCs

$$\mathcal{M}_n := \{S, A, (A(i)), (q_{ij}^n(a)), (r_n(i, a))\}$$

for $n = 1, 2, \ldots$. We note that, as opposed to the discounted case in Sec. 4.3, the control models $\mathcal{M}_n$ have the *same state space and action sets* as the original control model $\mathcal{M}$.

We suppose that the $\mathcal{M}_n$ satisfy the conditions stated in Sec. 2.2. In particular, the transition rates are conservative, i.e., for all $n \geq 1$,

$$\sum_{j \in S} q_{ij}^n(a) = 0 \quad \forall \, (i, a) \in K,$$

and stable, i.e., for all $n \geq 1$,

$$q_n(i) := \sup_{a \in A(i)} \{-q_{ii}^n(a)\} < \infty \quad \forall \, i \in S.$$

Let us also note that the sets Φ and $\mathbb{F}$ of Markov policies and deterministic stationary policies, respectively, are the same for the control models $\mathcal{M}$ and $\mathcal{M}_n$, for $n \geq 1$.

Next, we state the assumptions for the sequence of control models $\mathcal{M}_n$ (cf. Assumption 4.5).

Assumption 4.13.

(a) There exist a Lyapunov function w on S, constants $c > 0$ and $b \geq 0$, and finite subsets C and $\{C_n\}_{n \geq 1}$ of S such that

$$\sum_{j \in S} q_{ij}(a)w(j) \leq -cw(i) + b \cdot \mathbf{I}_C(i) \quad \forall \, (i, a) \in K$$

and, in addition, for all $n \geq 1$,

$$\sum_{j \in S} q_{ij}^n(a)w(j) \leq -cw(i) + b \cdot \mathbf{I}_{C_n}(i) \quad \forall \, (i, a) \in K.$$

(b) For each $i \in S$, we have $q(i) \leq w(i)$. Moreover, for every $n \geq 1$ and $i \in S$, $q_n(i) \leq w(i)$.

(c) There exist constants $\tilde{c} \in \mathbb{R}$ and $\tilde{b} \geq 0$ such that

$$\sum_{j \in S} q_{ij}(a)w^2(j) \leq -\tilde{c}w^2(i) + \tilde{b} \quad \forall \, (i, a) \in K$$

and, for all $n \geq 1$,

$$\sum_{j \in S} q_{ij}^n(a)w^2(j) \leq -\tilde{c}w^2(i) + \tilde{b} \quad \forall \, (i, a) \in K.$$

(d) Every $f \in \mathbb{F}$ is irreducible under each control model $\mathcal{M}$ and $\{\mathcal{M}_n\}_{n \geq 1}$.

(e) There exists a constant $M > 0$ such that $|r(i, a)| \leq Mw(i)$ and $|r_n(i, a)| \leq Mw(i)$ for every $(i, a) \in K$ and $n \geq 1$.

(f) For each $i \in S$, the set $A(i)$ is compact. Moreover, for every $i, j \in S$ and $n \geq 1$, the functions $q_{ij}(a)$, $r(i, a)$, $q_{ij}^n(a)$, and $r_n(i, a)$ are continuous in $a \in A(i)$.

The conditions in Assumption 4.13 are standard. Roughly speaking, they correspond to the average reward optimality Assumptions 3.1, 3.3, and 3.11, except that they are now supposed to hold "uniformly" in $n \geq 1$; for instance, the constants c and b in Assumption 4.13(a) are the *same* for all $n \geq 1$, although the finite sets $C_n \subset S$ are allowed to vary with n. We recall from Remark 4.6(ii) that, for each $i \in S$, the functions

$$\sum_{j \in S} q_{ij}(a)w(j) \quad \text{and} \quad \sum_{j \in S} q_{ij}^n(a)w(j)$$

are continuous in $a \in A(i)$ for every $n \geq 1$ (cf. Assumption 3.3(b)).

Now, we introduce some more notation. Given a control model $\mathcal{M}_n$ and a deterministic stationary policy $f \in \mathbb{F}$, we will denote by $P_{ij}^{n,f}(t)$ the corresponding transition probability function, for $i, j \in S$ and $t \geq 0$. Also, we will denote by $E_{n,i}^f$ the associated expectation operator when the initial state is $i \in S$.

It is important to note that, by Theorem 2.5, the Markov chain $\{x^f(t)\}$ has, under each control model $\mathcal{M}_n$, a unique invariant probability measure μ_f^n for which $\mu_f^n(w) < \infty$.

Finally, as a consequence of Theorem 2.11, the control models $\mathcal{M}$ and $\{\mathcal{M}\}_{n \geq 1}$ are w-exponentially ergodic uniformly on $\mathbb{F}$; that is, there exist positive constants δ, δ_n, R, and R_n such that

$$\sup_{f \in \mathbb{F}} |E_i^f[u(x(t))] - \mu_f(u)| \leq Re^{-\delta t}||u||_w w(i) \tag{4.18}$$

(here, μ_f denotes the unique invariant probability measure of $f \in \mathbb{F}$ under $\mathcal{M}$), and

$$\sup_{f \in \mathbb{F}} |E_{n,i}^f[u(x(t))] - \mu_f^n(u)| \leq R_n e^{-\delta_n t}||u||_w w(i) \tag{4.19}$$

for every $i \in S$, $t \geq 0$, and $u \in \mathcal{B}_w(S)$. An assumption that the constants δ_n and R_n in (4.19) do not depend on n would be too restrictive and, hence, we allow them to depend on n. However, if the control models $\mathcal{M}_n$ satisfy the monotonicity conditions in Assumption 2.7 then, by Theorem 2.8, the constants $\delta_n = c$ and $R_n = 2(1 + b/c)$ turn out to be independent of n.

Let $g_n(f)$, for $f \in \mathbb{F}$, denote the gain of the policy $f \in \mathbb{F}$ for the control model $\mathcal{M}_n$. By Assumptions 4.13(a) and (e), and (3.24), for all $n \geq 1$,

$$|g_n(f)| \leq \frac{bM}{c} \quad \forall \, f \in \mathbb{F}. \tag{4.20}$$

As a consequence of Assumption 4.13, each control model $\mathcal{M}_n$ satisfies the hypotheses of Theorem 3.20, and the corresponding AROEs are, for $n \geq 1$,

$$g_n^* = \max_{a \in A(i)} \left\{ r_n(i, a) + \sum_{j \in S} q_{ij}^n(a) h_n(j) \right\} \quad \forall \, i \in S, \tag{4.21}$$

where the constant g_n^* is the corresponding optimal gain, and h_n is in $\mathcal{B}_w(S)$.

Convergence of control models Our next step is to give a suitable definition of convergence for the control models.

Definition 4.14. Given the above defined control models $\mathcal{M}$ and $\mathcal{M}_n$, for $n \geq 1$, we say that the sequence $\{\mathcal{M}_n\}_{n \geq 1}$ converges to $\mathcal{M}$ (denoted by $\mathcal{M}_n \to \mathcal{M}$) if

(i) for each $i \in S$, $r_n(i, \cdot)$ converges to $r(i, \cdot)$ uniformly on $A(i)$ and, moreover,

(ii) for every $i, j \in S$, $q_{ij}^n(\cdot)$ converges to $q_{ij}(\cdot)$ uniformly on $A(i)$.

Observe that this definition is similar to that given in [3, Theorem 6.1(2)] for the convergence of reward functions.

Before stating our convergence results, we need two preliminary results. The first one shows that Definition 4.14(ii) implies the following, seemingly stronger, condition.

Lemma 4.15. *Suppose that the control models $\mathcal{M}$ and $\mathcal{M}_n$, for $n \geq 1$, satisfy Assumption 4.13 and that Definition 4.14(ii) holds. Then, for every $i \in S$,*

$$\lim_{n \to \infty} \sup_{a \in A(i)} \sum_{j \in S} |q_{ij}^n(a) - q_{ij}(a)| w(j) = 0. \tag{4.22}$$

Proof. We know from Remark 4.6 and Lemma 4.7 (in particular, we use Assumptions 4.13(b) and (c)) that, given $i \in S$ and $\varepsilon > 0$, there exists some K_0 large enough such that, for all $n \geq 1$ and $a \in A(i)$,

$$\sum_{j \geq K_0} q_{ij}(a) w(j) < \varepsilon/4 \quad \text{and} \quad \sum_{j \geq K_0} q_{ij}^n(a) w(j) < \varepsilon/4.$$

Therefore, for every $n \geq 1$,

$$\sup_{a \in A(i)} \sum_{j \geq K_0} |q_{ij}^n(a) - q_{ij}(a)| w(j) \leq \varepsilon/2.$$

Now, by Definition 4.14(ii), we can choose n sufficiently large so that

$$\sup_{a \in A(i)} \sum_{j < K_0} |q_{ij}^n(a) - q_{ij}(a)| w(j) \leq \varepsilon/2,$$

and the stated result follows. $\qquad\square$

Our next lemma is an extension of Lemma 3.8(i).

Lemma 4.16. *Suppose that the control models $\mathcal{M}$ and $\mathcal{M}_n$, for $n \geq 1$, satisfy Assumption 4.13 and that Definition 4.14(ii) holds. In addition, assume that $\{h_n\}_{n \geq 1}$ is a bounded sequence in $\mathcal{B}_w(S)$ that converges pointwise to $h_0 \in \mathcal{B}_w(S)$. Also, given $i \in S$, suppose that $\{a_n\}_{n \geq 1} \subseteq A(i)$ converges to $a_0 \in A(i)$. Then*

$$\lim_{n \to \infty} \sum_{j \in S} q_{ij}^n(a_n) h_n(j) = \sum_{j \in S} q_{ij}(a_0) h_0(j).$$

Proof.   Suppose that $\|h_n\|_w \leq H$ for all $n \geq 0$. First of all, we have

$$\sum_{j \in S} q_{ij}^n(a_n) h_n(j) = \sum_{j \in S} (q_{ij}^n(a_n) - q_{ij}(a_n)) h_n(j) + \sum_{j \in S} q_{ij}(a_n) h_n(j).$$

$$(4.23)$$

By hypothesis, we have

$$\left| \sum_{j \in S} (q_{ij}^n(a_n) - q_{ij}(a_n)) h_n(j) \right| \leq H \cdot \sup_{a \in A(i)} \sum_{j \in S} |q_{ij}^n(a) - q_{ij}(a)| w(j),$$

$$(4.24)$$

which, by Lemma 4.15, converges to zero as $n \to \infty$. On the other hand, by Lemma 3.8(i),

$$\lim_{n \to \infty} \sum_{j \in S} q_{ij}(a_n) h_n(j) = \sum_{j \in S} q_{ij}(a_0) h_0(j). \qquad (4.25)$$

Finally, the stated result follows from (4.24) and (4.25) by taking the limit as $n \to \infty$ in the right-hand side of (4.23). $\qquad\square$

Sufficient conditions for convergence Next, we propose two different sufficient conditions ensuring the convergence of the optimal gains and the optimal policies. The first one imposes a boundedness condition on the sequence $\{h_n\}$ in the AROEs (4.21); see Theorem 4.17 below. Our second condition does not assume such a boundedness condition but, instead, it strengthens the Lyapunov inequality in Assumption 4.13(c); see Theorem 4.18.

Theorem 4.17. *Suppose that the control models $\mathcal{M}$ and $\mathcal{M}_n$, for $n \geq 1$, satisfy Assumption 4.13, and that $\mathcal{M}_n \to \mathcal{M}$. Suppose also that there exist solutions h_n of (4.21) such that*

$$\sup_n \|h_n\|_w < \infty.$$

Then the following results hold:

(i) $\lim_{n \to \infty} g_n^* = g^*$, *where g^* is the optimal gain of the control model $\mathcal{M}$.*
(ii) *If $f_n \in \mathbb{F}$, for each $n \geq 1$, is an average reward optimal policy for $\mathcal{M}_n$, then any limit policy (in the sense of componentwise convergence) of $\{f_n\}$ is average reward optimal for $\mathcal{M}$.*

Proof. *Proof of (i).* By (4.20), the sequence $\{g_n^*\}_{n \geq 1}$ is bounded, and so we may consider a convergent subsequence n', that is, $\lim_{n'} g_{n'}^* = \bar{g}$ for some $\bar{g} \in \mathbb{R}$. Besides, since $\{h_{n'}\}$ is bounded in the w-norm, there exists a further subsequence n'' and $\bar{h} \in \mathcal{B}_w(S)$ such that

$$\lim_{n''} h_{n''}(i) = \bar{h}(i) \quad \forall \, i \in S.$$

Finally, let $f_{n''} \in \mathbb{F}$ be a gain optimal policy for the control model $\mathcal{M}_{n''}$, and let n''' be a further subsequence such that, for some $\bar{f} \in \mathbb{F}$,

$$\lim_{n'''} f_{n'''}(i) = \bar{f}(i) \quad \forall \, i \in S.$$

For notational ease, the subsequence n''' will be simply denoted by n.
    We fix $i \in S$ and $a \in A(i)$. We have, by (4.21),

$$g_n^* \geq r_n(i, a) + \sum_{j \in S} q_{ij}^n(a) h_n(j).$$

Taking the limit as $n \to \infty$ in this expression yields, by Definition 4.14(i) and Lemma 4.16,

$$\bar{g} \geq r(i, a) + \sum_{j \in S} q_{ij}(a) \bar{h}(j),$$

and so, since $a \in A(i)$ is arbitrary,

$$\bar{g} \geq \max_{a \in A(i)} \left\{ r(i,a) + \sum_{j \in S} q_{ij}(a)\bar{h}(j) \right\} \quad \forall\, i \in S. \tag{4.26}$$

We will next prove the reverse inequality. Fix $i \in S$. Since f_n is optimal for the control model $\mathcal{M}_n$, by Theorem 3.20(iii) we have

$$g_n^* = r_n(i, f_n) + \sum_{j \in S} q_{ij}^n(f_n)h_n(j).$$

Since the limit in Definition 4.14(i) is uniform, by Lemma 4.16 it follows that

$$\bar{g} = r(i, \bar{f}) + \sum_{j \in S} q_{ij}(\bar{f})\bar{h}(j) \quad \forall\, i \in S. \tag{4.27}$$

Therefore, from (4.26) and (4.27), we conclude that

$$\bar{g} = \max_{a \in A(i)} \left\{ r(i,a) + \sum_{j \in S} q_{ij}(a)\bar{h}(j) \right\} = r(i, \bar{f}) + \sum_{j \in S} q_{ij}(\bar{f})\bar{h}(j)$$

for all $i \in S$. Consequently, by Theorem 3.20, we see that $\bar{g} = g^*$ and that $\bar{f}$ is gain optimal for the control model $\mathcal{M}$. Hence, the unique limit point of the bounded sequence $\{g_n^*\}$ is g^*, and so the whole sequence converges to g^*, thus proving statement (i).

Proof of (ii). First of all, note that there is indeed a limit policy because $\mathbb{F}$ is compact. Let n' be a subsequence such that $f_{n'}$ converges pointwise to $\bar{f} \in \mathbb{F}$. Using the same arguments as in the proof of statement (i), we obtain

$$g^* = r(i, \bar{f}) + \sum_{j \in S} q_{ij}(\bar{f})\bar{h}(j) \quad \forall\, i \in S,$$

that is, $(g^*, \bar{h})$ is a solution of the Poisson equation for $\bar{f}$. Hence, by Proposition 3.14, we have $g^* = g(\bar{f})$, and $\bar{f}$ is a gain optimal policy for the control model $\mathcal{M}$. $\qquad\square$

As already mentioned, if the conditions in Assumption 2.7 are verified by each control model $\mathcal{M}_n$ then, by Theorem 2.8, the constants R_n and δ_n in (4.19) are $R_n = 2(1 + b/c)$ and $\delta_n = c$ for all $n \geq 1$ (the constants b and c are from Assumption 4.13(a)). Moreover, the function $h_n \in \mathcal{B}_w(S)$ in (4.21) can be chosen as the bias of a gain optimal policy, which satisfies (3.27), and so

$$\sup_{n \geq 1} \|h_n\|_w \leq \frac{2M(1 + b/c)}{c}.$$

Therefore, the conditions of Theorem 4.17 are satisfied.

Our next result replaces the boundedness condition in Theorem 4.17 with a suitable Lyapunov condition; see [136, Theorem 3.2].

Theorem 4.18. *Suppose that the control models* $\mathcal{M}$ *and* $\mathcal{M}_n$, *for* $n \geq 1$, *satisfy Assumption 4.13, and that* $\mathcal{M}_n \to \mathcal{M}$. *Suppose, moreover, that there exists* $\eta > 0$, *and constants* $\hat{c} > 0$ *and* $\hat{b} \geq 0$ *such that, for each* $n \geq 1$,

$$\sum_{j \in S} q_{ij}^n(a) w^{2+\eta}(j) \leq -\hat{c} w^{2+\eta}(i) + \hat{b} \quad \forall \ (i, a) \in K.$$

Then the following statements hold:

(i) $\lim_{n \to \infty} g_n^* = g^*$, *the optimal gain of* $\mathcal{M}$.
(ii) *If* $f_n \in \mathbb{F}$ *is, for each* $n \geq 1$, *an average reward optimal policy for* $\mathcal{M}_n$, *then any limit policy of* $\{f_n\}$ *is average reward optimal for* $\mathcal{M}$.

Proof. *Proof of (i).* This proof requires two preliminary facts. The first, which follows from (2.18), is that for each $n \geq 1$, $i \in S$, $h \in \mathcal{B}_w(S)$, and $(i, a) \in K$,

$$\left| \sum_{j \in S} q_{ij}^n(a) h(j) \right| \leq ||h||_w (2 + b) w^2(i). \tag{4.28}$$

The second fact is the following uniform integrability result: Given $\varepsilon > 0$, there exists K_0 such that, for each $n \geq 1$ and $f \in \mathbb{F}$,

$$\sum_{j \geq K_0} \mu_f^n(j) w^2(j) \leq \varepsilon. \tag{4.29}$$

Indeed, given $t \geq 0$, $n \geq 1$, $k \in S$, $f \in \mathbb{F}$, and an initial state $i \in S$,

$$\sum_{j \geq k} P_{ij}^{n,f}(t) w^2(j) \leq \frac{1}{w^\eta(k)} \sum_{j \geq k} P_{ij}^{n,f}(t) w^{2+\eta}(j)$$

$$\leq \frac{1}{w^\eta(k)} \sum_{j \in S} P_{ij}^{n,f}(t) w^{2+\eta}(j).$$

Now, by the inequality (2.13) for the Lyapunov function $w^{2+\eta}$,

$$\sum_{j \in S} P_{ij}^{n,f}(t) w^{2+\eta}(j) = E_{n,i}^f[w^{2+\eta}(x(t))] \leq e^{-\hat{c}t} w^{2+\eta}(i) + \hat{b}/\hat{c},$$

and thus

$$\sum_{j \geq k} P_{ij}^{n,f}(t) w^2(j) \leq \frac{1}{w^\eta(k)} (e^{-\hat{c}t} w^{2+\eta}(i) + \hat{b}/\hat{c}).$$

By ergodicity, $P_{ij}^{n,f}(t) \to \mu_f^n(j)$ as $t \to \infty$; hence, it is easily shown that

$$\sum_{j \geq k} \mu_f^n(j) w^2(j) \leq \frac{\hat{b}}{\hat{c} w^n(k)}, \qquad (4.30)$$

and (4.29) follows.

Our next step in the proof of Theorem 4.18 is to show that

$$\lim_{n \to \infty} \sup_{f \in \mathbb{F}} |g_n(f) - g(f)| = 0. \qquad (4.31)$$

To this end, fix $f \in \mathbb{F}$, and let $g(f)$ and $h_f \in \mathcal{B}_w(S)$ be its gain and bias, respectively, for the original control model $\mathcal{M}$. The Poisson equation (3.30) for f reads

$$g(f) = r(i, f) + \sum_{j \in S} q_{ij}(f) h_f(j) \quad \forall \, i \in S.$$

Multiplying both sides of this equation by $\mu_f^n(i)$, for arbitrary $n \geq 1$, and summing over $i \in S$ yields

$$g(f) = \sum_{i \in S} r(i, f) \mu_f^n(i) + \sum_{j \in S} \sum_{i \in S} \mu_f^n(i) q_{ij}(f) h_f(j); \qquad (4.32)$$

note that these sums are finite and interchangeable as a consequence of (4.28) and (4.30).

Observe that $g_n(f) = \sum_{i \in S} r_n(i, f) \mu_f^n(i)$ (recall (3.25)) and, by Theorem 2.5(ii),

$$\sum_{i \in S} \mu_f^n(i) q_{ij}^n(f) = 0 \quad \forall \, j \in S.$$

Therefore, (4.32) can be rewritten as

$$g(f) - g_n(f) = \sum_{i \in S} (r(i, f) - r_n(i, f)) \mu_f^n(i)$$

$$+ \sum_{i \in S} \mu_f^n(i) \sum_{j \in S} [q_{ij}(f) - q_{ij}^n(f)] h_f(j). \qquad (4.33)$$

Now, fix $K_0 \in S$. We have

$$\left| \sum_{i \geq K_0} (r(i, f) - r_n(i, f)) \mu_f^n(i) \right| \leq \sum_{i \geq K_0} 2M w(i) \mu_f^n(i)$$

$$\leq \frac{2M}{w(K_0)} \sum_{i \geq K_0} w^2(i) \mu_f^n(i),$$

which combined with (4.30) yields

$$\left| \sum_{i \geq K_0} (r(i,f) - r_n(i,f))\mu_f^n(i) \right| \leq \frac{2M\hat{b}}{\hat{c}w^{1+\eta}(K_0)}.$$

Furthermore, recalling (4.28) and the bound on $||h_f||_w$ in (3.27),

$$\left| \sum_{i \geq K_0} \mu_f^n(i) \sum_{j \in S} [q_{ij}(f) - q_{ij}^n(f)]h_f(j) \right| \leq \frac{2RM(2+b)}{\delta} \sum_{i \geq K_0} w^2(i)\mu_f^n(i)$$

$$\leq \frac{2RM(2+b)\hat{b}}{\hat{c}\delta w^\eta(K_0)},$$

where the constants $\delta > 0$ and $R > 0$ are taken from (4.18).

Therefore, given $\varepsilon > 0$, let $K_0 \in S$ (which does not depend on f) be such that

$$\max\left\{ \frac{2M\hat{b}}{\hat{c}w^{1+\eta}(K_0)}, \frac{2RM(2+b)\hat{b}}{\hat{c}\delta w^\eta(K_0)} \right\} \leq \frac{\varepsilon}{4}. \qquad (4.34)$$

By Definition 4.14, there exists N_0 (which does not depend on f, but it does depend on K_0) such that $n \geq N_0$ implies

$$\left| \sum_{i < K_0} (r(i,f) - r_n(i,f))\mu_f^n(i) \right| \leq \max_{0 \leq i < K_0} \sup_{a \in A(i)} |r(i,a) - r_n(i,a)| \leq \frac{\varepsilon}{4}$$

$$(4.35)$$

and also that, by Lemma 4.15,

$$\left| \sum_{i < K_0} \mu_f^n(i) \sum_{j \in S} [q_{ij}(f) - q_{ij}^n(f)]h_f(j) \right|$$

is bounded above by

$$\frac{RM}{\delta} \cdot \max_{0 \leq i < K_0} \sup_{a \in A(i)} \sum_{j \in S} |q_{ij}(a) - q_{ij}^n(a)|w(j) \leq \frac{\varepsilon}{4}. \qquad (4.36)$$

In conclusion, we have shown that, given $\varepsilon > 0$, there exists N_0 such that

$$\sup_{f \in \mathbb{F}} |g(f) - g_n(f)| \leq \varepsilon \quad \forall\, n \geq N_0, \qquad (4.37)$$

which proves (4.31).

We will now complete the proof of (i), that is, let us prove that $g_n^* \to g^*$. Given $\varepsilon > 0$, let N_0 be as in (4.37), and choose $n \geq N_0$. Let f_n be a gain optimal policy for $\mathcal{M}_n$. By (4.37), we have

$$g_n^* = g_n(f_n) \leq \varepsilon + g(f_n) \leq \varepsilon + g^*. \tag{4.38}$$

Besides, if $f \in \mathbb{F}$ is a gain optimal policy for $\mathcal{M}$, by (4.37) we obtain

$$g^* = g(f) \leq \varepsilon + g_n(f) \leq \varepsilon + g_n^*. \tag{4.39}$$

Combining (4.38) and (4.39), it follows that $|g_n^* - g^*| \leq \varepsilon$, which completes the proof of part (i).

Proof of (ii). Suppose that $\{f_n\}$ is a sequence of optimal policies for the control models $\{\mathcal{M}_n\}$, and let n' be a subsequence such that $f_{n'} \to f$ for some $f \in \mathbb{F}$.

Using (4.31), it follows that

$$\lim_{n'} [g_{n'}(f_{n'}) - g(f_{n'})] = 0.$$

On the other hand, by Theorem 3.17, $\lim_{n'} g(f_{n'}) = g(f)$, and, by part (i),

$$\lim_{n'} g_{n'}(f_{n'}) = \lim_{n'} g_{n'}^* = g^*.$$

Hence, we have proved that $g(f) = g^*$, that is, the limit policy f is gain optimal for the original control model $\mathcal{M}$. $\qquad\square$

The proofs of Theorems 4.17 and 4.18 are quite different. That of Theorem 4.17 takes the limit in the AROE (4.21) for the control model $\mathcal{M}_n$ to obtain the AROE for the original control model $\mathcal{M}$. In contrast, in Theorem 4.18, we use the Poisson equation for each $f \in \mathbb{F}$. In particular, this makes the boundedness hypothesis on $\{h_n\}$ of Theorem 4.17 unnecessary, because the biases h_f in the control model $\mathcal{M}$ are uniformly bounded in $\mathcal{B}_w(S)$; see (3.27).

Regarding the applications of the above results, we make the following important remark.

Remark 4.19. Suppose that the hypotheses of Theorem 4.18 hold and that the constants δ and R of the uniform ergodicity of $\mathcal{M}$ (see (4.18)) are known. Then, given $\varepsilon > 0$, we can *explicitly* determine n such that $|g_n^* - g^*| \leq \varepsilon$. Indeed, the constants K_0 and N_0 are determined in (4.34)–(4.36) from the initial data of the control model (and also from δ and R).

Finally, let us mention that the condition in Assumption 4.13(c) can be weakened in the following sense.

Remark 4.20. Suppose that the control models $\mathcal{M}$ and $\mathcal{M}_n$, for $n \geq 1$, satisfy Assumption 4.13, where, in Assumption 4.13(c), the constants $\tilde{c} \in \mathbb{R}$ and $\tilde{b} \geq 0$ are allowed to depend on n. If $\mathcal{M}_n \to \mathcal{M}$ and, in addition, (4.22) is satisfied for every $i \in S$, then Theorems 4.17 and 4.18 still hold.

Application to finite approximations In Theorems 4.17 and 4.18 we have analyzed the approximation of a denumerable state control model $\mathcal{M}$ by means of a sequence of denumerable state control models $\{\mathcal{M}_n\}_{n \geq 1}$. In practice, however, finite state approximations of a denumerable state control model are of particular interest.

Let $\mathcal{M}$ be a denumerable state CMC

$$\mathcal{M} := \{S, A, (A(i)), (q_{ij}(a)), (r(i, a))\}$$

and consider a sequence of control models $\mathcal{M}_n$

$$\mathcal{M}_n := \{S_n, A, (A(i)), (q_{ij}^n(a)), (r_n(i, a))\},$$

whose respective state spaces are

$$S_n := \{0, 1, \ldots, n\} \quad \forall\, n \geq 1.$$

We suppose that the control models $\mathcal{M}$ and $\mathcal{M}_n$, for $n \geq 1$, satisfy Assumption 4.13. Regarding the conditions in Assumption 4.13 for finite state control models, observe that they can be also formulated when we restrict ourselves to the states $i, j \in S_n$. Other features such as the w-norm of a function $u : S_n \to \mathbb{R}$ (recall (4.9)) or the convergence of a sequence of policies $f_n \in \mathbb{F}$ on S_n to a policy $f \in \mathbb{F}$ on S are given the obvious definitions.

Under our standing hypotheses, the properties of the finite state control models $\mathcal{M}_n$ are similar to those of the denumerable state control model $\mathcal{M}$. In particular, we can establish the corresponding finite state AROE

$$g_n^* = \max_{a \in A(i)} \left\{ r_n(i, a) + \sum_{j \in S_n} q_{ij}^n(a) h_n(j) \right\} \quad \forall\, i \in S_n, \tag{4.40}$$

where g_n^* is the optimal gain and h_n is in $\mathcal{B}_w(S_n)$; see [112].

Moreover, since the conditions in Definition 4.14 are stated for fixed states $i, j \in S$, the convergence of $\{\mathcal{M}_n\}_{n \geq 1}$ to $\mathcal{M}$ is given the same definition.

Theorem 4.21. *Suppose that the control models $\mathcal{M}$ and $\{\mathcal{M}_n\}_{n \geq 1}$, with respective state spaces S and S_n, for $n \geq 1$, verify Assumption 4.13. If $\mathcal{M}_n \to \mathcal{M}$, and either*

(a) *the functions h_n in (4.40) verify $\sup_{n \geq 1} \|h_n\|_w < \infty$, or*

(b) *there exist constants $\eta > 0$, $\hat{c} > 0$, and $\hat{b} \geq 0$ such that, for each $n \geq 1$,*

$$\sum_{j \in S_n} q_{ij}^n(a) w^{2+\eta}(j) \leq -\hat{c} w^{2+\eta}(i) + \hat{b} \quad \forall \, i \in S_n, \ a \in A(i),$$

then the conclusions of Theorem 4.17 (or Theorem 4.18) are satisfied.

The proof of this theorem is omitted because it is similar to that of Theorems 4.17 and 4.18.

So far, we have analyzed the approximation of a denumerable state CMC by means of a sequence of *pre-given* finite state CMCs. Now, we suppose that we are given a denumerable state control model $\mathcal{M}$ and we show how to construct a sequence of finite state control models $\mathcal{M}_n$ satisfying Theorem 4.21.

Consider a denumerable state control model

$$\mathcal{M} := \{S, A, (A(i)), (q_{ij}(a)), (r(i, a))\}$$

that satisfies the conditions in Assumption 4.13. For each $n \geq 1$, let us define the "truncated" control model $\mathcal{M}_n$ with state space

$$S_n := \{0, 1, \ldots, n\},$$

action sets $A(i)$ for $i \in S_n$, reward rates

$$r_n(i, a) := r(i, a) \quad \text{for } i \in S_n \text{ and } a \in A(i),$$

and transition rates, for $i, j \in S_n$ and $a \in A(i)$, given by

$$q_{ij}^n(a) := q_{ij}(a) \quad \text{for } i \in S_n \text{ and } 0 \leq j < n,$$

and

$$q_{in}^n(a) := \sum_{j \geq n} q_{ij}(a) \quad \text{for } i \in S_n$$

(cf. the finite state and action truncation defined for discounted CMCs in Sec. 4.3). The proof of our next result is omitted because it is similar to that of Proposition 4.12.

Proposition 4.22. *Suppose that the original control model $\mathcal{M}$ satisfies Assumption 4.13 and, for each $n \geq 1$, consider the truncated model $\mathcal{M}_n$. Then the control models $\mathcal{M}_n$ satisfy Assumption 4.13 (except perhaps (d)) and, furthermore, $\mathcal{M}_n \to \mathcal{M}$.*

To conclude, we state our main result in this section.

Theorem 4.23. *Suppose that the control model $\mathcal{M}$ verifies Assumption 4.13 and consider the truncated finite state approximations $\mathcal{M}_n$, for $n \geq 1$. In addition, suppose that each $f \in \mathbb{F}$ is irreducible for every $\mathcal{M}_n$, and that either*

(a) *the functions h_n in (4.40) verify $\sup_n \|h_n\|_w < \infty$, or*
(b) *there exist constants $\eta > 0$, $\hat{c} > 0$, and $\hat{b} \geq 0$ such that*

$$\sum_{j \in S} q_{ij}(a) w^{2+\eta}(j) \leq -\hat{c} w^{2+\eta}(i) + \hat{b} \quad \forall \, (i, a) \in K \qquad (4.41)$$

(cf. condition (b) in Theorem 4.21).

Then $\lim_n g_n^* = g^*$, and any limit policy of gain optimal policies for $\mathcal{M}_n$ is gain optimal for $\mathcal{M}$.

Proof. The proof of this theorem goes along the same lines as those of Theorems 4.17, 4.18, or 4.21, and we omit it. For further details the reader may consult [136, Theorem 3.4].

However, it is worth giving some details of the proof of the theorem in the case when (b) holds. We choose an arbitrary $f \in \mathbb{F}$ and, proceeding as in (4.33), we obtain

$$g(f) - g_n(f) = \sum_{i \in S_n} \mu_f^n(i) \sum_{j > n} (h_f(j) - h_f(n)) q_{ij}(f).$$

Pick $0 < K_0 < n$. Recalling (3.27), (4.28), and (4.30), we have

$$\left| \sum_{K_0 < i \leq n} \mu_f^n(i) \sum_{j > n} (h_f(j) - h_f(n)) q_{ij}(f) \right| \leq \frac{2\hat{b}(2 + b)RM}{\delta \hat{c} w^\eta(K_0)}. \qquad (4.42)$$

On the other hand, by Assumption 4.13(b) and (4.41),

$$\left| \sum_{0 \leq i \leq K_0} \mu_f^n(i) \sum_{j > n} (h_f(j) - h_f(n)) q_{ij}(f) \right|$$

is bounded above by

$$\frac{2RM}{\delta} \sum_{0 \leq i \leq K_0} \frac{\mu_f^n(i)}{w^{1+\eta}(n)} \sum_{j > n} q_{ij}(f) w^{2+\eta}(j)$$

$$\leq \frac{2RM}{\delta} \sum_{0 \leq i \leq K_0} \frac{\mu_f^n(i)(\hat{b} + w^{3+\eta}(i))}{w^{1+\eta}(n)},$$

and so

$$\left| \sum_{0 \le i \le K_0} \mu_f^n(i) \sum_{j>n} (h_f(j) - h_f(n)) q_{ij}(f) \right| \le \frac{2RM(\hat{b} + w^{3+\eta}(K_0))}{\delta w^{1+\eta}(n)}.$$

(4.43)

Therefore, since the bounds (4.42) and (4.43) are uniform on $\mathbb{F}$, given $\varepsilon > 0$ we can choose K_0 and n large enough so that $|g(f) - g_n(f)| \le \varepsilon$ for all $f \in \mathbb{F}$. Therefore, $|g_n^* - g^*| \le \varepsilon$, which completes the proof of the convergence $g_n^* \to g^*$.

The rest of the proof is skipped. $\square$

As a consequence of the above proof, we have the following important fact (cf. Remark 4.19).

Remark 4.24. Suppose that condition (b) in Theorem 4.23 holds. The bounds (4.42) and (4.43) show that, given ε, we can explicitly determine n such that $|g_n^* - g^*| \le \varepsilon$ provided that all the involved constants are known.

(In what follows, recall the notation $\sim$ and O introduced in Sec. 1.4.) In addition, if $w(n) \sim n^\beta$, for some $\beta > 0$, which is the case in many applications such as queueing systems or controlled birth-and-death processes (see, e.g., [45, 131], or the examples in Chapter 9 below), we can choose $K_0 \sim n^\alpha$ for some $0 < \alpha < \frac{\beta(1+\eta)}{3+\eta}$; recall (4.43). In particular, we can let $K_0 \sim n^{\beta/3}$, so that

$$|g_n^* - g^*| = O(n^{-2\beta\eta/3}).$$

Therefore, the convergence is at least of order $2\beta\eta/3$ as the truncation size n grows. Obviously, the larger η in (4.41), the larger the convergence order. Hence, we obtain an explicit bound on the convergence order, in contrast to what is mentioned in [100, Sec. 1]: that the approximation methods do not have an error estimate and they are "asymptotic in nature."

4.5.   Conclusions

In this chapter, we have presented the policy iteration algorithm for discounted and average reward CMCs, and we have proved its convergence. We have also proposed approximation theorems for discounted and average reward CMCs, which are very much like "continuity theorems" for control models. An application of practical interest of such approximation results is the truncation technique studied for both discounted and average models.

It is worth noting that, in some cases, we can explicitly give the order of convergence of these approximations.

Concerning the approximation theorems in Secs 4.3 and 4.4, let us mention some open issues. In Sec. 4.3 we have not obtained the convergence rate of V_α^{*n} to V_α^*, whereas such convergence rates have been established, for some particular average reward models, in Remark 4.24. Therefore, giving bounds on $|V_\alpha^{*n}(i) - V_\alpha^*(i)|$ remains an open issue.

We note also that, in Sec. 4.3, the action sets of the approximating control models $\mathcal{M}_n$ can vary with n. On the other hand, for average reward problems in Sec. 4.4, the control models $\mathcal{M}$ and $\mathcal{M}_n$ have the same state and action spaces (though in Theorem 4.21 the state space depends on n). It would be useful to allow the action sets $A(i)$ of the approximating models $\mathcal{M}_n$ to vary with n; hence, an interesting open issue is to establish Theorems 4.17 and 4.18 when the approximating control models are of the form

$$\mathcal{M}_n = \{S_n, A, (A_n(i)), (q_{ij}^n(a)), (r_n(i, a))\}.$$

Let us mention that the policy iteration algorithm (for discrete-time finite state and finite action spaces CMCs) can be modified [138] in order to account for the so-called advanced optimality criteria (such as bias optimality or sensitive discount optimality), which we will study in the subsequent chapters. Hence, an issue worth studying is whether the above defined finite state truncation technique can be used to define a policy iteration algorithm to approximate, say, bias optimal policies.

In Chapter 9 below, we apply the finite state truncation technique described in Secs 4.3 and 4.4 to two controlled population models, namely, a controlled birth-and-death process in Sec. 9.3 and a controlled population system with catastrophes in Sec. 9.4. In particular, we will show that this approach can be used to make explicit numerical approximations of the optimal rewards and the optimal policy of a denumerable state control model.

Chapter 5

Overtaking, Bias, and Variance Optimality

5.1. Introduction

It is a well-known fact that the average reward optimality criterion is underselective because it does not take into account the behavior of the reward process $r(t, x(t), \varphi)$ on finite time intervals. For instance, there might exist two policies $\varphi, \varphi' \in \Phi$ such that $J_T(i, \varphi) > J_T(i, \varphi')$ for some $i \in S$ and every $T > 0$, but with the same long-run average reward $J(i, \varphi) = J(i, \varphi')$. Thus, φ and φ' have the same long-run average reward although, clearly, φ should be preferred to φ'. Moreover, as mentioned in the conclusions of Chapter 3, one way to find an intermediate optimality criterion between discount and average optimality is by maximizing the growth of the finite horizon total expected reward $J_T(i, \varphi)$.

Let us mention that, starting from (3.29), we will prove that the total expected reward $J_T(i, \varphi)$ of a stationary policy is, asymptotically as the time horizon $T \to \infty$, a straight line with slope $g(f)$ — the gain — and ordinate $h_f(i)$ — the bias. Therefore, one should try to find a policy with maximal bias (i.e., with maximal ordinate) within the class of average reward optimal policies (i.e., with maximal slope). This is what bias optimality is about.

The concept of bias optimality was introduced by Blackwell [17] and also by Veinott [160] as a tool to analyze the so-called near optimality criterion defined by Blackwell [17], subsequently known as the strong 0-discount optimality criterion (see Chapter 6). Denardo and Veinott studied some other characterizations of bias optimality [27]. Bias optimality has been extensively studied for discrete-time CMCs. See, for instance, the papers by Denardo [26], Lewis and Puterman [102, 103], or Haviv and Puterman [62], the latter showing an application of bias optimality to queueing systems. Bias optimality for discrete-time models is also analyzed by Hernández-Lerma and Lasserre [70], and Hilgert and Hernández-Lerma [74]. For continuous-time control models, there have been recent developments in the study of bias optimality. For continuous-time denumerable state CMCs, see the paper by Prieto-Rumeau and Hernández-Lerma [129]. Bias optimality for multichain CMCs with finite state space is analyzed in [56]. The case of jump Markov control processes was studied by Zhu and Prieto-Rumeau [174], while Jasso-Fuentes and Hernández-Lerma deal with multidimensional controlled diffusions [82, 84].

The notion of overtaking optimality is particularly important in economic growth problems. It was first mentioned by Ramsey [140], and later used by Gale [41], von Weizsäcker [162] — from which we borrow the concept of weak overtaking optimality; see our definition below — Carlson, Haurie, and Leizarowitz [20], and Leizarowitz [99], among many other authors.

There is still another way of improving the average reward optimality criterion. It is to select a gain optimal policy with minimal limiting average variance

$$\lim_{T \to \infty} \frac{1}{T} E_i^f \left[\left(\int_0^T r(x(t), f) dt - J_T(i, f) \right)^2 \right].$$

This optimality criterion selects a policy with the maximal average reward and whose finite horizon expected reward, in addition, has the "smallest deviation" from the line with slope g^*.

This criterion is related to the Markowitz allocation problem, in which an investor tries to maximize a certain expected return with the minimum risk (or variance); see, e.g., [10]. The discrete-time variance criterion was studied by Puterman [138], Hernández-Lerma and Lasserre [70], and also by Hernández-Lerma, Vega-Amaya, and Carrasco [73]. For continuous-time CMCs, we refer to Guo and Hernández-Lerma [52], and Prieto-Rumeau and Hernández-Lerma [132].

The rest of the chapter is organized as follows. In Section 5.2 we study the bias and the overtaking optimality criteria: we show that they are equivalent and derive the corresponding optimality equations. This section is mainly based on [129] and it also relies on results given in Section 3.4 on the bias and the Poisson equation. In Section 5.3 we analyze the variance optimality criterion, following [132]. In Section 5.4 we show by means of a simple example that bias and variance optimality are, in fact, opposite criteria (meaning that bias optimality assumes a large risk, as opposed to variance optimality). We propose a multicriteria approach that identifies the Pareto solutions for the intermediate positions between bias and variance optimality. We state our conclusions in Section 5.5.

In this chapter, we suppose that Assumptions 3.1, 3.3, *and* 3.11 hold.

5.2. Bias and overtaking optimality

To account for the fact that the average reward optimality criterion is underselective (recall the discussion in Sec. 5.1 above), we introduce next the *overtaking optimality criterion* for which we will restrict ourselves to the class $\mathbb{F}$ of deterministic stationary policies. Indeed, it will be evident that overtaking optimal policies are, necessarily, average optimal, and, thus, we will be working with a subset of $\mathbb{F}_{ao} = \mathbb{F}_{ca}$. (See Remark 3.22.)

We say that a policy $f \in \mathbb{F}$ *overtakes* the policy $f' \in \mathbb{F}$ if, for every $i \in S$ and $\varepsilon > 0$, there exists $T_0 > 0$ such that

$$J_T(i, f) \geq J_T(i, f') - \varepsilon \quad \forall\, T \geq T_0, \tag{5.1}$$

or, equivalently,

$$\liminf_{T \to \infty} [J_T(i, f) - J_T(i, f')] \geq 0.$$

(In the literature, (5.1) is sometimes referred to as *weak* overtaking, whereas overtaking refers to the case $\varepsilon = 0$.)

Therefore, f overtakes f' when the total expected reward of f is larger (except for a "small" constant) than the total expected reward of f' over any finite horizon $[0, T]$, provided that T is large enough.

Definition 5.1. A policy $f \in \mathbb{F}$ is said to be *overtaking optimal* in $\mathbb{F}$ if f overtakes every $f' \in \mathbb{F}$.

Multiplying both sides of (5.1) by T^{-1} and letting $T \to \infty$, it is evident that an overtaking optimal policy is necessarily gain optimal (recall (3.25)). Therefore, overtaking optimality is indeed a refinement of the average reward optimality criterion.

An obvious question is whether the family of overtaking optimal policies is nonempty. To answer this question we need to analyze the asymptotic behavior of $J_T(i, f)$ as $T \to \infty$. To this end, we recall equation (3.29) in Remark 3.13:

$$J_T(i, f) = g(f)T + h_f(i) - E_i^f [h_f(x(T))]$$

for every $f \in \mathbb{F}$, $i \in S$, and $T \geq 0$. Since $\mu_f(h_f) = 0$ (by (3.28)), Assumption 3.11(c), together with (3.27), gives

$$|E_i^f[h_f(x(T))]| \leq R^2 M e^{-\delta T} w(i)/\delta.$$

Therefore,

$$J_T(i, f) = g(f)T + h_f(i) + U(i, f, T), \qquad (5.2)$$

where the residual term $U(i, f, T)$ verifies that

$$\sup_{f \in \mathbb{F}} \|U(\cdot, f, T)\|_w \leq R^2 M e^{-\delta T}/\delta.$$

Thus, we have the following interpretation of the bias h_f. Given a stationary policy $f \in \mathbb{F}$ and an initial state $i \in S$, the total expected reward on $[0, T]$ is, approximately as $T \to \infty$, a straight line whose slope is the gain $g(f)$ of f, and whose ordinate is the bias $h_f(i)$. The error of this approximation decreases, as $T \to \infty$, exponentially in the w-norm uniformly in $f \in \mathbb{F}$.

Bias optimality   The above interpretation of the bias gives a clear intuition of how to obtain overtaking optimal policies. Indeed, if $g(f) > g(f')$ then, by (5.2), f overtakes f'. Now, if $g(f) = g(f')$ and $h_f \geq h'_f$ componentwise, then f overtakes f' as well. Hence, to obtain an overtaking optimal policy, one should try to find a policy with maximal bias within the class of average reward optimal policies. This leads to our next definition, which uses the notation $\mathbb{F}_{\mathrm{ca}}$ for the set of canonical policies (recall Remark 3.22).

Definition 5.2. We define the optimal bias function $\hat{h} \in \mathcal{B}_w(S)$ as

$$\hat{h}(i) = \sup_{f \in \mathbb{F}_{\mathrm{ca}}} h_f(i).$$

(By (3.27), $\hat{h}$ is indeed in $\mathcal{B}_w(S)$.) A policy $\hat{f} \in \mathbb{F}$ is said to be bias optimal if it is gain optimal and, in addition, $h_{\hat{f}} = \hat{h}$.

Our next theorem is a direct consequence of the previous arguments, and it is stated without proof.

Theorem 5.3. *Suppose that Assumptions* 3.1, 3.3, *and* 3.11 *hold. A policy* $f \in \mathbb{F}$ *is overtaking optimal in* $\mathbb{F}$ *if and only if it is bias optimal.*

Of course, this theorem is for the moment useless because we do not yet know whether there exist bias optimal policies. Thus, the rest of the section is devoted to proving the existence of bias optimal policies.

Let $(g^*, h) \in \mathbb{R} \times \mathcal{B}_w(S)$ be a solution of the AROE

$$g^* = \max_{a \in A(i)} \left\{ r(i, a) + \sum_{j \in S} q_{ij}(a) h(j) \right\} \quad \forall \, i \in S,$$

and recall (from (3.38)) that, for each $i \in S$, $A^*(i) \subseteq A(i)$ is the subset of actions that attain the maximum in the AROE.

We say that $(g, h, h') \in \mathbb{R} \times \mathcal{B}_w(S) \times \mathcal{B}_w(S)$ is a solution of the *bias optimality equations* if

$$g = \max_{a \in A(i)} \left\{ r(i, a) + \sum_{j \in S} q_{ij}(a) h(j) \right\} \quad \forall \, i \in S, \tag{5.3}$$

$$h(i) = \max_{a \in A^*(i)} \left\{ \sum_{j \in S} q_{ij}(a) h'(j) \right\} \quad \forall \, i \in S. \tag{5.4}$$

We now state our main result in this section.

Theorem 5.4. *Under Assumptions 3.1, 3.3, and 3.11, the following results hold:*

(i) *There exist solutions $(g, h, h') \in \mathbb{R} \times \mathcal{B}_w(S) \times \mathcal{B}_w(S)$ to the bias optimality equations and, necessarily, $g = g^*$ and $h = \hat{h}$, the optimal bias function.*

(ii) *A policy $f \in \mathbb{F}$ is bias optimal if and only if $f(i)$ attains the maximum in (5.3) and (5.4) for every $i \in S$.*

(iii) *A canonical policy $f \in \mathbb{F}_{\mathrm{ca}}$ is bias optimal if and only if $\mu_f(\hat{h}) = 0$.*

Proof. Proof of (i). Obviously, (5.3) corresponds to the AROE (3.36) and then $g = g^*$, by Theorem 3.20. Given any canonical policy f, its bias h_f and h differ by a constant (see Corollary 3.23). As a consequence, recalling that $\mu_f(h_f) = 0$, we have

$$h_f(i) = h(i) - \mu_f(h) \quad \forall \, i \in S. \tag{5.5}$$

Therefore, by Definition 5.2,

$$\hat{h}(i) = h(i) + \max_{f \in \mathbb{F}_{\mathrm{ca}}} \{-\mu_f(h)\} \quad \forall \, i \in S. \tag{5.6}$$

Now observe that $\sigma^* := \max_{f \in \mathbb{F}_{\mathrm{ca}}} \{-\mu_f(h)\}$ is the optimal gain of a control problem with reward rate function $-h(i)$ and action sets $A^*(i)$, for $i \in S$; see (3.25). Hence, consider a new control model with state space S, compact action sets $A^*(i)$, transition rates $q_{ij}(a)$, and reward rate $-h(i)$. It is clear that this control model satisfies Assumptions 3.1, 3.3, and 3.11 and, therefore, the corresponding AROE is

$$\sigma^* = \max_{a \in A^*(i)} \left\{ -h(i) + \sum_{j \in S} q_{ij}(a) h'(j) \right\} \quad \forall \, i \in S$$

for some $h' \in \mathcal{B}_w(S)$. Consequently, $(g^*, \hat{h}, h')$ is a solution of the bias optimality equations.

To complete the proof of statement (i), it remains to show that if (g^*, h, h') is a solution of the bias optimality equations, then $h = \hat{h}$. To this end, note that (5.4) implies that $\sigma^* = 0$ and thus $h = \hat{h}$ follows from (5.6).

Proof of (ii). Suppose that $f \in \mathbb{F}$ is bias optimal. Then, f is canonical and so it attains the maximum in (5.3). In addition, h_f and $\hat{h}$ differ by a constant. More precisely,

$$h_f(i) = \hat{h}(i) - \mu_f(\hat{h}) \quad \forall \, i \in S.$$

But, since f is bias optimal, we must have $\mu_f(\hat{h}) = 0$, that is, f is optimal for the control model with action sets $A^*(i)$ and reward rate $-\hat{h}$. Hence, f attains the maximum in the corresponding AROE, i.e., (5.4). This proves the "only if" part.

Suppose now that $f \in \mathbb{F}$ attains the maximum in the bias optimality equations, and fix an arbitrary $f' \in \mathbb{F}_{ca}$. By (5.6) and recalling that $h = \hat{h}$ by statement (i), we have $\mu_{f'}(\hat{h}) \geq 0$, with equality if $f' = f$. Now, from (5.5) we derive that $h_{f'} \leq \hat{h}$ with equality if $f' = f$, which shows that f is bias optimal.

Proof of (iii). The last statement is now a direct consequence of parts (i) and (ii). $\square$

Theorem 5.4 thus proves the existence of bias optimal policies and characterizes them by means of two nested average reward optimality equations. Combined with Theorem 5.3, it also establishes the existence of overtaking optimal policies in $\mathbb{F}$.

5.3.   Variance minimization

In this section we propose an alternative refinement of the average reward optimality criterion. Namely, we want to obtain, within the class of gain optimal policies, a policy with minimal *limiting average variance*, which is defined as

$$\sigma^2(i,f) := \lim_{T\to\infty} \frac{1}{T} E_i^f \left[\left(\int_0^T r(x(t),f)dt - J_T(i,f) \right)^2 \right]$$

for each deterministic stationary policy $f \in \mathbb{F}$ and every initial state i in S. Equivalent expressions of the limiting average variance are given in Proposition 5.8 and (5.15) below.

For the variance minimization problem to be well posed, we introduce the following assumptions on the control model.

Assumption 5.5. There exists a Lyapunov function w on S such that

(a) For all $i \in S$, $q(i) \leq w(i)$.
(b) There exist constants $c > 0$ and $b \geq 0$, and a finite set $C \subset S$ such that

$$\sum_{j\in S} q_{ij}(a)w^2(j) \leq -cw^2(i) + b \cdot \mathbf{I}_C(i) \quad \forall \, (i,a) \in K$$

(cf. (2.25) or Assumption 3.11(a)).
(c) There exists a constant $M > 0$ such that $|r(i,a)| \leq Mw(i)$ for each $(i,a) \in K$ (this is the same as Assumption 3.1(c)).

A condition similar to Assumption 5.5(a) was imposed in Assumptions 4.5(b) and 4.13(b). By Assumption 5.5(b) and Theorem 2.13, the condition in Assumption 3.11(a) is also satisfied. It is useful to note that, as in (2.13), we deduce from Assumption 5.5(b) that

$$E_i^\varphi[w^2(x(t))] \leq e^{-ct}w^2(i) + b(1 - e^{-ct})/c \tag{5.7}$$

for all $\varphi \in \Phi$, $t \geq 0$, and $i \in S$.

We also need the standard continuity and compactness conditions in Assumption 3.3, which we restate here in a slightly different form, as follows.

Assumption 5.6.

(a) For each $i \in S$, $A(i)$ is a compact Borel space.
(b) For each fixed $i, j \in S$, the functions $a \mapsto q_{ij}(a)$ and $a \mapsto r(i,a)$ are continuous on $A(i)$.

The continuity of $a \mapsto \sum_{j\in S} q_{ij}(a)w(j)$ imposed in Assumption 3.3(b) is obtained (recall Remark 4.6(ii)) from Assumption 5.5(b) and is, therefore, omitted.

Assumption 5.7. Every policy in $\mathbb{F}$ is irreducible.

The above assumptions imply, in particular, that our control model is w-exponentially ergodic uniformly on $\mathbb{F}$ (see Theorem 2.11).

In our next result we obtain an explicit expression for $\sigma^2(i, f)$ that involves the bias $h_f \in \mathcal{B}_w(S)$ of f and its invariant probability measure μ_f. In particular, we show that the limiting average variance does not depend on the initial state $i \in S$.

Proposition 5.8. *Suppose that Assumptions* 5.5, 5.6, *and* 5.7 *are satisfied. Given a policy $f \in \mathbb{F}$ and an initial state $i \in S$, the corresponding limiting average variance*

$$\sigma^2(f) := 2 \sum_{j \in S} (r(j, f) - g(f)) h_f(j) \mu_f(j).$$

Proof. Pick an arbitrary policy $f \in \mathbb{F}$ and define, for $T \geq 0$,

$$U_T := \int_0^T [r(x(t), f) - g(f)] dt \quad \text{and} \quad M_T := U_T + h_f(x(T)) - h_f(i). \quad (5.8)$$

From the Poisson equation (Proposition 3.14), Dynkin's formula (Proposition 2.3), and (5.7), it follows that $\{M_T\}_{T \geq 0}$ is a square-integrable P_i^f-martingale.

The proof of this proposition now proceeds in several steps. First of all, let us prove that

$$\lim_{n \to \infty} \frac{1}{n} E_i^f [U_n^2] = \sigma^2(f). \quad (5.9)$$

Indeed, recalling (5.8),

$$\frac{1}{n} E_i^f [U_n^2] = \frac{1}{n} E_i^f [M_n^2] + \frac{1}{n} E_i^f [(h_f(x(n)) - h_f(i))^2]$$
$$- \frac{2}{n} E_i^f [M_n (h_f(x(n)) - h_f(i))]. \quad (5.10)$$

By (5.7), $\sup_n E_i^f [(h_f(x(n)) - h_f(i))^2] < \infty$, and so

$$\lim_{n \to \infty} \frac{1}{n} E_i^f [(h_f(x(n)) - h_f(i))^2] = 0. \quad (5.11)$$

Similarly, it follows from (5.7) that $\sup_n E_i^f [(M_{n+1} - M_n)^2] < \infty$ and, by the orthogonality property of martingale differences, $E_i^f [M_n^2] = O(n)$. Using the Cauchy–Schwartz inequality,

$$\lim_{n \to \infty} \frac{2}{n} E_i^f [M_n (h_f(x(n)) - h_f(i))] = 0. \quad (5.12)$$

On the other hand,

$$\frac{1}{n}E_i^f[M_n^2] = \frac{1}{n}\sum_{k=1}^n E_i^f[(M_k - M_{k-1})^2] = \frac{1}{n}\sum_{k=1}^n E_i^f[\Psi(x(k-1))],$$

where Ψ is the function defined as

$$\Psi(i) := E_i^f\left[\left(\int_0^1 r(x(s), f)ds + h(x(1)) - h(i) - g(f)\right)^2\right] \quad \forall\, i \in S.$$

It is clear that Ψ is in $\mathcal{B}_{w^2}(S)$ (recall (2.10)), and by w^2-ergodicity (use Theorem 2.5 (iii) in connection with Assumption 5.5(b)), we obtain

$$\lim_{n\to\infty}\frac{1}{n}E_i^f[M_n^2] = \int_S \Psi d\mu_f = 2\sum_{j\in S}(r(j,f) - g(f))h_f(j)\mu_f(j), \quad (5.13)$$

where the latter equality follows from [15, Equation (2.12)]. Combining (5.10) with (5.11)–(5.13) proves (5.9). (An alternative proof of (5.13) is given in [52, Theorem 10.3].)

Our next step in the proof of Proposition 5.8 is to show that (5.9) holds when the integer $n \geq 0$ is replaced with $T \geq 0$, i.e.,

$$\lim_{T\to\infty}\frac{1}{T}E_i^f[U_T^2] = \sigma^2(f), \quad (5.14)$$

or, equivalently,

$$\lim_{T\to\infty}\left[\frac{1}{T}E_i^f[U_T^2] - \frac{1}{[T]}E_i^f[U_{[T]}^2]\right] = 0,$$

where $[T]$ denotes the integer part of T. Indeed, note that

$$\left|\frac{1}{T}E_i^f[U_T^2] - \frac{1}{[T]}E_i^f[U_{[T]}^2]\right| \leq \left(\frac{1}{[T]} - \frac{1}{T}\right)E_i^f[U_{[T]}^2] + \frac{1}{T}E_i^f|U_T^2 - U_{[T]}^2|.$$

Hence, as a consequence of (5.9), $(\frac{1}{[T]} - \frac{1}{T})E_i^f[U_{[T]}^2]$ converges to zero. On the other hand,

$$\frac{1}{T}E_i^f[|U_T^2 - U_{[T]}^2|] \leq \frac{1}{T}\sqrt{E_i^f[(U_T - U_{[T]})^2]E_i^f[(U_T + U_{[T]})^2]},$$

which, by arguments similar to those used to obtain (5.11) and (5.12), also converges to zero. Therefore, (5.14) holds.

Finally, using (5.2) and (5.7), it is easily seen that

$$\sigma^2(i, f) = \lim_{T \to \infty} \frac{1}{T} E_i^f [U_T^2].$$

By (5.14), this completes the proof.   □

In the proof of Proposition 5.8 we have also shown that

$$\sigma^2(f) = \lim_{T \to \infty} \frac{1}{T} E_i^f \left[\left(\int_0^T r(x(t), f) dt - J_T(i, f) \right)^2 \right]$$

$$= \lim_{T \to \infty} \frac{1}{T} E_i^f \left[\left(\int_0^T [r(x(t), f) - g(f)] dt \right)^2 \right]. \qquad (5.15)$$

Remark 5.9. As a consequence of [15, Theorem 2.1],

$$\frac{1}{\sqrt{T}} \left(\int_0^T r(x(t), f) dt - J_T(i, f) \right) \quad \text{and} \quad \frac{1}{\sqrt{T}} \int_0^T [r(x(t), f) - g(f)] dt$$

converge weakly to a normal distribution with zero mean and variance $\sigma^2(f)$. Since weak convergence does not imply the convergence of moments (in particular, second order moments), we cannot deduce Proposition 5.8 from the above mentioned weak convergence.

Next we give the definition of a variance optimal policy. We recall that $\mathbb{F}_{ca}$ denotes the set of canonical policies of the average reward CMC. We say that a policy $f^* \in \mathbb{F}_{ca}$ is *variance optimal* if, for every $f \in \mathbb{F}_{ca}$, $\sigma^2(f^*) \leq \sigma^2(f)$.

Our next theorem shows that variance minimization optimal policies can be found by solving two nested *variance optimality equations* (as is the case for the bias optimality criterion; see (5.3) and (5.4)).

Theorem 5.10. *Suppose that Assumptions* 5.5, 5.6, *and* 5.7 *hold, and that, in addition, there exist constants* $\tilde{c} \in \mathbb{R}$ *and* $\tilde{b} \geq 0$ *such that*

$$\sum_{j \in S} q_{ij}(a) w^3(j) \leq -\tilde{c} w^3(i) + \tilde{b} \quad \forall \, (i, a) \in K, \qquad (5.16)$$

where the Lyapunov function w is taken from Assumption 5.5. Then:

(i) *There exists a solution $(g^*, h, \sigma^2, u) \in \mathbb{R} \times \mathcal{B}_w(S) \times \mathbb{R} \times \mathcal{B}_{w^2}(S)$ to the system of equations*

$$g^* = \max_{a \in A(i)} \left\{ r(i,a) + \sum_{j \in S} q_{ij}(a) h(j) \right\}, \tag{5.17}$$

$$\sigma^2 = \min_{a \in A^*(i)} \left\{ 2(r(i,a) - g)h(i) + \sum_{j \in S} q_{ij}(a)u(j) \right\}, \tag{5.18}$$

for every $i \in S$, where $A^*(i)$ is defined in (3.38).

(ii) *A policy f in $\mathbb{F}$ is variance optimal if and only if $f(i)$ attains the maximum and the minimum in (5.17) and (5.18), respectively, for every $i \in S$. The minimal limiting average variance $\min_{f \in \mathbb{F}_{ca}} \sigma^2(f)$ equals σ^2 in (5.18).*

Proof. First of all, let us mention that (5.16) ensures the application of the Dynkin formula (Proposition 2.3) for functions in $\mathcal{B}_{w^2}(S)$, and also, by Theorem 2.11, w^2-exponential ergodicity uniformly on $\mathbb{F}$.

Now we proceed to prove statement (i). The existence of solutions to the AROE (5.17) follows from Theorem 3.20. Suppose now that $f \in \mathbb{F}$ is a canonical policy. By Proposition 3.14, $g(f) = g^*$ and $h_f = h - \mu_f(h)$. Thus, by Proposition 5.8, the limiting average variance of f verifies that

$$\sigma^2(f) = 2 \sum_{i \in S} (r(i,f) - g(f)) h_f(i) \mu_f(i) = 2 \sum_{i \in S} (r(i,f) - g^*) h(i) \mu_f(i),$$

where the right-hand equality is derived from (3.25). This shows that $\sigma^2(f)$ is the *expected average cost* (which is to be minimized) of the policy f when the cost rate function is

$$r'(i,a) := 2(r(i,a) - g^*) h(i) \quad \forall \; (i,a) \in K.$$

Note that for some constant $M' > 0$ we have $|r'(i,a)| \leq M' w^2(i)$ for every $(i,a) \in K$.

Hence, the control model with state space S, compact action sets $A^*(i)$ for $i \in S$, transition rates $q_{ij}(a)$, and cost rate given by r' above verifies the hypotheses in Theorem 3.20 when the Lyapunov function w is replaced with w^2. Thus, under our assumptions, there exist solutions to the corresponding average cost optimality equation (5.18), in analogy with the average reward optimality equation (AROE).

The proof of (ii) is now straightforward. $\qquad\qquad\square$

Note that, in (5.16), we have to go one degree further in the Lyapunov conditions, and impose them on the function w^3. This is because the Lyapunov condition ensuring the application of the Dynkin formula is, typically, one degree larger than the reward (or cost) rate function. In this case, the cost rate function r' being in $\mathcal{B}_{w^2}(K)$, we need a Lyapunov condition on w^3.

5.4.   Comparison of variance and overtaking optimality

At this point, the situation is that a decision-maker who wishes to improve the average reward optimality criterion has two different possibilities: either choose a policy with minimal variance or a policy with maximal bias. Two questions arise. Is either of these two criteria "better" than the other one? Can we optimize both criteria simultaneously? More explicitly, can we find, in general control models, a canonical policy with minimal limiting average variance and maximal bias at the same time?

As already pointed out (see (5.2)), the expected reward $J_T(i, f)$ lines up with a straight line with slope $g(f)$ (for the asymptotic variance, the ordinate $h_f(i)$ is not relevant). Therefore, when minimizing within $\mathbb{F}_{\mathrm{ca}}$ (recall (5.15))

$$\sigma^2(f) = \lim_{T \to \infty} \frac{1}{T} E_i^f \left[\left(\int_0^T [r(x(t), f) - g^*] dt \right)^2 \right],$$

the decision-maker chooses a policy with minimal limiting average deviation from $T \mapsto g^* T$. On the other hand, recalling (5.2), when choosing a bias optimal policy, the decision-maker selects, among the family of parallel lines $g^* T + h_f(i)$, for $f \in \mathbb{F}_{\mathrm{ca}}$, the one with the largest ordinate. Hence, intuitively, it seems that the two criteria are, in fact, opposed because the variance criterion assumes minimal risk (or variance), whereas the bias criterion tries to maximize the expected growth rate whatever the risk is. This is made clear in the following example.

Example 5.11. Consider the control model with state space $S = \{1, 2\}$, action sets $A(1) = [4, 5]$ and $A(2) = \{a_0\}$, transition rates $q_{11}(a) = \frac{1}{2}(3 - a)$ and $q_{22}(a_0) = -1$, and reward rates $r(1, a) = a$ and $r(2, a_0) = 1$. We identify each policy $f \in \mathbb{F}$ with its value $f(1) = a \in [4, 5]$.

Simple calculations show that the invariant probability measures are

$$\mu(a) = \frac{1}{a - 1} (2, a - 3),$$

so that every policy $f \in \mathbb{F}$ is gain optimal with $g(a) = 3 = g^*$. Similarly, the corresponding bias and limiting average variance are

$$h(a) = \frac{1}{a-1} \cdot \begin{pmatrix} 2a-6 \\ -4 \end{pmatrix} \quad \text{and} \quad \sigma^2(a) = \frac{8(a-3)}{a-1},$$

respectively. The maximal bias is attained at $a = 5$ (which corresponds to the *maximal* variance), whereas the minimal variance is attained at $a = 4$ (which corresponds to the *minimal* bias).

This example shows that the average variance and the bias cannot, in general, be simultaneously minimized and maximized, respectively. Therefore, we end up with a *multicriteria control problem* in which we want to minimize and maximize, respectively, the components of the vector criterion $(\sigma^2(f), h_f)$ for f in $\mathbb{F}_{\mathrm{ca}}$. We are interested in determining the corresponding set of *nondominated* (or *Pareto*) *solutions*, that is, the set of policies $f \in \mathbb{F}_{\mathrm{ca}}$ such that, for any other canonical policy f', either $\sigma^2(f') > \sigma^2(f)$ or $h_{f'} < h_f$.

This is achieved in the next result by using the usual "scalarization approach" (see, e.g., [72]), which consists in considering a control model whose cost rate function is a convex combination of the corresponding cost rates for variance minimization and overtaking optimality (recall Theorems 5.4 and 5.10). Observe that this result introduces a "risk parameter" β, which is interpreted as the decision-maker's attitude towards risk between the two extremes $\beta = 0$ (bias optimality) and $\beta = 1$ (variance optimality).

Let us first introduce some notation. Let $\hat{h}$ be the optimal bias function (Definition 5.2). By Corollary 3.23, we can write

$$\hat{h}(i) = h_f(i) + c_f \quad \forall\, i \in S, \tag{5.19}$$

where the constant $c_f \geq 0$ equals 0 for a bias optimal policy.

The proof of Theorem 5.12 below is similar to that of Theorem 5.10 and is, therefore, omitted.

Theorem 5.12. *Suppose that Assumptions 5.5, 5.6, and 5.7, as well as (5.16), hold, and fix a risk parameter $0 \leq \beta \leq 1$. Let $g^* \in \mathbb{R}$ and $\hat{h} \in \mathcal{B}_w(S)$ be the optimal gain and the optimal bias function, respectively. Then:*

(i) *There exist solutions $(v(\beta), u) \in \mathbb{R} \times \mathcal{B}_w(S)$ to the optimality equation*

$$v(\beta) = \min_{a \in A^*(i)} \left\{ 2\beta(r(i,a) - g^*)\hat{h}(i) + (1-\beta)\hat{h}(i) + \sum_{j \in S} q_{ij}(a)u(j) \right\} \tag{5.20}$$

for all $i \in S$, with $A^*(i)$ as in (5.18).

(ii) *Using the notation of* (5.19),

$$v(\beta) = \min_{f \in \mathbb{F}_{\mathrm{ca}}} \{\beta \sigma^2(f) + (1 - \beta)c_f\}.$$

(iii) *As β varies in $[0, 1]$, the set of policies attaining the minimum in* (5.20) *for every $i \in S$ ranges over the set of nondominated policies of the vector criterion $\{(\sigma^2(f), h_f)\}_{f \in \mathbb{F}_{\mathrm{ca}}}$.*

Hence, the set of Pareto optimal policies is determined by, first, choosing an arbitrary $0 \leq \beta \leq 1$ and, second, solving the optimality equation (5.20).

5.5. Conclusions

In this chapter we have analyzed the bias (or overtaking) and the variance optimality criteria, both of which are refinements of the average reward optimality criterion. The structure of both analyses is similar. Indeed, it is proved that bias optimal and variance optimal policies are obtained by solving an average reward or cost problem (with a suitably defined reward or cost rate function) within the class of gain optimal policies. The corresponding optimality equations are then a pair of nested AROE-like equations. In Chapter 6 below, when dealing with sensitive discount optimality criteria, we will obtain as well nested average optimality equations, which are extensions of the bias optimality equations.

An interesting open issue is to define a policy iteration algorithm, as we did in Chapter 4 for average reward CMCs, but this time for the bias optimality criterion. Let us mention that, for discrete-time CMCs, a policy iteration for the bias optimality criterion is described in [138].

Technically, the difference between bias and variance optimality is that, in the first case, the reward rate of the nested control model is in $\mathcal{B}_w(K)$, while the cost rate function for variance optimality is in $\mathcal{B}_{w^2}(K)$. This means that the hypotheses for variance optimality are stronger, but there is no substantial difference.

Also, we note that some generalizations of overtaking optimality have been proposed. For instance, in [27] and [160], an averaged overtaking optimality criterion is discussed. Finally, let us mention that the concepts of bias and overtaking optimality can be extended to two-person stochastic games; see Chapter 11. In this case, however, the equivalence between bias and overtaking optimality does not hold in general.

Chapter 6

Sensitive Discount Optimality

6.1. Introduction

In Chapter 5 we studied two refinements of the average reward optimality criterion, namely, bias (or, equivalently, overtaking) optimality and variance minimization. The common feature of both refined criteria is that they deal with the asymptotic behavior of the controlled process as the time parameter goes to infinity (either by analyzing the growth of the total expected reward on $[0, T]$ or the average variance of the returns). In this chapter we follow a different approach: we want to analyze α-discounted CMCs as the discount rate $\alpha \downarrow 0$ (recall the discussion in Sec. 3.8), that is, we try to obtain a policy $\varphi \in \Phi$ such that

$$V_\alpha(i, \varphi) \geq V_\alpha(i, \varphi') \tag{6.1}$$

"asymptotically as $\alpha \downarrow 0$" for every $\varphi' \in \Phi$. Of course, the ideal situation would be to find a policy φ that is α-discount optimal for all sufficiently small discount rates $\alpha > 0$. This is precisely what Blackwell optimality is about (see Chapter 7). But before studying Blackwell optimality, we must study some "intermediate" criteria, which are the sensitive discount optimality criteria.

The main tool used to analyze the α-discounted total expected reward, as $\alpha \downarrow 0$, is the Laurent series (from functional analysis; see, e.g., [89, 168]), which usually refers to a power series with a finite number of negative powers. The Laurent series expansion for the discounted reward of a stationary policy $\varphi \in \Phi_s$ is of the form

$$V_\alpha(\cdot, \varphi) = \frac{1}{\alpha} \cdot g + h_0 + \alpha h_1 + \alpha^2 h_2 + \ldots,$$

for sufficiently small α, where $g \in \mathbb{R}$ is a constant and the h_i, for $i \geq 0$, are functions on S. Note that the Laurent series expansion is related to the series expansion of the resolvent of the Markov process $\{x^\varphi(t)\}$; see, e.g., [30, Section 4] or [64].

For technical reasons, to "asymptotically maximize" $V_\alpha(i, \varphi)$ as $\alpha \downarrow 0$, one has to consider a finite truncation of the above power series, say,

$$\frac{1}{\alpha} \cdot g + h_0 + \alpha h_1 + \alpha^2 h_2 + \cdots + \alpha^n h_n.$$

A careful inspection of the above expression shows that this "asymptotic maximization" can be carried out equivalently either by maximizing

$$\alpha^{-n} V_\alpha(i, \varphi)$$

as $\alpha \downarrow 0$, or by lexicographically maximizing the coefficients of the Laurent series

$$(g, h_0, \ldots, h_n).$$

This is precisely the n-discount optimality criterion (see Definition 6.1 below).

Blackwell [17] introduced the 0-discount optimality criterion, which was also studied by Veinott [161], and Denardo and Veinott [27]. They used a Laurent series expansion with two terms plus a residual. The n-discount optimality criteria and the complete Laurent series expansion for control models were introduced by Veinott [161] as tools to analyze Blackwell optimality, and they were also used by Miller and Veinott [114]. The Laurent series expansion has been extensively used for discrete-time CMCs; see, e.g., Puterman [138] or Kadota [85]. For continuous-time control models, the Laurent series expansion for the expected discounted reward has been proposed by Veinott [161], Taylor [157], and Sladký [151]. For controlled diffusions on compact intervals, Puterman [137] proposed such a series expansion, which Jasso-Fuentes and Hernández-Lerma extended to the case of multidimensional diffusions [83, 84]. For the case of a continuous-time CMC with denumerable state space, we refer the reader to Prieto-Rumeau and Hernández-Lerma [127].

Another optimality criterion that naturally arises when dealing with n-discount optimality is *strong* n-discount optimality. This optimality criterion has been analyzed for discrete-time CMCs by Cavazos-Cadena and Lasserre [21], Nowak [116], Jaśkiewicz [79], and Hilgert and Hernández-Lerma [74] under various recurrence and boundedness hypotheses. The continuous-time, denumerable state case was developed by Prieto-Rumeau and Hernández-Lerma [127]; Zhu dealt with this optimality criterion for continuous-time jump processes [172], while strong

n-discount optimality for controlled diffusions was studied by Jasso-Fuentes and Hernández-Lerma in [84].

The formal definition of the sensitive discount optimality criteria, in which we use the notation (3.6) and (3.8), is now given.

Definition 6.1. Let $\Phi' \subseteq \Phi$ be a class of policies, and fix an integer $n \geq -1$. We say that $\varphi^* \in \Phi'$ is n-*discount optimal* in Φ' if

$$\liminf_{\alpha \downarrow 0} \alpha^{-n}(V_\alpha(i, \varphi^*) - V_\alpha(i, \varphi)) \geq 0 \quad \forall\, i \in S,\ \varphi \in \Phi'.$$

Moreover, $\varphi^* \in \Phi$ is said to be *strong n-discount optimal* if

$$\lim_{\alpha \downarrow 0} \alpha^{-n}(V_\alpha(i, \varphi^*) - V_\alpha^*(i)) = 0 \quad \forall\, i \in S.$$

Obviously, if φ^* is strong n-discount optimal, then it is n-discount optimal in Φ. Also note that the above defined criteria become more restrictive as n grows, in the sense that if φ^* is (strong) n-discount optimal for some $n \geq 0$, then it is (strong) k-discount optimal for all $-1 \leq k \leq n$.

The rest of this chapter, in which we mainly follow [53] and [127], is organized as follows. In Sec. 6.2 we propose a Laurent series expansion for the discounted reward of stationary policies, and we prove that the coefficients of the Laurent series expansion can be determined by solving a sequence of Poisson-like equations. We make a small digression in Sec. 6.3, in which we prove some important results for the *vanishing discount approach* to average optimality (recall Sec. 3.5). In Sec. 6.4 we prove the existence of n-discount optimal policies in the class $\mathbb{F}$ of deterministic stationary policies and we characterize them by means of a sequence of nested AROE-like optimality equations. Strong discount optimality and its relations with average and bias optimality are analyzed in Sec. 6.5. Finally, n-discount optimality is extended, under some additional hypotheses, to the class Φ_s of stationary policies in Sec. 6.6. We conclude with some final remarks in Sec. 6.7.

Throughout this chapter, we suppose that Assumptions 3.1, 3.3, and 3.11 hold.

6.2. The Laurent series expansion

In what follows, we fix an arbitrary $f \in \mathbb{F}$. We define the operator $H_f : \mathcal{B}_w(S) \to \mathcal{B}_w(S)$ as

$$(H_f u)(i) := \int_0^\infty E_i^f[u(x(t)) - \mu_f(u)]dt \quad \forall\, i \in S. \tag{6.2}$$

The operator H_f is usually referred to as the *bias operator* (cf. the definition of the bias in (3.26)). Note that H_f is well defined as a consequence of Assumption 3.11(c). Besides, $H_f u$ is uniformly bounded on $\mathbb{F}$ in the w-norm; more precisely

$$\sup_{f \in \mathbb{F}} \|H_f u\|_w \leq \frac{R}{\delta} \|u\|_w \quad \forall \, u \in \mathcal{B}_w(S).$$

Also note that taking the μ_f-expectation in (6.2) gives

$$\mu_f(H_f u) = 0 \quad \forall \, f \in \mathbb{F}, \ u \in \mathcal{B}_w(S). \tag{6.3}$$

The successive compositions of H_f with itself will be denoted by H_f^k, for $k \geq 1$.

Let us recall the notation introduced in (3.10): for $i \in S$, $f \in \mathbb{F}$, and $u \in \mathcal{B}_w(S)$,

$$V_\alpha(i, f, u) := E_i^f \left[\int_0^\infty e^{-\alpha t} u(x(t)) dt \right],$$

where $\alpha > 0$ is an arbitrary discount rate. The following proposition expresses $V_\alpha(i, f, u)$ as a *Laurent series*.

Proposition 6.2. *Given a discount rate $0 < \alpha < \delta$, where δ is as in Assumption 3.11(c), and arbitrary $f \in \mathbb{F}$ and $u \in \mathcal{B}_w(S)$, we have*

$$V_\alpha(i, f, u) = \frac{1}{\alpha} \mu_f(u) + \sum_{k=0}^\infty (-\alpha)^k (H_f^{k+1} u)(i) \quad \forall \, i \in S,$$

and the series converges in the w-norm.

Proof.   Since $f \in \mathbb{F}$ remains fixed in this proof, for ease of notation we will write H instead of H_f. Let $(G_t u)(i)$ be the integrand in (6.2), i.e.,

$$(G_t u)(i) := E_i^f [u(x(t))] - \mu_f(u)$$

for all $t \geq 0$ and $i \in S$. By the uniform exponential ergodicity in Assumption 3.11(c),

$$\|G_t u\|_w \leq R e^{-\delta t} \|u\|_w. \tag{6.4}$$

With this notation, we may express $V_\alpha(i, f, u)$ as

$$V_\alpha(i, f, u) = \frac{1}{\alpha} \mu_f(u) + \int_0^\infty e^{-\alpha t} (G_t u)(i) dt \quad \forall \, i \in S. \tag{6.5}$$

Our next step in this proof is to show that, for each $i \in S$,

$$\int_0^\infty e^{-\alpha t}(G_t u)(i)dt = \sum_{k=0}^\infty (-\alpha)^k \int_0^\infty \frac{t^k}{k!}(G_t u)(i)dt, \qquad (6.6)$$

where the convergence of the series is in the w-norm. To this end, for each integer $k \geq 0$ and $|\alpha| < \delta$, let

$$M_{k,\alpha}(u, i) := \int_0^\infty \frac{(-\alpha t)^k}{k!}(G_t u)(i)dt \quad \forall\, i \in S.$$

Recalling (6.4), we have

$$\|M_{k,\alpha}(u, \cdot)\|_w \leq \frac{R|\alpha|^k}{\delta^{k+1}} \cdot \|u\|_w. \qquad (6.7)$$

It follows that

$$\sum_{k=0}^j M_{k,\alpha}(u, \cdot) = \int_0^\infty \sum_{k=0}^j \frac{(-\alpha t)^k}{k!}(G_t u)(\cdot)dt$$

is a Cauchy sequence in the Banach space $\mathcal{B}_w(S)$ that converges, as $j \to \infty$, to the left-hand side of (6.6) — indeed, the integrand is bounded by the integrable function $R\|u\|_w w(i)e^{(|\alpha|-\delta)t}$. This completes the proof of (6.6).

The next step in the proof of Proposition 6.2 is to show that

$$\int_0^\infty \frac{t^k}{k!}(G_t u)(i)dt = (H^{k+1}u)(i) \quad \forall\, i \in S, \; k \geq 0. \qquad (6.8)$$

Let

$$(C_k u)(i) := \int_0^\infty \frac{t^k}{k!}(G_t u)(i)dt \quad \forall\, i \in S, \; k = 0, 1, \dots.$$

We will use induction to prove that $C_k u = H^{k+1}u$ for each $k \geq 0$. By the definition of H in (6.2), $C_0 = H$. Suppose now that $C_{k-1} = H^k$ for some $k \geq 1$. Then, for any fixed $i \in S$,

$$(C_k u)(i) = \int_0^\infty \left(\int_0^t \frac{s^{k-1}}{(k-1)!}ds \right)(G_t u)(i)dt$$

$$= \int_0^\infty \frac{s^{k-1}}{(k-1)!} \left(\int_s^\infty (G_t u)(i)dt \right) ds \qquad \text{[by Fubini's theorem]}$$

$$= \int_0^\infty \frac{s^{k-1}}{(k-1)!} \left(\int_0^\infty (G_{s+t}u)(i)dt \right) ds.$$

Now observe that

$$\int_0^\infty (G_{s+t}u)(i)dt = \int_0^\infty [E_i^f[u(x(s+t))] - \mu_f(u)]dt$$

$$= \int_0^\infty \left[\sum_{j \in S} P_{ij}^f(s)(E_j^f[u(x(t))] - \mu_f(u))\right] dt$$

$$= \sum_{j \in S} P_{ij}^f(s) \int_0^\infty [E_j^f[u(x(t))] - \mu_f(u)]dt$$

$$= \sum_{j \in S} P_{ij}^f(s)(Hu)(j) = (G_s Hu)(i),$$

where the last equality follows from (6.3). Therefore,

$$C_k u = \int_0^\infty \frac{s^{k-1}}{(k-1)!}(G_s Hu)ds = C_{k-1}Hu = H^{k+1}u,$$

by the induction hypothesis.

Finally, combining (6.5) with (6.6) and (6.8), we complete the proof of the proposition.   □

From (6.7) we derive the following bound on the residuals of the Laurent series expansion.

Corollary 6.3. *Given* $f \in \mathbb{F}$, $u \in \mathcal{B}_w(S)$, *and* $0 < \alpha < \delta$, *define the* k-*residual of the Laurent series expansion in Proposition 6.2 as*

$$R_k(f, u, \alpha) := \sum_{j=k}^\infty (-\alpha)^j H_f^{j+1}u.$$

Then

$$\sup_{f \in \mathbb{F}} \|R_k(f, u, \alpha)\|_w \le R\|u\|_w \frac{\alpha^k}{\delta^k(\delta - \alpha)} \quad \forall\, k \ge 0.$$

If our irreducibility and ergodicity assumptions are extended to Φ_s, the class of randomized stationary policies, then the Laurent series expansion and the subsequent results are still valid. This is stated in the next remark.

Remark 6.4. If, in addition to Assumptions 3.1, 3.3, and 3.11, each randomized stationary policy is irreducible and the control model is w-exponentially ergodic uniformly on Φ_s, then the series expansion in Proposition 6.2 holds for policies in Φ_s, and the bound given in Corollary 6.3 is uniform on Φ_s.

The generalized Poisson equations Given $f \in \mathbb{F}$ and $k \geq 0$ we define

$$h_f^k(\cdot) := (-1)^k H_f^{k+1} r(\cdot, f).$$

In particular, from the definition (3.26) of the bias h_f and the definition (6.2) of H_f, it follows that $h_f^0 = H_f u = h_f \in \mathcal{B}_w(S)$. Therefore, for $k \geq 1$, the function $h_f^k \in \mathcal{B}_w(S)$ can be interpreted as the bias of the policy f when the reward rate is $-h_f^{k-1}$. The μ_f-expectation of the bias of f (whatever the reward rate) being zero (recall (3.28)), we obtain

$$\mu_f(h_f^k) = 0 \quad \forall \, k \geq 0. \tag{6.9}$$

Moreover, from Proposition 6.2 with $u(\cdot) = r(\cdot, f)$, the Laurent series of the discounted reward of the policy f can be written as

$$V_\alpha(i, f) = \frac{1}{\alpha} g(f) + \sum_{k=0}^{\infty} \alpha^k h_f^k(i) \quad \forall \, i \in S, \ f \in \mathbb{F}, \ 0 < \alpha < \delta. \tag{6.10}$$

In our next result we prove that the coefficients $\{h_f^k\}_{k \geq 0}$ are the solution of a system of linear equations, which are generalizations of the Poisson equation (3.30). Given $f \in \mathbb{F}$ and an integer $n \geq 0$, we consider the equations

$$g = r(i, f) + \sum_{j \in S} q_{ij}(f) h^0(j) \tag{6.11}$$

$$h^0(i) = \sum_{j \in S} q_{ij}(f) h^1(j) \tag{6.12}$$

$$\cdots \quad \cdots$$

$$h^n(i) = \sum_{j \in S} q_{ij}(f) h^{n+1}(j) \tag{6.13}$$

for every $i \in S$, where $g \in \mathbb{R}$ and $h^0, \ldots, h^{n+1} \in \mathcal{B}_w(S)$. The equations (6.11), (6.12),...,(6.13) are called the $-1, 0, \ldots, n$th *Poisson equations for f*, respectively. Note that the -1th Poisson equation corresponds to the usual Poisson equation (3.30). In the next proposition, we characterize the solutions of the Poisson equations.

Proposition 6.5. *Fix $f \in \mathbb{F}$ and an integer $n \geq -1$. The solutions of the $-1, 0, \ldots, n$th Poisson equations for f are*

$$g(f), h_f^0, \ldots, h_f^n \quad \text{and} \quad h_f^{n+1} + z\mathbf{1},$$

for arbitrary $z \in \mathbb{R}$.

Proof.   We will proceed by induction. The case $n = -1$ has been proved in Proposition 3.14. Suppose now that the stated result holds for some n, and let us prove it for $n + 1$.

First of all, we will show that h_f^{n+1} and h_f^{n+2} verify the $(n+1)$th Poisson equation. To this end, consider a control model with reward rate $-h_f^{n+1}$. The gain of f is zero (by (6.9)) and, therefore, h_f^{n+2} being the bias of f for the reward rate $-h_f^{n+1}$ (by the definition of H_f), the corresponding Poisson equation is precisely the $(n+1)$th Poisson equation. Obviously, for any $z \in \mathbb{R}$, $(h_f^{n+1}, h_f^{n+2} + z\mathbf{1})$ is also a solution of the $(n+1)$th Poisson equation. So far, we have shown that

$$(g(f), h_f^0, \ldots, h_f^n, h_f^{n+1}, h_f^{n+2} + z\mathbf{1}) \tag{6.14}$$

are solutions of the $-1, 0, \ldots, (n+1)$th Poisson equations, respectively.

It remains to show that the solutions of the $-1, 0, \ldots, (n+1)$th Poisson equations are necessarily of the form (6.14). To prove this, fix a solution

$$(g, h^0, \ldots, h^n, h^{n+1}, h^{n+2})$$

of the $-1, 0, \ldots, (n+1)$th Poisson equations. By the induction hypothesis,

$$g = g(f), \ h^0 = h_f^0, \ldots, h^n = h_f^n, \quad \text{and} \quad h^{n+1} = h_f^{n+1} + z\mathbf{1}$$

for some $z \in \mathbb{R}$. Now, the $(n+1)$th Poisson equation, i.e.,

$$0 = -h^{n+1}(i) + \sum_{j \in S} q_{ij}(f) h^{n+2}(j) \quad \forall\, i \in S,$$

implies that, for a control model with reward rate $-h^{n+1}$, the gain of f is zero and its bias equals h^{n+2} up to an additive constant; see Proposition 3.14. Therefore,

$$0 = \mu_f(h^{n+1}) = \mu_f(h_f^{n+1} + z\mathbf{1}) = z,$$

that is, $h^{n+1} = h_f^{n+1}$, and $h^{n+2} = h_f^{n+2} + z'\mathbf{1}$ for some $z' \in \mathbb{R}$. This completes the proof. $\square$

In connection with Remark 6.4, we have the following.

Remark 6.6. If Assumptions 3.1, 3.3, and 3.11 hold and, moreover, each randomized stationary policy is irreducible and the control model is w-exponentially ergodic uniformly on Φ_s, then the result of Proposition 6.5 holds for randomized stationary policies $\varphi \in \Phi_s$.

From Proposition 6.5, the next result easily follows.

Corollary 6.7. *Given $f \in \mathbb{F}$, the unique solution of the infinite set of Poisson equations is*

$$(g(f), h_f^0, h_f^1, \ldots).$$

In this section, we have proved that the total α-discounted expected reward $V_\alpha(i, \varphi)$ of a stationary policy $\varphi \in \Phi_s$ can be expanded as a power series of α, for small $\alpha > 0$; recall (6.10) and Remark 6.4. Let us mention that the α-discounted reward of a Markov policy $\varphi \in \Phi$ admits as well such a series expansion (see the conclusions of this chapter in Sec. 6.7).

6.3. The vanishing discount approach (revisited)

Let us recall that, in this chapter, we are supposing that Assumptions 3.1, 3.3, and 3.11 are satisfied.

The Laurent series expansion provides interesting results showing the relation existing between average reward optimality and the α-discounted optimality criterion as $\alpha \downarrow 0$. Our first result is a straightforward consequence of Corollary 6.3 and (6.10), and it should be linked with the Abelian theorem mentioned in Sec. 3.5; recall (3.43).

Lemma 6.8. *Given $f \in \mathbb{F}$ and $u \in \mathcal{B}_w(S)$, we have $\lim_{\alpha\downarrow0} \alpha V_\alpha(i, f, u) = \mu_f(u)$ for every $i \in S$. In particular, letting $u(\cdot) := r(\cdot, f)$,*

$$\lim_{\alpha\downarrow0} \alpha V_\alpha(i, f) = g(f) \quad \forall\, i \in S.$$

Similarly, if the conditions in Assumptions 3.11(b) and (c) are extended to the class Φ_s of randomized stationary policies, then for every $\varphi \in \Phi_s$ and $u \in \mathcal{B}_w(S)$

$$\lim_{\alpha\downarrow0} \alpha V_\alpha(i, \varphi, u) = \mu_\varphi(u) \quad and \quad \lim_{\alpha\downarrow0} \alpha V_\alpha(i, \varphi) = g(\varphi)$$

for all $i \in S$.

Proof. Given $f \in \mathbb{F}$, $u \in \mathcal{B}_w(S)$, and $i \in S$, by Proposition 6.2 and Corollary 6.3 we have

$$\alpha V_\alpha(i, f, u) = \mu_f(u) + \alpha R_0(f, u, \alpha)$$

for $0 < \alpha < \delta$, where

$$|R_0(f, u, \alpha)| \leq R\|u\|_w w(i)/(\delta - \alpha).$$

Hence, letting $\alpha \downarrow 0$, we obtain $\lim_{\alpha\downarrow0} \alpha V_\alpha(i, f, u) = \mu_f(u)$. The other statements in this lemma are proved similarly. $\square$

Lemma 6.8 above establishes the equality (3.43) for stationary policies. As a consequence of Theorem 3.17 and Corollary 3.18, we obtain the results for the vanishing discount approach to average optimality mentioned at the end of Sec. 3.5.

Theorem 6.9. *Suppose that Assumptions 3.1, 3.3, and 3.11 hold, and let $\{\alpha_n\}_{n\geq 1} \downarrow 0$ be a sequence of discount rates.*

If the sequence $\{f_n\}$ in $\mathbb{F}$ converges componentwise to $f \in \mathbb{F}$, then

$$\lim_{n\to\infty} \alpha_n V_{\alpha_n}(i, f_n) = g(f) \quad \forall\, i \in S.$$

Suppose, in addition, that irreducibility in Assumption 3.11(b) holds for every $\varphi \in \Phi_s$, and w-exponential ergodicity in Assumption 3.11(c) is uniform on Φ_s. If the sequence $\{\varphi_n\}$ in Φ_s converges to $\varphi \in \Phi_s$, then

$$\lim_{n\to\infty} \alpha_n V_{\alpha_n}(i, \varphi_n) = g(\varphi) \quad \forall\, i \in S.$$

Proof. We only prove the first part (the proof for randomized stationary policies is similar and it is based on Remark 6.4).

To simplify the proof, we suppose without loss of generality that $\alpha_n < \delta$ (recall Assumption 3.11(c)) for all $n \geq 1$. By Corollary 6.3 and (6.10), for every $i \in S$

$$\alpha_n V_{\alpha_n}(i, f_n) = g(f_n) + \alpha_n R_0(f_n, r(\cdot, f_n), \alpha_n)(i), \tag{6.15}$$

where

$$\sup_n |R_0(f_n, r(\cdot, f_n), \alpha_n)(i)| \leq RMw(i)/(\delta - \alpha_1).$$

Recalling that $g(f_n)$ converges to $g(f)$, by Theorem 3.17, and taking the limit as $n \to \infty$ in (6.15) yields the desired result. $\square$

Finally, we prove the generalized Abelian theorem and the gain optimality of vanishing discount optimal policies. At this point, recall that the set of deterministic stationary policies $\mathbb{F}$ is compact (see Sec. 2.2). Therefore, any sequence $\{f_n\}_{n\geq 1}$ in $\mathbb{F}$ has limit policies, meaning that there exists a subsequence $\{n'\}$ and a policy $f \in \mathbb{F}$ such that $f_{n'} \to f$.

Theorem 6.10. *Suppose that Assumptions 3.1, 3.3, and 3.11 are satisfied. Then the following statements hold:*

(i) *For every $i \in S$, $\lim_{\alpha\downarrow 0} \alpha V_\alpha^*(i) = g^*$.*

(ii) *Suppose that $\{\alpha_n\}$ is a sequence of discount rates converging to 0 and that, for every n, the policy $f_n \in \mathbb{F}$ is α_n-discount optimal. Then, any limit policy of $\{f_n\}_{n\geq 1}$ is average reward optimal.*

Proof. *Proof of (i).* Given a discount rate $0 < \alpha < \delta$, let $f_\alpha \in \mathbb{F}$ be an α-discount optimal policy. Then, by Corollary 6.3 and (6.10),

$$V_\alpha^*(i) = V_\alpha(i, f_\alpha) = \frac{g(f_\alpha)}{\alpha} + R_0(f_\alpha, r(\cdot, f_\alpha), \alpha)(i) \quad \forall\, i \in S,$$

where the residual verifies

$$|R_0(f_\alpha, r(\cdot, f_\alpha), \alpha)(i)| \le RM \frac{w(i)}{\delta - \alpha} \quad \forall\, 0 < \alpha < \delta,\ i \in S.$$

Similarly, if $f^* \in \mathbb{F}$ is an average reward optimal policy, then

$$V_\alpha(i, f^*) = \frac{g^*}{\alpha} + R_0(f^*, r(\cdot, f^*), \alpha)(i) \quad \forall\, i \in S,$$

and the residual verifies

$$|R_0(f^*, r(\cdot, f^*), \alpha)(i)| \le RM \frac{w(i)}{\delta - \alpha} \quad \forall\, 0 < \alpha < \delta,\ i \in S.$$

Therefore, for each $i \in S$,

$$\liminf_{\alpha \downarrow 0} \alpha(V_\alpha(i, f^*) - V_\alpha^*(i)) = \liminf_{\alpha \downarrow 0} [g^* - g(f_\alpha)] \ge 0. \tag{6.16}$$

On the other hand, since $V_\alpha(i, f^*) \le V_\alpha^*(i)$, we obtain

$$\limsup_{\alpha \downarrow 0} \alpha(V_\alpha(i, f^*) - V_\alpha^*(i)) \le 0.$$

Consequently,

$$\lim_{\alpha \downarrow 0} \alpha(V_\alpha(i, f^*) - V_\alpha^*(i)) = 0 \quad \forall\, i \in S.$$

This yields, by Lemma 6.8, statement (i).

Proof of (ii). Suppose that, for some subsequence $\{n'\}$ and policy $f \in \mathbb{F}$, we have $f_{n'} \to f$. Then, by part (i) and Theorem 6.9, for each $i \in S$,

$$g^* = \lim_{n' \to \infty} \alpha_{n'} V_{\alpha_{n'}}^*(i) = \lim_{n' \to \infty} \alpha_{n'} V_{\alpha_{n'}}(i, f_{n'}) = g(f).$$

Hence, f is average reward optimal, which completes the proof. □

A result similar to Theorem 6.10(i) for discrete-time Markov control processes and stochastic games is proposed in [74] and [14, 115], respectively.

6.4.   The average reward optimality equations

In the same way that the AROE (3.36) is obtained by "maximization" of the Poisson equation, the family of average reward optimality equations is obtained by "maximization" of the Poisson equations defined in Sec. 6.2.

Indeed, consider the system of equations

$$g = \max_{a \in A(i)} \left\{ r(i, a) + \sum_{j \in S} q_{ij}(a) h^0(j) \right\} \qquad (6.17)$$

$$h^0(i) = \max_{a \in A^0(i)} \left\{ \sum_{j \in S} q_{ij}(a) h^1(j) \right\} \qquad (6.18)$$

$$\cdots \quad \cdots$$

$$h^n(i) = \max_{a \in A^n(i)} \left\{ \sum_{j \in S} q_{ij}(a) h^{n+1}(j) \right\} \qquad (6.19)$$

for each $i \in S$, with $g \in \mathbb{R}$ and $h^0, \ldots, h^{n+1} \in \mathcal{B}_w(S)$, and where $A^0(i)$ is the set of actions $a \in A(i)$ attaining the maximum in (6.17), $A^1(i)$ is the set of $a \in A^0(i)$ attaining the maximum in (6.18), and so forth. The equations (6.17), (6.18), \ldots,(6.19) are referred to as the $-1, 0, \ldots, n$th *average reward optimality equations*, respectively, or optimality equations, for short, when there is no risk of confusion.

Observe that our assumptions (in particular, Assumption 3.3) ensure that the sets $A^k(i)$, for $0 \leq k \leq n$, form a monotone nonincreasing sequence of nonempty compact subsets of $A(i)$ (of course, provided that solutions to the average reward optimality equations exist, which will be proved later).

It should also be noted that the -1th optimality equation corresponds to the AROE (3.36), whereas the -1 and 0th optimality equations correspond to the bias optimality equations (5.3) and (5.4). Therefore, $g = g^*$, the optimal gain, and $h^0 = \hat{h}$, the optimal bias function.

Next, we prove that solutions to the average reward optimality equations indeed exist.

Proposition 6.11. *There exist solutions $(g, h^0, \ldots, h^{n+1})$ to the -1, $0, \ldots, n$th average reward optimality equations. Besides, for every $n \geq 0$, the solutions $g, h^0, \ldots, h^n$ are unique, whereas h^{n+1} is unique up to additive constants.*

Proof. The proof is by induction on $n \geq -1$. The case $n = -1$ is proved in Theorem 3.20. Suppose now that the stated result holds for some $n \geq -1$, and let us prove it for $n + 1$.

First of all, we show that there exists a solution to the $(n+1)$th optimality equation. Let $(g, h^0, \dots, h^{n+1})$ be a solution of the $-1, 0, \dots, n$th average reward optimality equations. Consider now a control model with compact action sets $A^{n+1}(i)$ (the set of $a \in A^n(i)$ attaining the maximum in (6.19)) for each $i \in S$, and reward rate $-h^{n+1}$. This control model satisfies our standing hypotheses, namely, Assumptions 3.1, 3.3, and 3.11. Therefore, there exists a constant $g_{n+1} \in \mathbb{R}$ and a function $h^{n+2} \in \mathcal{B}_w(S)$ that satisfy the corresponding AROE:

$$g_{n+1} = \max_{a \in A^{n+1}(i)} \left\{ -h^{n+1}(i) + \sum_{j \in S} q_{ij}(a) h^{n+2}(j) \right\} \quad \forall \, i \in S.$$

We conclude that $(g, h^0, \dots, h^{n+1} + g_{n+1}\mathbf{1}, h^{n+2})$ solves the $-1, 0, \dots,$ $(n+1)$th average reward optimality equations.

To prove uniqueness, suppose that

$$(g, h^0, \dots, h^n, h^{n+1}, h^{n+2}) \quad \text{and} \quad (g, h^0, \dots, h^n, \bar{h}^{n+1}, \bar{h}^{n+2})$$

are solutions of the $-1, 0, \dots, (n+1)$th average reward optimality equations. By the induction hypothesis, we know that $(g, h^0, \dots, h^n)$ is unique and that $h^{n+1} - \bar{h}^{n+1} = z\mathbf{1}$ for some $z \in \mathbb{R}$. The $(n+1)$th optimality equation shows that, for the control model with compact action sets $A^{n+1}(i)$ for $i \in S$ and reward rate $-h^{n+1}$ or $-\bar{h}^{n+1}$, the optimal gain is zero. Since both reward rates differ by a constant, the set of gain optimal policies coincide. Therefore, there exists some $f \in \mathbb{F}$ such that $\mu_f(h^{n+1}) = \mu_f(\bar{h}^{n+1}) = 0$ and so, necessarily, $z = 0$; hence, $h^{n+1} = \bar{h}^{n+1}$. The fact that h^{n+2} and $\bar{h}^{n+2}$ differ by a constant is derived from Theorem 3.20(ii). The proof is now complete. □

Let us mention that, since h^{n+1} in the nth optimality equation (6.19) is unique up to additive constants, the set $A^{n+1}(i) \subseteq A^n(i)$ of maxima is well defined and does not depend on the particular solution h^{n+1}.

The next result should be linked with Corollary 6.7.

Corollary 6.12. *The solution to the infinite set of average reward optimality equations is unique.*

In the remainder of this chapter, we will denote by $(g^*, h^0, h^1, \dots)$ the unique solution of the infinite system of average reward optimality equations.

Characterizations of sensitive discount optimality   We begin with two definitions.

For an integer $n \geq -1$, let $\mathbb{F}^n$ be the family of policies $f \in \mathbb{F}$ such that $f(i) \in A^{n+1}(i)$ for every $i \in S$. That is, $\mathbb{F}^n$ is the family of deterministic stationary policies attaining the maximum in the $-1, 0,\ldots,n$th average reward optimality equations. In particular, $\mathbb{F}^{-1} = \mathbb{F}_{ca}$ and $\mathbb{F}^0$ is the set of bias optimal policies.

As already mentioned, the sets $A^n(i)$, for $n \geq 0$, form a monotone nonincreasing sequence of nonempty compact sets. Therefore, $\cap A^n(i)$ is nonempty for each $i \in S$, and so is

$$\mathbb{F}^\infty := \bigcap_{n=-1}^{\infty} \mathbb{F}^n. \tag{6.20}$$

Given two vectors $\mathbf{u}$ and $\mathbf{v}$ in $\mathbb{R}^d$ (with $1 \leq d \leq \infty$), we say that $\mathbf{u}$ is *lexicographically greater than or equal to* $\mathbf{v}$ if the first nonzero component of $\mathbf{u} - \mathbf{v}$ (if it exists) is positive. This will be denoted by $\mathbf{u} \succeq \mathbf{v}$. If $\mathbf{u} \succeq \mathbf{v}$ and $\mathbf{u} \neq \mathbf{v}$, we will write $\mathbf{u} \succ \mathbf{v}$.

Our next theorem characterizes the family of n-discount optimal policies in $\mathbb{F}$ and the lexicographic maximization of the coefficients of the Laurent series (6.10).

Theorem 6.13.   *We suppose that Assumptions 3.1, 3.3, and 3.11 are satisfied. Given $n \geq -1$ and $f \in \mathbb{F}$, the following statements are equivalent.*

(i) *f is in $\mathbb{F}^n$.*
(ii) *f is n-discount optimal in $\mathbb{F}$.*
(iii) *For every $f' \in \mathbb{F}$, $(g(f), h_f^0, \ldots, h_f^n) \succeq (g(f'), h_{f'}^0, \ldots, h_{f'}^n)$, meaning that the lexicographic inequality holds for every $i \in S$.*

Proof.   *Proof of (i) $\Rightarrow$ (iii).* Fix $f \in \mathbb{F}^n$, and let $f' \in \mathbb{F}$ be arbitrary.

If $f' \notin \mathbb{F}^{-1}$ then $g(f) > g(f')$ and then (iii) holds.

If $f' \in \mathbb{F}^{-1}$ and $f' \notin \mathbb{F}^n$, then there exists a maximal $-1 \leq k < n$ such that $f' \in \mathbb{F}^k$ and $f' \notin \mathbb{F}^{k+1}$. Since f' is in $\mathbb{F}^k$, then

$$(g(f), h_f^0, \ldots, h_f^k) = (g(f'), h_{f'}^0, \ldots, h_{f'}^k),$$

and so both f and f' are average reward optimal for the control problem with action sets $A^k(i)$, for $i \in S$, and reward rate $-h^k$. Besides, f is bias optimal for this control model, whereas f' is not. Since the bias of two gain optimal policies differ by a constant, then $h_f^{k+1}(i) > h_{f'}^{k+1}(i)$ for every $i \in S$, which proves (iii).

Finally, if $f' \in \mathbb{F}^n$ then, by Proposition 6.5,

$$(g(f), h_f^0, \ldots, h_f^n) = (g(f'), h_{f'}^0, \ldots, h_{f'}^n)$$

and (iii) also holds in this case.

Proof of (iii) $\Rightarrow$ (i). Suppose that (iii) holds for some $f \in \mathbb{F}$. In particular, f is gain optimal, that is, $f \in \mathbb{F}^{-1}$. If $f \notin \mathbb{F}^n$ then, as before, there exists a maximal $-1 \le k < n$ such that $f \in \mathbb{F}^k$ and $f \notin \mathbb{F}^{k+1}$. But the set $\mathbb{F}^{k+1}$ being nonempty, given $\bar{f} \in \mathbb{F}^{k+1}$ and using the same argument as in the previous proof, we obtain

$$(g(\bar{f}), h_{\bar{f}}^0(i), \ldots, h_{\bar{f}}^{k+1}(i)) \succ (g(f), h_f^0(i), \ldots, h_f^{k+1}(i)) \quad \forall \, i \in S,$$

which is a contradiction. Therefore f is in $\mathbb{F}^n$, which completes the proof.

Proof of (ii) $\Leftrightarrow$ (iii). First of all, observe that for $0 < \alpha < \delta$, arbitrary $f \in \mathbb{F}$, and $i \in S$, the Laurent series (6.10) can be expressed as

$$V_\alpha(i, f) = \frac{g(f)}{\alpha} + \sum_{k=0}^{n} \alpha^k h_f^k(i) + R_{n+1}(f, r(\cdot, f), \alpha)(i),$$

where, by Corollary 6.3,

$$|R_{n+1}(f, r(\cdot, f), \alpha)(i)| \le RM \frac{\alpha^{n+1} w(i)}{\delta^{n+1}(\delta - \alpha)}.$$

This shows that, given f and f' in $\mathbb{F}$,

$$\liminf_{\alpha \downarrow 0} \alpha^{-n}(V_\alpha(i, f) - V_\alpha(i, f'))$$

$$= \lim_{\alpha \downarrow 0} \left[\frac{g(f) - g(f')}{\alpha^{n+1}} + \sum_{k=0}^{n} \frac{1}{\alpha^{n-k}}(h_f^k(i) - h_{f'}^k(i)) \right].$$

From this expression, it is clear that $\liminf_{\alpha \downarrow 0} \alpha^{-n}(V_\alpha(i, f) - V_\alpha(i, f')) \ge 0$ for every $i \in S$ and $f' \in \mathbb{F}$ if and only if (iii) holds. This completes the proof of the theorem. $\qquad \square$

We note that, for the case $n = 0$, the definition of bias optimality given in Chapter 5 — that is, maximize the bias within the class of average reward optimal policies — indeed corresponds to the lexicographic maximization of the vector criterion $(g(f), h_f)$ for $f \in \mathbb{F}$.

6.5. Strong discount optimality

Our goal in this section is to prove that a policy $f \in \mathbb{F}$ that is average reward optimal (respectively, bias optimal) is strong -1-discount optimal (respectively, strong 0-discount optimal).

Strong −1-discount optimality   We have the following result.

Theorem 6.14. *Suppose that Assumptions 3.1, 3.3, and 3.11 hold. A policy $f \in \mathbb{F}$ is gain optimal if and only if it is strong −1-discount optimal.*

Proof.   If $f \in \mathbb{F}$ is strong −1-discount optimal, then it is −1-discount optimal and, by Theorem 6.13, f is average reward optimal.

Conversely, suppose that f is average reward optimal. It follows from the inequality (6.16) that f is strong −1-discount optimal.   □

Strong 0-discount optimality   To analyze strong 0-discount optimality we need to strengthen our hypotheses. More precisely, in addition to Assumptions 3.1, 3.3, and 3.11, *we also suppose that the constant c' in Assumption 3.1(b) is nonnegative: $c' \geq 0$.*

Next we introduce the so-called *discrepancy functions*. We recall that $(g^*, h^0, h^1, \dots)$ denotes the unique solution of the infinite system of average reward optimality equations (see Corollary 6.12). Define the Mandl discrepancy function $\Delta_0 : K \to \mathbb{R}$ as

$$\Delta_0(i,a) := r(i,a) - g^* + \sum_{j \in S} q_{ij}(a) h^0(j) \quad \forall\, (i,a) \in K.$$

Clearly, $\Delta_0(i,a) \leq 0$ for every $(i,a) \in K$ and $\Delta_0(i,a) = 0$ if and only if $a \in A^0(i)$ (see the notation introduced at the beginning of Sec. 6.4). Similarly, for $n \geq 1$, the nth discrepancy function is

$$\Delta_n(i,a) := -h^{n-1}(i) + \sum_{j \in S} q_{ij}(a) h^n(j) \quad \forall\, (i,a) \in K.$$

Obviously, if $a \in A^{n-1}(i)$ then $\Delta_n(i,a) \leq 0$, and $\Delta_n(i,a) = 0$ if $a \in A^n(i)$.

Remark 6.15. We note that, strictly speaking, Mandl's discrepancy function refers to Δ_0. By an abuse of terminology, however, the functions Δ_n, for $n \geq 1$, are also referred to as Mandl's discrepancy functions.

It follows from arguments similar to those used in (2.18) that

$$\Delta_n \in \mathcal{B}_{w+w'}(K) \quad \forall\, n \geq 0, \tag{6.21}$$

(the notation $\mathcal{B}_{w+w'}(K)$ is defined in (2.10)) where w' is the function in Assumption 3.1(b). It is also easily seen that, for $n \geq 0$ and each $i \in S$, the mapping $a \mapsto \Delta_n(i,a)$ is continuous on $A(i)$.

The notation $\Delta_n(t,i,\varphi)$, for $t \geq 0$, $i \in S$, and $\varphi \in \Phi$, and also $\Delta_n(i,\varphi)$, for $i \in S$ and $\varphi \in \Phi_s$, are given the obvious definitions (recall (3.1)).

In our next result, we use the notation in (3.10) and (3.11).

Lemma 6.16. *Given* $\varphi \in \Phi$, $\alpha > 0$, $i \in S$, *and* $n \geq 0$,

$$V_\alpha(i, \varphi) = \frac{g^*}{\alpha} + \sum_{k=0}^{n} \alpha^k h^k(i) + \alpha^{n+1} V_\alpha(i, \varphi, -h^n) + \sum_{k=0}^{n} \alpha^k V_\alpha(i, \varphi, \Delta_k).$$

Proof. The proof is by induction on $n \geq 0$. We will use the notation introduced in (2.14).

Let $n = 0$. From the definition of Δ_0 we easily deduce that

$$V_\alpha(i, \varphi, \Delta_0) = V_\alpha(i, \varphi) - \frac{g^*}{\alpha} + E_i^\varphi \left[\int_0^\infty e^{-\alpha t} (L^{t,\varphi} h^0)(x(t)) dt \right].$$

Now, by (2.19), for any given $T > 0$,

$$E_i^\varphi [e^{-\alpha T} h^0(x(T))] - h^0(i) = E_i^\varphi \left[\int_0^T e^{-\alpha t} (-\alpha h^0 + L^{t,\varphi} h^0)(x(t)) dt \right].$$

Letting $T \to \infty$ we obtain:

- $E_i^\varphi [e^{-\alpha T} h^0(x(T))] \to 0$, by (2.13), because $h^0 \in \mathcal{B}_w(S)$.
- $E_i^\varphi [\int_0^T e^{-\alpha t} h^0(x(t)) dt] \to V_\alpha(i, \varphi, h^0)$ also by (2.13).
- By (2.16), (6.21), and our hypothesis that $c' \geq 0$, the interchange of the limit $T \to \infty$ and expectation is allowed and gives

$$E_i^\varphi \left[\int_0^T e^{-\alpha t} (L^{t,\varphi} h^0)(x(t)) dt \right] \to E_i^\varphi \left[\int_0^\infty e^{-\alpha t} (L^{t,\varphi} h^0)(x(t)) dt \right].$$

Summarizing, we have shown that

$$E_i^\varphi \left[\int_0^\infty e^{-\alpha t} (L^{t,\varphi} h^0)(x(t)) dt \right] = \alpha V_\alpha(i, \varphi, h^0) - h^0(i),$$

and the desired result for $n = 0$ follows.

Suppose now that the result holds for some $n \geq 0$. An argument similar to that in the case $n = 0$ gives

$$V_\alpha(i, \varphi, -h^n) = h^{n+1}(i) - \alpha V_\alpha(i, \varphi, h^{n+1}) + V_\alpha(i, \varphi, \Delta_{n+1}), \qquad (6.22)$$

from which the stated result for $n + 1$ easily follows. $\qquad \square$

The following is our main result in this section: it shows the equivalence between bias optimality and strong 0-discount optimality. (Compare with the equivalence between *bias* and *overtaking* optimality in Theorem 5.3.)

Theorem 6.17. *Let us suppose that Assumptions 3.1, 3.3, and 3.11 hold and that, in addition, the constant c' in Assumption 3.1(b) is nonnegative. Then a policy $f \in \mathbb{F}$ is bias optimal if and only if it is strong 0-discount optimal.*

Proof.   Clearly, if $f \in \mathbb{F}$ is strong 0-discount optimal, then it is 0-discount optimal in $\mathbb{F}$ and, by Theorem 6.13, it is bias optimal.

Conversely, let us prove that a bias optimal policy $f \in \mathbb{F}$ is strong 0-discount optimal. We will proceed by contradiction. Thus, if $f \in \mathbb{F}$ is not strong 0-discount optimal, then there exist $i \in S$, $\varepsilon > 0$, and a sequence $\{\alpha_n\}$ of discount rates converging to 0 such that

$$V_{\alpha_n}(i, f) - V_{\alpha_n}^*(i) + \varepsilon \le 0. \tag{6.23}$$

If f_n is an α_n-discount optimal policy, we can suppose without loss of generality that $\{f_n\}$ converges componentwise to some $\bar{f} \in \mathbb{F}$ (this is by the compactness of $\mathbb{F}$). As a consequence of Theorem 6.10(ii), the policy $\bar{f}$ is average reward optimal.

Now we use Lemma 6.16 for $n = 0$. Since $f \in \mathbb{F}$ is gain optimal, $\Delta_0(j, f) = 0$ for every $j \in S$, and, thus,

$$V_{\alpha_n}(i, f) = \frac{g^*}{\alpha_n} + h^0(i) - \alpha_n V_{\alpha_n}(i, f, h^0).$$

For each f_n, we have $\Delta_0(j, f_n) \le 0$, and so

$$V_{\alpha_n}^*(i) = V_{\alpha_n}(i, f_n) \le \frac{g^*}{\alpha_n} + h^0(i) - \alpha_n V_{\alpha_n}(i, f_n, h^0).$$

Hence, from (6.23),

$$-\alpha_n V_{\alpha_n}(i, f, h^0) + \alpha_n V_{\alpha_n}(i, f_n, h^0) + \varepsilon \le 0.$$

The result in Lemma 6.8 shows that $\alpha_n V_{\alpha_n}(i, f, h^0) \to \mu_f(h^0)$, and $\mu_f(h^0)$ equals 0 because f is bias optimal (recall Theorem 5.4(iii) and the equations (6.17) and (6.18)). By Theorem 6.9, we also have $\alpha_n V_{\alpha_n}(i, f_n, h^0) \to \mu_{\bar{f}}(h^0)$. This shows that

$$\mu_{\bar{f}}(-h^0) > 0,$$

which is a contradiction because, $\bar{f}$ being gain optimal, $\mu_{\bar{f}}(-h^0) \le 0$ (see, e.g., (6.18)). This completes the proof of the theorem. $\square$

Remark 6.18 (An open problem). *We have shown that the policies in* $\mathbb{F}^{-1}$ *and* $\mathbb{F}^0$ *are strong* -1*- and strong 0-discount optimal, respectively. Proving that a policy in* $\mathbb{F}^n$*, for* $n \ge 1$*, is strong n-discount optimal remains an open issue.*

6.6. Sensitive discount optimality in the class of stationary policies

In Theorem 6.13 we proved that a policy in $\mathbb{F}^n$ is n-discount optimal in $\mathbb{F}$. In this section we show that, under additional conditions, a policy in $\mathbb{F}^n$ is also n-discount optimal in Φ_s, the family of randomized stationary policies. This should be related to Theorem 7.1 below, in which we will establish the existence of a Blackwell optimal policy in Φ_s.

We suppose that Assumption 3.1 holds, where the constant c' is nonnegative. We also suppose that Assumptions 3.3 and 3.11 hold where, in Assumption 3.11(b), we suppose irreducibility of every $\varphi \in \Phi_s$.

We introduce some notation. Let $\Phi^{-1} := \Phi_s$ and, for $n \geq 0$, let Φ^n be the family of policies $\varphi \in \Phi_s$ such that

$$\varphi(A^n(i)|i) = 1 \quad \forall\, i \in S,$$

that is, Φ^n is the class of stationary policies taking random actions in the set of solutions of the $-1, 0, \ldots, (n-1)$th average reward optimality equations.

Our next result uses the discrepancy functions Δ_n defined in Sec. 6.5.

Lemma 6.19. *Given an arbitrary stationary policy $\varphi \in \Phi_s$ and an integer $n \geq 0$, if $\varphi \in \Phi^{n-1}$ and $\varphi \notin \Phi^n$ then*

$$\int_0^\infty E_i^\varphi[\Delta_n(x(t), \varphi)]dt = -\infty \quad \forall\, i \in S.$$

Proof. For every $i \in S$, $\Delta_n(i, a) \leq 0$ with $\varphi(\cdot|i)$-probability one because $\varphi \in \Phi^{n-1}$. Hence, $\Delta_n(i, \varphi) \leq 0$ for each $i \in S$ with strict inequality for some $i_0 \in S$ (because $\varphi \notin \Phi^n$). Now, for arbitrary $i \in S$,

$$\int_0^\infty E_i^\varphi[\Delta_n(x(t), \varphi)]dt \leq \int_0^\infty E_i^\varphi[\Delta_n(i_0, \varphi) \cdot \mathbf{I}_{\{x(t)=i_0\}}]dt$$

$$= \Delta_n(i_0, \varphi) \int_0^\infty E_i^\varphi[\mathbf{I}_{\{x(t)=i_0\}}]dt$$

$$= \Delta_n(i_0, \varphi) \cdot E_i^\varphi[\zeta_{i_0}],$$

where ζ_{i_0} denotes the total time spent by the process $\{x(t)\}_{t \geq 0}$ at state $i_0 \in S$. By irreducibility and recurrence of $\{x^\varphi(t)\}$, $E_i^\varphi[\zeta_{i_0}] = \infty$, and the lemma follows. $\qquad\square$

We now state our main result in this section.

Theorem 6.20. *Suppose that Assumptions 3.1 (with $c' \geq 0$), 3.3, and 3.11 (where irreducibility in (b) is supposed for every $\varphi \in \Phi_s$) are satisfied. Then a policy $f \in \mathbb{F}^n$ is n-discount optimal in Φ_s.*

Proof. Fix $f \in \mathbb{F}^n$ and $\varphi \in \Phi_s$. Then there are three possibilities:

(a) For some $-1 \le m < n$, $\varphi \in \Phi^m$ and $\varphi \notin \Phi^{m+1}$, or
(b) $\varphi \in \Phi^n$ and $\varphi \notin \Phi^{n+1}$, or
(c) $\varphi \in \Phi^{n+1}$.

In the case that (a) holds, then, by Lemma 6.16,

$$V_\alpha(i, f) = \frac{g^*}{\alpha} + \sum_{k=0}^{m+1} \alpha^k h^k(i) + \alpha^{m+2} V_\alpha(i, f, -h^{m+1}) \quad \forall i \in S$$

because $\Delta_0(i, f) = \cdots = \Delta_{n+1}(i, f) = 0$. Also, for each $i \in S$,

$$V_\alpha(i, \varphi) = \frac{g^*}{\alpha} + \sum_{k=0}^{m+1} \alpha^k h^k(i) + \alpha^{m+2} V_\alpha(i, \varphi, -h^{m+1}) + \alpha^{m+1} V_\alpha(i, \varphi, \Delta_{m+1})$$

because

$$\Delta_0(i, \varphi) = \cdots = \Delta_m(i, \varphi) = 0 \quad \forall i \in S.$$

Therefore, for each $i \in S$,

$$\alpha^{-(m+1)}(V_\alpha(i, f) - V_\alpha(i, \varphi)) = \alpha V_\alpha(i, f, -h^{m+1}) - \alpha V_\alpha(i, \varphi, -h^{m+1})$$
$$- V_\alpha(i, \varphi, \Delta_{m+1}). \tag{6.24}$$

Now we observe that $\alpha V_\alpha(i, f, -h^{m+1})$ and $\alpha V_\alpha(i, \varphi, -h^{m+1})$ are bounded as $\alpha \downarrow 0$; see, for instance, (3.7). On the other hand

$$V_\alpha(i, \varphi, \Delta_{m+1}) = E_i^\varphi \left[\int_0^\infty [e^{-\alpha t} \Delta_{m+1}(x(t), \varphi)] dt \right]$$
$$= \int_0^\infty E_i^\varphi [e^{-\alpha t} \Delta_{m+1}(x(t), \varphi)] dt$$

because $\Delta_{m+1}(\cdot, \varphi) \le 0$. By monotone convergence and Lemma 6.19,

$$\lim_{\alpha \downarrow 0} V_\alpha(i, \varphi, \Delta_{m+1}) = \int_0^\infty E_i^\varphi [\Delta_{m+1}(x(t), \varphi)] dt = -\infty.$$

Finally, we conclude from (6.24) that

$$\lim_{\alpha \downarrow 0} \alpha^{-(m+1)}(V_\alpha(i, f) - V_\alpha(i, \varphi)) = \infty \quad \forall i \in S,$$

and since $m + 1 \le n$,

$$\lim_{\alpha \downarrow 0} \alpha^{-n}(V_\alpha(i, f) - V_\alpha(i, \varphi)) = \infty \quad \forall i \in S. \tag{6.25}$$

Let us suppose now that (b) holds. As in (6.24) we obtain, for each $i \in S$,

$$\alpha^{-n}(V_\alpha(i, f) - V_\alpha(i, \varphi)) = \alpha^2 V_\alpha(i, f, -h^{n+1}) - \alpha^2 V_\alpha(i, \varphi, -h^{n+1})$$
$$- \alpha V_\alpha(i, \varphi, \Delta_{n+1}).$$

Now, by (3.7) for the reward rate $-h^{n+1}$,

$$\lim_{\alpha \downarrow 0} \alpha^2 V_\alpha(i, f, -h^{n+1}) = \lim_{\alpha \downarrow 0} \alpha^2 V_\alpha(i, \varphi, -h^{n+1}) = 0. \tag{6.26}$$

Since $\varphi \in \Phi^n$, it follows that $\Delta_n(\cdot, \varphi) = 0$ and $\Delta_{n+1}(\cdot, \varphi) \leq 0$, and, thus, $\alpha V_\alpha(i, \varphi, \Delta_{n+1}) \leq 0$. Therefore,

$$\liminf_{\alpha \downarrow 0} \alpha^{-n}(V_\alpha(i, f) - V_\alpha(i, \varphi)) \geq 0.$$

In fact, from (3.7) and (6.22), we deduce that

$$0 \leq \liminf_{\alpha \downarrow 0} \alpha^{-n}(V_\alpha(i, f) - V_\alpha(i, \varphi)) < \infty \quad \forall\, i \in S. \tag{6.27}$$

Finally, let us assume that condition (c) holds. The beginning of the proof is similar to that of (a). Lemma 6.16 implies that

$$V_\alpha(i, f) = \frac{g^*}{\alpha} + \sum_{k=0}^{n+1} \alpha^k h^k(i) + \alpha^{n+2} V_\alpha(i, f, -h^{n+1}) \quad \forall\, i \in S$$

because $\Delta_0(i, f) = \cdots = \Delta_{n+1}(i, f) = 0$. Similarly,

$$V_\alpha(i, \varphi) = \frac{g^*}{\alpha} + \sum_{k=0}^{n+1} \alpha^k h^k(i) + \alpha^{n+2} V_\alpha(i, \varphi, -h^{n+1}) \quad \forall\, i \in S$$

because, this time,

$$\Delta_0(i, \varphi) = \cdots = \Delta_{n+1}(i, \varphi) = 0 \quad \forall\, i \in S.$$

Hence,

$$\alpha^{-n}(V_\alpha(i, f) - V_\alpha(i, \varphi)) = \alpha^2 V_\alpha(i, f, -h^{n+1}) - \alpha^2 V_\alpha(i, \varphi, -h^{m+1})$$

for every $i \in S$. Using (6.26), we obtain

$$\lim_{\alpha \downarrow 0} \alpha^{-n}(V_\alpha(i, f) - V_\alpha(i, \varphi)) = 0 \quad \forall\, i \in S. \tag{6.28}$$

From (6.25), (6.27), and (6.28), we conclude that $f \in \mathbb{F}^n$ is n-discount optimal in Φ_s. $\square$

6.7. Conclusions

In this chapter, we have made a thorough analysis of the sensitive discount optimality criteria. The key result for this is the Laurent series expansion in Proposition 6.2 (see also Remark 6.4). This result follows from the standard series expansion of the resolvent of a Markov chain, though the proof herein allowed us to derive useful bounds on the residuals of this series expansion; see Corollary 6.3.

The bounds on the residuals of the Laurent series expansion yield the result for the vanishing discount approach to average optimality in Theorem 6.10. This theorem could have been proved in Chapter 3, but the techniques developed in Chapter 6 allowed us to make a simpler proof. It is interesting to mention that the vanishing discount approach is also valid for *constrained* CMCs (see Sec. 8.5).

The Laurent series expansion holds for stationary policies. Lemma 6.16 shows that there exists as well a series expansion for the discounted reward of a nonstationary policy $\varphi \in \Phi$ in terms of Mandl's discrepancy functions Δ_k, for $k \geq 0$. More precisely, by taking the limit as $n \to \infty$ in the equation in Lemma 6.16, we obtain for small $\alpha > 0$

$$V_\alpha(i, \varphi) = \frac{g^*}{\alpha} + \sum_{k=0}^{\infty} \alpha^k h^k(i) + \sum_{k=0}^{\infty} \alpha^k V_\alpha(i, \varphi, \Delta_k)$$

$$= V_\alpha(i, f^*) + \sum_{k=0}^{\infty} \alpha^k V_\alpha(i, \varphi, \Delta_k),$$

where f^* is any policy in $\mathbb{F}^\infty$. The problem when dealing with this series is that, as opposed to the Laurent series for stationary policies, its coefficients are in $\mathcal{B}_{w+w'}(S)$ since the Δ_k are in $\mathcal{B}_{w+w'}(K)$.

As established in Sec. 6.5, the strong -1- and strong 0-discount optimality criteria respectively correspond to the average and the bias optimality criteria. An issue that, to the best of our knowledge, has not been yet analyzed is the existence of strong n-discount optimal policies for $n \geq 1$; recall Remark 6.18. In such a case, it would be very interesting to provide as well an interpretation of such optimal policies.

As already mentioned in the conclusions of Chapter 5, it would be interesting to know whether the policy iteration algorithm for discrete-time CMCs under the sensitive discount optimality criteria, developed in [138], can be adapted to our continuous-time CMC framework.

Chapter 7

Blackwell Optimality

7.1. Introduction

The motivation for the Blackwell optimality criterion comes from the idea of obtaining a policy $\varphi^* \in \Phi$ that is asymptotically α-discount optimal as the discount rate $\alpha \downarrow 0$. Blackwell [17] proposed the following definition: A policy $\varphi^* \in \Phi$ is optimal if there exists a discount rate $\alpha_0 > 0$ such that φ^* is α-discount optimal for all $0 < \alpha < \alpha_0$ (recall (6.1)), that is,

$$V_\alpha(i, \varphi^*) \geq V_\alpha(i, \varphi) \ \forall i \in S, \ 0 < \alpha < \alpha_0, \ \varphi \in \Phi. \tag{7.1}$$

This optimality criterion is known in the literature as *strong Blackwell optimality*.

The existence of such optimal policies for a discrete-time CMC with finite state and action spaces was established by Blackwell [17]. This existence result was also proved by Miller and Veinott [114, 161] by using the Laurent series technique; see Sec. 6.2. More precisely, it is shown in [161] that a policy that is n-discount optimal for all sufficiently large n is, necessarily, strong Blackwell optimal.

This optimality criterion turned out to be too restrictive when trying to generalize it to nonfinite state and action spaces CMCs. The weaker notion of *Blackwell optimality* was then introduced. Roughly speaking, it allowed α_0 in (7.1) to depend on both the initial state $i \in S$ and the policy $\varphi \in \Phi$ (a precise definition is given in (7.2) below). The existence of Blackwell optimal policies for discrete-time CMCs with denumerable state space and compact action space was proved by Dekker and Hordijk: this was achieved by imposing conditions on the deviation matrix [23], and also by making some recurrence assumptions [24]; see also the paper by Lasserre [95]. Blackwell optimality for discrete-time CMCs with countable state space has also been treated by Kadota [85].

In subsequent papers, Blackwell optimality was generalized by Yushkevich to discrete-time CMCs with Borel state space and denumerable action space [169], and compact action space [170]. A thorough study of Blackwell optimality for discrete-time CMCs with Borel state space and compact action space was made by Hordijk and Yushkevich [75, 76] by imposing drift and uniform geometric ergodicity conditions. As we shall see, our assumptions for studying Blackwell optimality for continuous-time denumerable state CMCs also include such drift and uniform exponential ergodicity conditions.

The continuous-time counterpart of Blackwell optimality is, by far, less developed. Indeed, only a few papers analyze this issue. Veinott [161] and Sladký [151] study continuous-time CMCs with finite state and action spaces. In some recent papers, we have studied Blackwell optimality for denumerable state and compact action sets; see [126, 127] and also [52, 53]. For a continuous state space, Puterman [137] dealt with one-dimensional controlled diffusions on compact intervals, while Jasso-Fuentes and Hernández-Lerma [83] made a thorough analysis of Blackwell optimality for multidimensional controlled diffusions.

All the aforementioned papers on Blackwell optimality have some common features, which are worth mentioning. First of all, we already know that n-discount optimality is related to the lexicographic maximization of the first terms of the Laurent series expansion (recall Theorem 6.13). In this way, Blackwell optimality corresponds to the lexicographic maximization of the whole sequence of coefficients of the Laurent series expansion. Secondly, let us mention that, usually, proving the existence of a Blackwell optimal policy within the class of stationary policies is rather straightforward (once the n-discount optimality criterion has been studied). Proving the existence of a Blackwell optimal policy in the class of Markov (nonstationary) policies is a more complicated problem (see, e.g., [76] or [126]), and additional hypotheses are needed.

Our analysis of Blackwell optimality is mainly based on [126, 127]. In Sec. 7.2 we study Blackwell optimality in the class of stationary policies. This result is an almost direct consequence of Theorem 6.13. Proving the existence of Blackwell optimal policies in the class Φ of Markov policies requires supplementary hypotheses. This is achieved in Sec. 7.3. Our conclusions are stated in Sec. 7.4.

Now, let us give a formal definition of Blackwell optimality. Let $\Phi' \subseteq \Phi$ be a class of control policies. We say that $\varphi^* \in \Phi'$ is *Blackwell optimal in* Φ' if for each $\varphi \in \Phi'$ and every $i \in S$ there exists a discount rate

$\alpha_0 = \alpha_0(i, \varphi) > 0$, that may depend on both i and φ, such that

$$V_\alpha(i, \varphi^*) \geq V_\alpha(i, \varphi) \quad \forall\, 0 < \alpha < \alpha_0. \tag{7.2}$$

7.2. Blackwell optimality in the class of stationary policies

The existence of a Blackwell optimal policy in the class of stationary (either deterministic or not) policies is a direct consequence of the n-discount optimality criteria analyzed in Chapter 6. We recall that the set $\mathbb{F}^\infty \subseteq \mathbb{F}$ (see (6.20)) of the policies that are n-discount optimal for every $n \geq -1$ is nonempty.

Theorem 7.1. *Suppose that Assumptions* 3.1, 3.3, *and* 3.11 *hold. The following results hold*:

(i) *Every $f \in \mathbb{F}^\infty$ is Blackwell optimal in $\mathbb{F}$.*
(ii) *In addition, suppose that in Assumption* 3.1(b) *the constant c' is non-negative and that, in Assumption* 3.11(b), *irreducibility holds for every policy in Φ_s. Then every $f \in \mathbb{F}^\infty$ is Blackwell optimal in Φ_s.*

Proof. *Proof of* (i). Let $f \in \mathbb{F}^\infty$ and fix an arbitrary $f' \in \mathbb{F}$. If f' is in $\mathbb{F}^\infty$, then the Laurent series of $V_\alpha(i, f)$ and $V_\alpha(i, f')$ coincide for small α and every $i \in S$. Thus $V_\alpha(i, f) = V_\alpha(i, f')$, so that (7.2) holds.

Otherwise, let $n \geq -1$ be the minimal n such that $f \notin \mathbb{F}^n$. The notation in this proof implicitly assumes that $n \geq 0$; the proof for the case $n = -1$ is similar. For each $i \in S$ (recall the proof of Theorem 6.13)

$$g(f) = g(f'), \ h_f^0(i) = h_{f'}^0(i), \ldots, h_f^{n-1}(i) = h_{f'}^{n-1}(i), \quad \text{and} \quad h_f^n(i) > h_{f'}^n(i).$$

Therefore, by the same argument used in the proof of Theorem 6.13,

$$\liminf_{\alpha \downarrow 0} \alpha^{-n}(V_\alpha(i, f) - V_\alpha(i, f')) = h_f^n(i) - h_{f'}^n(i) > 0,$$

from which $V_\alpha(i, f) > V_\alpha(i, f')$ for small α follows. This completes the proof of statement (i).

Proof of (ii). We will use the notation introduced in Secs 6.5 and 6.6. We fix $f \in \mathbb{F}^\infty$ and an arbitrary $\varphi \in \Phi_s$. If φ is in Φ^n for every $n \geq -1$ then $V_\alpha(i, \varphi, \Delta_n) = 0$ for every $i \in S$, $n \geq 0$, and $\alpha > 0$; therefore, by Lemma 6.16,

$$V_\alpha(i, \varphi) = \frac{g^*}{\alpha} + \sum_{k=0}^n \alpha^k h^k(i) + \alpha^{n+1} V_\alpha(i, \varphi, -h^n).$$

Now, h^0 being the bias of $f \in \mathbb{F}^\infty$, we have $\|h^0\|_w \leq RM/\delta$ (recall (3.27)). Recursively, since h^n is the bias of f for the reward rate $-h^{n-1}$, we deduce that $\|h^n\|_w \leq M(R/\delta)^{n+1}$. Hence, as a consequence of (3.7) for the reward rate $-h^n$, it follows that for $\alpha > 0$ sufficiently small,

$$\lim_{n \to \infty} \alpha^{n+1} V_\alpha(i, \varphi, -h^n) = 0.$$

In particular, for small $\alpha > 0$ and all $i \in S$,

$$V_\alpha(i, \varphi) = \frac{g^*}{\alpha} + \sum_{k=0}^{\infty} \alpha^k h^k(i) = V_\alpha(i, f),$$

and (7.2) holds.

Now suppose that there exists $n \geq 0$ such that $\varphi \in \Phi^{n-1}$ and $\varphi \notin \Phi^n$. We fix a state $i \in S$. To prove that $V_\alpha(i, f) \geq V_\alpha(i, \varphi)$ for small $\alpha > 0$, it suffices to prove that

$$\liminf_{\alpha \downarrow 0} \alpha^{-n} (V_\alpha(i, f) - V_\alpha(i, \varphi)) > 0.$$

Now, we have $V_\alpha(i, \varphi, \Delta_m) = 0$ for $0 \leq m < n$ and $\alpha > 0$. As a consequence,

$$V_\alpha(i, \varphi) = \frac{g^*}{\alpha} + \sum_{k=0}^{n} \alpha^k h^k(i) + \alpha^{n+1} V_\alpha(i, \varphi, -h^n) + \alpha^n V_\alpha(i, \varphi, \Delta_n). \quad (7.3)$$

Similarly, since $\Delta_n(i, f) = 0$ for every $n \geq 0$,

$$V_\alpha(i, f) = \frac{g^*}{\alpha} + \sum_{k=0}^{n} \alpha^k h^k(i) + \alpha^{n+1} V_\alpha(i, f, -h^n). \quad (7.4)$$

Therefore,

$$\alpha^{-n}(V_\alpha(i, f) - V_\alpha(i, \varphi)) = \alpha(V_\alpha(i, f, -h^n) - V_\alpha(i, \varphi, -h^n)) - V_\alpha(i, \varphi, \Delta_n).$$

As a consequence of (3.7), $\alpha(V_\alpha(i, f, -h^n) - V_\alpha(i, \varphi, -h^n))$ remains bounded as $\alpha \downarrow 0$. On the other hand,

$$V_\alpha(i, \varphi, \Delta_n) = E_i^\varphi \left[\int_0^\infty e^{-\alpha t} \Delta_n(x(t), \varphi) dt \right]$$

$$= \int_0^\infty e^{-\alpha t} E_i^\varphi [\Delta_n(x(t), \varphi)] dt$$

(the interchange of integral and expectation is a consequence of Assumption 3.1(b)). Now, $E_i^\varphi[\Delta_n(x(t), \varphi)] \leq 0$ (because $\varphi \in \Phi^{n-1}$) and, thus, by monotone convergence

$$\lim_{\alpha \downarrow 0} V_\alpha(i, \varphi, \Delta_n) = \int_0^\infty E_i^\varphi[\Delta_n(x(t), \varphi)] dt = -\infty,$$

by Lemma 6.19. Hence, we have shown that

$$\liminf_{\alpha \downarrow 0} \alpha^{-n}(V_\alpha(i, f) - V_\alpha(i, \varphi)) = \infty,$$

which completes the proof of (ii). □

7.3. Blackwell optimality in the class of all policies

Proving the existence of a Blackwell optimal policy in the class of all control policies is quite involved, and it needs several additional hypotheses. Instead of stating all these additional hypotheses, we prefer to make some assumptions that, though not in their weaker form, are more easily verifiable in practice, and yield the existence of a Blackwell optimal policy.

It is worth noting that some of the conditions in Assumption 7.2 below have already been used (see, for instance, Assumptions 3.1 and 3.11, or Assumption 5.5).

Assumption 7.2. There exists a Lyapunov function w on S such that:

(a) There are constants $c > 0$ and $b \geq 0$, and a finite set $C \subset S$ such that

$$\sum_{j \in S} q_{ij}(a)w(j) \leq -cw(i) + b \cdot \mathbf{I}_C(i) \quad \forall \, (i, a) \in K.$$

(b) For each $i \in S$, $q(i) \leq w(i)$.
(c) There exist constants $c', c'' > 0$ and $b', b'' \geq 0$ such that

$$\sum_{j \in S} q_{ij}(a)w^2(j) \leq -c'w^2(i)+b' \quad \text{and} \quad \sum_{j \in S} q_{ij}(a)w^3(j) \leq -c''w^3(i)+b''$$

for all $(i, a) \in K$.
(d) There exists a constant $M > 0$ such that $|r(i, a)| \leq Mw(i)$ for every $(i, a) \in K$.

The condition in (b) is usually verified in practice, and it is not a restrictive requirement. The Assumption 7.2(c) is needed for the application of the Dynkin formula (as we shall see in Chapter 9, despite its appearance this condition is not restrictive at all).

Assumption 7.3.

(a) For each $i \in S$, the set $A(i)$ is compact.
(b) For every $i, j \in S$, the functions $a \mapsto q_{ij}(a)$ and $a \mapsto r(i, a)$ are continuous on $A(i)$.

Note that the continuity of $a \mapsto \sum_{j \in S} q_{ij}(a)w(j)$ in Assumption 3.3(b) follows from Remark 4.6(ii) and Assumption 7.2, and is, therefore, omitted in Assumption 7.3.

Assumption 7.4. Each $f \in \mathbb{F}$ is irreducible.

The uniform ergodicity condition in Assumption 3.11(c), with the new Assumptions 7.2, 7.3, and 7.4, follows from Theorem 2.11 and is omitted as well.

Our main result in this section is the following.

Theorem 7.5. *If Assumptions* 7.2, 7.3, *and* 7.4 *are satisfied, then every policy in* $\mathbb{F}^\infty$ *is Blackwell optimal in* Φ.

The proof of Theorem 7.5 requires several preliminary results. We begin with the following simple version of an Abelian theorem.

Lemma 7.6. *Let* $v : [0, \infty) \to \mathbb{R}$ *be a measurable function that is bounded on compact intervals, and suppose also that the integral* $\int_0^\infty e^{-\alpha t} v(t)dt$ *is finite for each* $\alpha > 0$. *Then*

$$\limsup_{\alpha \downarrow 0} \alpha \int_0^\infty e^{-\alpha t} v(t)dt \leq \limsup_{T \to \infty} v(T).$$

Proof. Let $C := \limsup_{T \to \infty} v(T)$. The stated result is obvious if $C = -\infty$ or $C = +\infty$. Therefore, suppose that C is finite. For each $\varepsilon > 0$ there exists T_0 such that $t \geq T_0$ implies $v(t) \leq C + \varepsilon$, and thus

$$\alpha \int_0^\infty e^{-\alpha t} v(t)dt = \alpha \int_0^{T_0} e^{-\alpha t} v(t)dt + \alpha \int_{T_0}^\infty e^{-\alpha t} v(t)dt$$

$$\leq \alpha \int_0^{T_0} e^{-\alpha t} v(t)dt + (C + \varepsilon)e^{-\alpha T_0}.$$

Now we take the lim sup as $\alpha \downarrow 0$ to obtain the stated result.   □

Stronger versions of Lemma 7.6 involve $\limsup_{T \to \infty} \frac{1}{T} \int_0^T v(t)dt$; see [64, Lemma 3.6], for instance. Lemma 7.6, however, suffices for our purposes.

We also need the following Fatou-like lemma, which in turn requires an inequality similar to (2.13) or (2.16); namely, for every $\varphi \in \Phi$, $i \in S$, and $t \geq 0$,

$$E_i^\varphi[w^3(x(t))] \leq e^{-c''t}w^3(i) + b''/c'', \tag{7.5}$$

where $c'' > 0$ is as in Assumption 7.2(c). In addition, it should be noted that the proof of Lemma 7.7 uses Assumption 7.2(c), whereas in [126] we used a uniform integrability condition [126, Assumption D].

Lemma 7.7. *Given a policy* $\varphi \in \Phi$, *an initial state* $i \in S$, *a function* $v \in \mathcal{B}_{w^2}(K)$ *(recall the notation introduced in (2.10)), and a nonnegative sequence* $\{t_k\}_{k \geq 1}$ *increasing to* ∞,

$$\limsup_{k \to \infty} E_i^\varphi[v(t_k, x(t_k), \varphi)] \leq \sum_{j \in S} \limsup_{k \to \infty} P_{ij}^\varphi(0, t_k) v(t_k, j, \varphi),$$

where $v(t, i, \varphi)$ *is as in* (3.11).

Proof. We first note that, given $j_0 \in S$ and $t \geq 0$,

$$\sum_{j \geq j_0} P_{ij}^\varphi(0, t) w^2(j) \leq \frac{1}{w(j_0)} \sum_{j \geq j_0} P_{ij}^\varphi(0, t) w^3(j)$$

$$\leq \frac{1}{w(j_0)} \sum_{j \in S} P_{ij}^\varphi(0, t) w^3(j)$$

$$\leq \frac{1}{w(j_0)} (e^{-c''t} w^3(i) + b''/c''),$$

where the latter inequality is obtained from (7.5). So, for fixed $\varepsilon > 0$ and $i \in S$, there exists a finite set $K_\varepsilon \subset S$ such that

$$\sup_{t \geq 0} \sum_{j \notin K_\varepsilon} P_{ij}^\varphi(0, t) w^2(j) \leq \varepsilon. \tag{7.6}$$

Now we turn to the proof of this lemma. Fix $\varepsilon > 0$ and note that

$$\limsup_{k \to \infty} E_i^\varphi[v(t_k, x(t_k), \varphi)]$$

is less than or equal to

$$\sum_{j \in K_\varepsilon} \limsup_{k \to \infty} P_{ij}^\varphi(0, t_k) v(t_k, j, \varphi) + \limsup_{k \to \infty} \sum_{j \notin K_\varepsilon} P_{ij}^\varphi(0, t_k) v(t_k, j, \varphi),$$

where K_ε is taken from (7.6). Since $v \in \mathcal{B}_{w^2}(K)$, there exists a constant $C > 0$ verifying $|v(t_k, j, \varphi)| \leq C w^2(j)$ for every $j \in S$ and k. Thus, we deduce from (7.6) that

$$\limsup_{k\to\infty} E_i^\varphi[v(t_k, x(t_k), \varphi)] \le \sum_{j\in K_\varepsilon} \limsup_{k\to\infty} P_{ij}^\varphi(0, t_k)v(t_k, j, \varphi) + C\varepsilon. \quad (7.7)$$

On the other hand, we have

$$-C\varepsilon \le \limsup_{k\to\infty} \sum_{j\notin K_\varepsilon} -CP_{ij}^\varphi(0, t_k)w^2(j)$$

$$\le \sum_{j\notin K_\varepsilon} \limsup_{k\to\infty} [-CP_{ij}^\varphi(0, t_k)w^2(j)]$$

$$\le \sum_{j\notin K_\varepsilon} \limsup_{k\to\infty} P_{ij}^\varphi(0, t_k)v(t_k, j, \varphi),$$

where the first inequality comes from (7.6), the second one is derived from the standard Fatou lemma for nonpositive functions, and, finally, the last inequality is a consequence of the fact that $v \in \mathcal{B}_{w^2}(K)$. Hence, from (7.7) we deduce

$$\limsup_{k\to\infty} E_i^\varphi[v(t_k, x(t_k), \varphi)] \le \sum_{j\in S} \limsup_{k\to\infty} P_{ij}^\varphi(0, t_k)v(t_k, j, \varphi) + 2C\varepsilon.$$

But $\varepsilon > 0$ being arbitrary, we reach the desired result.   □

Continuing with the preliminaries to prove Theorem 7.5, we now state the following continuity result.

Lemma 7.8. *For every $u \in \mathcal{B}_{w^2}(K)$, $\varphi \in \Phi$, and $i \in S$, the mapping*

$$t \mapsto E_i^\varphi[u(t, x(t), \varphi)]$$

is Lipschitz continuous on $[0, \infty)$; hence, it is uniformly continuous.

Proof. Given $u \in \mathcal{B}_{w^2}(K)$ and $\varphi \in \Phi$, we define

$$(L^{s,\varphi}u)(s, i, \varphi) := \sum_{j\in S} q_{ij}(s, \varphi)u(s, j, \varphi)$$

for $i \in S$ and $s \ge 0$; see the notation of (2.14) and (3.11). By standard arguments we deduce that, for some constant C,

$$|(L^{s,\varphi}u)(s, i, \varphi)| \le Cw^3(i) \quad \forall i \in S, \ s \ge 0.$$

Then, from (7.5) we obtain, for every initial state $i \in S$,

$$E_i^\varphi \left[\int_0^t |(L^{s,\varphi}u)(s, x(s), \varphi)| ds \right] < \infty \quad \forall\, t \geq 0.$$

Consequently (see the comment after Proposition 2.3), the Dynkin formula applies; that is, given $t \geq 0$,

$$E_i^\varphi [u(t, x(t), \varphi)] - u(0, i, \varphi) = E_i^\varphi \left[\int_0^t (L^{s,\varphi}u)(s, x(s), \varphi) ds \right].$$

On the other hand, it also follows from (7.5) that, given $0 \leq s < t$,

$$\left| E_i^\varphi \left[\int_s^t (L^{v,\varphi}u)(v, x(v), \varphi) dv \right] \right| \leq C'(t - s)$$

where the constant C' depends on $i \in S$. We conclude that

$$|E_i^\varphi [u(t, x(t), \varphi)] - E_i^\varphi [u(s, x(s), \varphi)]| \leq C'(t - s),$$

and so $t \mapsto E_i^\varphi [u(t, x(t), \varphi)]$ is Lipschitz continuous on $[0, \infty)$. $\qquad \square$

Now observe that, under our standing hypotheses, the discrepancy function Δ_n, defined in Sec. 6.5, is in $\mathcal{B}_{w^2}(K)$ for each $n \geq 0$ (cf. (6.21)). Given a policy $\varphi \in \Phi$ and a state $i \in S$, we define

$$\mathbf{u}_n(i, \varphi) := \int_0^\infty E_i^\varphi [\Delta_n(t, x(t), \varphi)] dt \quad \text{for } n \geq 0,$$

and let

$$\mathbf{u}(i, \varphi) := \{\mathbf{u}_n(i, \varphi)\}_{n \geq 0}.$$

Our next result shows that this sequence is lexicographically nonpositive. We denote by $\mathbf{0}$ the zero sequence $\{0\}_{n \geq 0}$.

Lemma 7.9. *For each $\varphi \in \Phi$ and $i \in S$, $\mathbf{u}(i, \varphi) \preceq \mathbf{0}$.*

Proof. Proving this lemma is equivalent to showing that

(i) $\mathbf{u}_0(i, \varphi) \leq 0$, and
(ii) $\mathbf{u}_k(i, \varphi) = 0$ for $0 \leq k \leq n$ implies $\mathbf{u}_{n+1}(i, \varphi) \leq 0$.

Proof of (i). This statement is obvious since $\Delta_0 \leq 0$. (Note that we do not exclude the possibility $\mathbf{u}_0(i, \varphi) = -\infty$.)

Proof of (ii). If $\mathbf{u}_0(i, \varphi) = 0$, then using Lemma 7.8 and the fact that Δ_0 is nonpositive,

$$E_i^\varphi[\Delta_0(t, x(t), \varphi)] = 0 \quad \forall\, t \geq 0.$$

In particular, for all $t \geq 0$, $\Delta_0(j, a) = 0$ with $\varphi_t(\cdot|j)$-probability one for each $j \in S$ such that $P_{ij}^\varphi(0, t) > 0$. Hence, for such j, $\Delta_1(j, a) \leq 0$ with $\varphi_t(\cdot|j)$-probability one, and then $E_i^\varphi[\Delta_1(t, x(t), \varphi)] \leq 0$ for all $t \geq 0$. So, arguing as before, if

$$\mathbf{u}_1(i, \varphi) = \int_0^\infty E_i^\varphi[\Delta_1(t, x(t), \varphi)]dt = 0,$$

then for all $t \geq 0$ and each $j \in S$ such that $P_{ij}^\varphi(0, t) > 0$, we have $\Delta_1(j, a) = 0$ and $\Delta_2(j, a) \leq 0$ with $\varphi_t(\cdot|j)$-probability one.

Recursively, to prove (ii), we can show that for all $t \geq 0$ and every $j \in S$ such that $P_{ij}^\varphi(0, t) > 0$,

$$\Delta_0(j, a) = \cdots = \Delta_n(j, a) = 0 \quad \text{and} \quad \Delta_{n+1}(j, a) \leq 0 \tag{7.8}$$

with $\varphi_t(\cdot|j)$-probability one. This completes the proof of (ii). $\qquad\square$

Now, fix a policy $\varphi \in \Phi$ and a state $i \in S$. If $m \geq 0$ is the minimal m for which $\mathbf{u}_m(i, \varphi) < 0$, then we say that $\varphi \in \Phi_m^*(i)$. If $\mathbf{u}(i, \varphi) = \mathbf{0}$ then we say that $\varphi \in \Phi_\infty^*(i)$. At this point, note that when restricting ourselves to stationary policies, using the notation introduced in Sec. 6.6 and recalling Lemma 6.19, we obtain $\Phi_s \cap \Phi_m^*(i) = \Phi^{m-1} - \Phi^m$. With this notation and as a direct consequence of (7.8), we derive our next result, which we state without proof.

Lemma 7.10. *Given* $\varphi \in \Phi$ *and* $i \in S$, *suppose that* $\varphi \in \Phi_m^*(i)$ *for some* $m \geq 0$. *Then, for all* $t \geq 0$ *and each* $j \in S$ *such that* $P_{ij}^\varphi(0, t) > 0$,

$$\Delta_0(t, j, \varphi) = \cdots = \Delta_{m-1}(t, j, \varphi) = 0 \quad \text{and} \quad \Delta_m(t, j, \varphi) \leq 0.$$

Now we are ready to prove our main result.

Proof of Theorem 7.5. Let f be a deterministic stationary policy in $\mathbb{F}^\infty$, and fix a policy $\varphi \in \Phi$ and a state $i \in S$. We want to prove that $V_\alpha(i, f) \geq V_\alpha(i, \varphi)$ for all sufficiently small $\alpha > 0$. To this end, we shall distinguish three cases:

(a) $\varphi \in \Phi_\infty^*(i)$.
(b) For some integer $m \geq 0$, $\varphi \in \Phi_m^*(i)$ and $\mathbf{u}_m(i, \varphi) = -\infty$.
(c) For some integer $m \geq 0$, $\varphi \in \Phi_m^*(i)$ and $-\infty < \mathbf{u}_m(i, \varphi) < 0$.

The proof of cases (a) and (b) is a verbatim copy of the proof of Theorem 7.1(ii). Indeed, that proof is mainly based on Lemma 6.16, which holds for nonstationary policies. Therefore, we omit these proofs.

Case (c). It suffices to prove that

$$\liminf_{\alpha\downarrow 0} \alpha^{-m}(V_\alpha(i,f) - V_\alpha(i,\varphi)) > 0.$$

To this end, we proceed one step further in the expansions (7.3) and (7.4) and we obtain

$$V_\alpha(i,\varphi) = \frac{g^*}{\alpha} + \sum_{k=0}^{m+1} \alpha^k h^k(i) + \alpha^{m+2}V_\alpha(i,\varphi,-h^{m+1})$$

$$+ \alpha^m V_\alpha(i,\varphi,\Delta_m) + \alpha^{m+1}V_\alpha(i,\varphi,\Delta_{m+1})$$

and

$$V_\alpha(i,f) = \frac{g^*}{\alpha} + \sum_{k=0}^{m+1} \alpha^k h^k(i) + \alpha^{m+2}V_\alpha(i,f,-h^{m+1}).$$

As a consequence of (3.7),

$$\liminf_{\alpha\downarrow 0} \alpha^{-m}(V_\alpha(i,f)-V_\alpha(i,\varphi)) = -\limsup_{\alpha\downarrow 0} [V_\alpha(i,\varphi,\Delta_m)+\alpha V_\alpha(i,\varphi,\Delta_{m+1})]$$

and, thus, we must prove that

$$\limsup_{\alpha\downarrow 0} [V_\alpha(i,\varphi,\Delta_m) + \alpha V_\alpha(i,\varphi,\Delta_{m+1})] < 0.$$

We will proceed by contradiction, and henceforth we suppose that

$$\limsup_{\alpha\downarrow 0} [V_\alpha(i,\varphi,\Delta_m) + \alpha V_\alpha(i,\varphi,\Delta_{m+1})] \geq 0. \tag{7.9}$$

As in the proof of Theorem 7.1(ii),

$$\lim_{\alpha\downarrow 0} V_\alpha(i,\varphi,\Delta_m) = \int_0^\infty E_i^\varphi[\Delta_m(t,x(t),\varphi)]dt = \mathbf{u}_m(i,\varphi) < 0.$$

Combined with (7.9), this yields

$$\limsup_{\alpha\downarrow 0} \alpha V_\alpha(i,\varphi,\Delta_{m+1}) > 0$$

and, by Lemma 7.6,

$$\limsup_{t\to\infty} E_i^\varphi[\Delta_{m+1}(t,x(t),\varphi)] > 0.$$

Therefore, there exists $\varepsilon > 0$ and a sequence t' increasing to ∞ such that $\limsup_{t'} E_i^\varphi[\Delta_{m+1}(t', x(t'), \varphi)] > \varepsilon$. Using Lemma 7.7 we deduce that

$$\sum_{j \in S} \limsup_{t'} P_{ij}^\varphi(0, t')\Delta_{m+1}(t', j, \varphi) > \varepsilon,$$

and, thus, there exists $j \in S$ and a further subsequence t'' such that, for each t'',

$$P_{ij}^\varphi(0, t'')\Delta_{m+1}(t'', j, \varphi) \geq \varepsilon/2^{j+1}.$$

In particular,

$$\Delta_{m+1}(t'', j, \varphi) \geq \varepsilon/2^{j+1}$$

for every t''. Also, for every t'' and some $\tilde{\varepsilon} > 0$,

$$P_{ij}^\varphi(0, t'') \geq \tilde{\varepsilon}, \tag{7.10}$$

because $\Delta_{m+1}(t'', j, \varphi)$ is bounded for fixed j.

Taking into account Lemma 7.10 and (7.10), we have shown the existence of a sequence $\{t''\}$ and a state $j \in S$ such that, for $0 \leq n < m$,

$$\Delta_n(t'', j, \varphi) = 0, \quad \Delta_m(t'', j, \varphi) \leq 0, \quad \text{and} \quad \Delta_{m+1}(t'', j, \varphi) \geq \varepsilon/2^{j+1}. \tag{7.11}$$

Now, observe that $\varphi_{t''}(\cdot|j)$ is a probability measure on the compact Borel space $A(j)$. Then, by the results in Sec. 1.4, there exist a probability measure on $A(j)$, say φ^*, and a subsequence of $\{t''\}$, which we still denote by $\{t''\}$, such that

$$\varphi_{t''}(\cdot|j) \xrightarrow{w} \varphi^*.$$

By Assumptions 7.2 and 7.3, the functions $a \mapsto \Delta_n(j, a)$ are continuous and bounded on $A(j)$, and, therefore,

$$\Delta_n(t'', j, \varphi) = \int_{A(j)} \Delta_n(j, a)\varphi_{t''}(da|j) \to \int_{A(j)} \Delta_n(j, a)\varphi^*(da) = \Delta_n(j, \varphi^*)$$

as $t'' \to \infty$ for all $n \geq 0$. Hence, from (7.11), for $0 \leq n < m$,

$$\Delta_n(j, \varphi^*) = 0, \quad \Delta_m(j, \varphi^*) \leq 0, \quad \text{and} \quad \Delta_{m+1}(j, \varphi^*) \geq \varepsilon/2^{j+1}.$$

Let us now prove that, in fact, $\Delta_m(j, \varphi^*) = 0$. Indeed, by the hypothesis in (c),

$$-\infty < \int_0^\infty E_i^\varphi[\Delta_m(t, x(t), \varphi)]dt < 0$$

and, $t \mapsto E_i^\varphi[\Delta_m(t, x(t), \varphi)]$ being a uniformly continuous (Lemma 7.8) and nonpositive (Lemma 7.10) function, we deduce that

$$\lim_{t \to \infty} E_i^\varphi[\Delta_m(t, x(t), \varphi)] = 0.$$

Since

$$E_i^\varphi[\Delta_m(t'', x(t''), \varphi)] \le P_{ij}^\varphi(0, t'')\Delta_m(t'', j, \varphi) \le 0,$$

letting $t'' \to \infty$ and recalling (7.10), we derive $\Delta_m(t'', j, \varphi) \to 0$, and so $\Delta_m(j, \varphi^*) = 0$.

So far we have proved that

$$\Delta_n(j, \varphi^*) = 0 \ \text{ for } 0 \le n \le m, \quad \text{and} \quad \Delta_{m+1}(j, \varphi^*) \ge \varepsilon/2^{j+1},$$

which combined with the fact that $\{\Delta_n(i, a)\}_{n \ge 0} \preceq \mathbf{0}$ for every $(i, a) \in K$ shows that, with φ^*-probability one, we have

$$\Delta_0(j, a) = \cdots = \Delta_m(j, a) = 0,$$

and so $\Delta_{m+1}(j, a) \le 0$. This contradicts that $\Delta_{m+1}(j, \varphi^*) \ge \varepsilon/2^{j+1}$. The proof of Theorem 7.5 is now complete. $\qquad \square$

It is worth noting that, when dealing with stationary policies, the case (c), stated at the beginning of this proof, cannot happen (recall the proof of Theorem 7.1). Hence, the difficulty in the proof of Theorem 7.5 relies precisely in dealing with this case.

7.4. Conclusions

The results in this chapter show the existence of a deterministic stationary policy that is Blackwell optimal in the class of stationary policies (Sec. 7.2), or in the class of all Markov policies (Sec. 7.3). The proof of the latter result is more involved and, in addition, it requires supplementary hypotheses. These hypotheses are the most restrictive imposed so far (recall, for instance, Assumption 7.2(c)) on our control model. As we shall see, however, these assumptions are not restrictive in practice and they are satisfied by a number of models of interest (see Chapter 9).

The Blackwell optimality criterion is the strongest undiscounted optimality criterion, since it implies gain, bias, and n-discount optimality (though it does not imply, in general, variance optimality). Proving the existence of Blackwell optimal policies requires, however, conditions on the CMC model that are stronger than those for gain, bias, and n-discount optimal policies.

Therefore, when dealing with average reward and related criteria, one should always choose a Blackwell optimal policy (provided that the corresponding sufficient conditions are satisfied). In many cases of interest, however, there exists a unique gain optimal policy, which is necessarily Blackwell optimal.

For the two possibilities for improving the average reward optimality criterion mentioned in Sec. 3.8, either by letting the finite time horizon $T \to \infty$ or the discount rate $\alpha \downarrow 0$, we conclude from our analysis in Chapters 5 to 7 that the best and richest refinements are obtained by letting $\alpha \downarrow 0$. Indeed, the 0-discount optimality criterion is equivalent to bias optimality and, therefore, n-discount optimality for $n \geq 1$ is stronger than bias optimality. In the limit as $n \to \infty$, we even reach Blackwell optimality.

Chapter 8

Constrained Controlled Markov Chains

8.1. Introduction

In this chapter, we are concerned with the control model introduced in Chapter 2, namely,

$$\mathcal{M} := \{S, A, (A(i)), (q_{ij}(a)), r, u\}.$$

In addition, we consider a function $u : K \to \mathbb{R}$, which is interpreted as a *cost rate*. The function u is such that $a \mapsto u(i, a)$ is measurable on $A(i)$ for each fixed $i \in S$. In general, we will assume that u satisfies the same conditions as the reward rate function r (see, e.g., Assumptions 3.1(c) and 3.3(b)).

Our goal in this chapter is to find policies that are *constrained optimal* for a given optimality criterion. More precisely, for the discounted reward optimality criterion, given an initial state $i_0 \in S$ we will find policies that maximize the total expected discounted reward $V_\alpha(i_0, \varphi)$ within the class of policies φ that satisfy the constraint

$$V_\alpha(i_0, \varphi, u) \leq \theta$$

(see the notation in (3.10)), where $\theta \in \mathbb{R}$ is a given constraint constant.

Similarly, for the average reward optimality criterion, we will find policies that maximize $J(i, \varphi)$ over the set of policies such that a given long-run expected average cost satisfies a certain constraint (a precise definition is given in (8.8) below). To deal with these constrained problems, we will use the *Lagrange multipliers* approach.

Our results in this chapter can be easily extended to control problems with any finite number of constraints. For instance, for the discounted reward optimality criterion, given a family $u_1, \ldots, u_m$ of cost rate functions with corresponding constraint constants $\theta_1, \ldots, \theta_m$, we must find a policy that maximizes $V_\alpha(i_0, \varphi)$ within the class of policies φ such that

$$V_\alpha(i_0, \varphi, u_j) \le \theta_j \quad \forall\, j \in \{1, \dots, m\}.$$

For an average reward control problem with a finite number of constraints, a similar definition can be given. For ease of exposition, however, we will restrict ourselves to constrained problems with a single constraint. The extension to any finite number of constraints is straightforward.

Constrained control problems are an active area of research, and they have an obvious motivation: one usually has to maximize a reward or a benefit, but subject to some, say, budgetary restrictions. The discrete-time models have been extensively studied. We can mention, for instance, Altman [4], Feinberg and Shwartz [34], Hernández-Lerma and González-Hernández [65], Hernández-Lerma, González-Hernández, and López-Martínez [66], Piunovskiy [121], Puterman [138], and Sennott [145, 148]. For the continuous-time case, however, there exist just a few references. Constrained CMCs were studied by Guo and Hernández-Lerma [46] under the discounted optimality criterion, and by Prieto-Rumeau and Hernández-Lerma [131] and Zhang and Guo [171] for the average optimality criterion.

Usually, in a constrained average reward CMC, we impose a constraint on an *expected* average cost. For applications, however, this might not make sense. In fact, Haviv [61] shows an explicit application in which expected constraints lead to erroneous conclusions. The latter are avoided when using *pathwise* or *hard* constraints, that is, constraints that must be satisfied with probability one. These have been studied by Haviv [61], Ross and Varadarajan [141, 142], and also by Wu, Arapostathis, and Shakkottai [166]. The reference [131] also deals with CMCs with pathwise constraints.

In the previous chapters, when dealing with optimality criteria such as discounted reward, average reward, sensitive discount optimality, or Blackwell optimality, we have proved that there exist optimal policies that are *deterministic* and stationary. To solve a constrained control problem, however, we have to consider *randomized* policies. The reason for this is, roughly speaking, that the set Φ_s of randomized stationary policies is the convex hull generated by the set $\mathbb{F}$ of deterministic stationary policies (for a precise statement, see [42]), and that, in general, when solving a constrained problem, an optimal solution does not need to be a vertex (that is, a deterministic policy) of the unconstrained feasible set.

The contents of this chapter are the following. In Sec. 8.2, we study constrained CMCs under the discounted reward optimality criterion. Expected average constrained CMCs are analyzed in Sec. 8.3, while Sec. 8.4 deals with CMCs with pathwise average constraints. The main result in Sec. 8.4, Theorem 8.17, states that the facts in Sec. 8.3 for *average* constrained problems are also valid in the *pathwise* case. The relations between discounted and average constrained problems, as the discount rate vanishes, are explored in Sec. 8.5. Let us mention that these relations are not as direct as in the unconstrained case (recall Secs 3.5 and 6.3). Actually, Example 8.19 shows that the standard result for the vanishing discount approach to average optimality is *not true* in general. We conclude with some final remarks in Sec. 8.6.

8.2. Discounted reward constrained CMCs

This section is based on [46]. We consider a control model that satisfies Assumptions 3.1 and 3.3. For ease of reference, we restate them here.

Assumption 8.1. There exists a Lyapunov function w on S such that:

(a) For some constants $c \in \mathbb{R}$ and $b \geq 0$,

$$\sum_{j \in S} q_{ij}(a)w(j) \leq -cw(i) + b \quad \forall\, (i,a) \in K.$$

(b) There exists a nonnegative function $w' : S \to \mathbb{R}$ such that $q(i)w(i) \leq w'(i)$ for each $i \in S$ and, in addition, there exist constants $c' \in \mathbb{R}$ and $b' \geq 0$ such that

$$\sum_{j \in S} q_{ij}(a)w'(j) \leq -c'w'(i) + b' \quad \forall\, (i,a) \in K.$$

(c) There exists a constant $M > 0$ such that $|r(i,a)| \leq Mw(i)$ for every $(i,a) \in K$.

We next state the usual continuity-compactness hypotheses.

Assumption 8.2.

(a) For each $i \in S$, the set $A(i)$ is compact.
(b) For every $i, j \in S$, the functions $a \mapsto q_{ij}(a)$, $a \mapsto r(i,a)$, and $a \mapsto \sum_{j \in S} q_{ij}(a)w(j)$ are continuous on $A(i)$, where w is the Lyapunov function in Assumption 8.1.

Let $\alpha > 0$ be a given discount rate and, in addition, suppose that $\alpha + c > 0$, where $c \in \mathbb{R}$ is the constant in Assumption 8.1(a). We also consider a nonnegative *cost* rate function u, an initial state $i_0 \in S$, and a constraint constant $\theta \in \mathbb{R}$ that satisfy the following conditions (note that in (c) below we use the notation of (3.10)).

Assumption 8.3.

(a) The cost rate u verifies $0 \le u(i, a) \le M'w(i)$ for all $(i, a) \in K$ and some constant $M' > 0$.

(b) For each fixed $i \in S$, $a \mapsto u(i, a)$ is continuous on $A(i)$.

(c) There exists a policy $\varphi \in \Phi$ such that $V_\alpha(i_0, \varphi, u) < \theta$ for some state $i_0 \in S$.

The *strict* inequality $V_\alpha(i_0, \varphi, u) < \theta$ in Assumption 8.3(c) states a lot more than just the feasibility of the constrained problem. It is in fact a Slater-like condition ensuring the existence of $\varphi \in \Phi$ in the "interior" of the feasible set. The reader interested in the Slater condition for constrained optimization problems on $\mathbb{R}^n$ can consult [125, 153].

Statement of the problem   Let $i_0 \in S$ be a given *fixed* initial state. Our goal in this section is to maximize $V_\alpha(i_0, \varphi)$ in the class of "constrained" policies, that is, the policies $\varphi \in \Phi$ such that $V_\alpha(i_0, \varphi, u) \le \theta$. More precisely, we say that a policy $\varphi^* \in \Phi$ such that $V_\alpha(i_0, \varphi^*, u) \le \theta$ is *discounted constrained optimal* if, for each policy $\varphi \in \Phi$ such that $V_\alpha(i_0, \varphi, u) \le \theta$, we have

$$V_\alpha(i_0, \varphi^*) \ge V_\alpha(i_0, \varphi).$$

Our results here can be readily extended to the case when the initial state of the system has a given distribution μ on S (see Theorem 8.7 below). To fix ideas, however, we will analyze the case of a fixed initial state $i_0 \in S$.

The Lagrange multipliers approach   To deal with the constrained optimization problem, we introduce a "Lagrange multiplier" $\lambda \ge 0$. Given $(i, a) \in K$ and $\lambda \ge 0$, let

$$v_\lambda(i, a) := r(i, a) - \lambda u(i, a).$$

Obviously, the function v_λ verifies the conditions in Assumptions 8.1(c) and 8.2(b) (or, equivalently, Assumptions 3.1(c) and 3.3(b)). Therefore, interpreting v_λ as a reward rate function and letting

$$V^*(i, \lambda) := \sup_{\varphi \in \Phi} V_\alpha(i, \varphi, v_\lambda) \quad \forall \, i \in S,$$

the corresponding DROE (by Theorem 3.7) is

$$\alpha V^*(i, \lambda) = \max_{a \in A(i)} \left\{ v_\lambda(i, a) + \sum_{j \in S} q_{ij}(a) V^*(j, \lambda) \right\} \quad \forall \, i \in S.$$

Besides, let

$$\mathbb{F}^*_\lambda := \left\{ f \in \mathbb{F} \ : \ \alpha V^*(i, \lambda) = v_\lambda(i, f) + \sum_{j \in S} q_{ij}(f) V^*(j, \lambda) \quad \forall \, i \in S \right\},$$

with $v_\lambda(i, f) := v_\lambda(i, f(i))$. Then, by Theorem 3.7 again, $\mathbb{F}^*_\lambda$ is nonempty and any $f \in \mathbb{F}^*_\lambda$ is α-discount optimal for the control problem with reward rate function v_λ.

For each $\lambda \geq 0$, we choose a *fixed* policy $f^*_\lambda \in \mathbb{F}^*_\lambda$ and define

$$V_u(\lambda) := V_\alpha(i_0, f^*_\lambda, u).$$

By [46, Lemma 3.4], $\lambda \mapsto V_u(\lambda)$ is a monotone nonincreasing function (in particular, this lemma makes use of the fact that the cost rate u is nonnegative). Therefore, the constant

$$\tilde{\lambda} := \inf_{\lambda \geq 0} \{V_u(\lambda) \leq \theta\} \tag{8.1}$$

is well defined, and it can take the values $0 \leq \tilde{\lambda} \leq \infty$. Our next result shows that, in fact, it is finite.

Lemma 8.4. *Under Assumptions 8.1, 8.2, and 8.3, the constant $\tilde{\lambda} \geq 0$ in (8.1) is finite.*

Proof. We proceed by contradiction. Hence, we suppose that $\tilde{\lambda} = \infty$, that is, $V_u(\lambda) > \theta$ for all $\lambda \geq 0$.

We choose a policy $\varphi \in \Phi$ with $V_\alpha(i_0, \varphi, u) < \theta$ (such a policy indeed exists by Assumption 8.3(c)), and let $\delta := \theta - V_\alpha(i_0, \varphi, u) > 0$.

Fix an arbitrary $\lambda > 0$. We have

$$V_\alpha(i_0, \varphi, v_\lambda) = V_\alpha(i_0, \varphi) - \lambda V_\alpha(i_0, \varphi, u) = V_\alpha(i_0, \varphi) - \lambda(\theta - \delta). \tag{8.2}$$

On the other hand,

$$\begin{aligned}
V^*(i_0, \lambda) &= V_\alpha(i_0, f^*_\lambda) - \lambda V_\alpha(i_0, f^*_\lambda, u) \\
&= V_\alpha(i_0, f^*_\lambda) - \lambda V_u(\lambda) \\
&< V_\alpha(i_0, f^*_\lambda) - \lambda \theta \leq V^*_\alpha(i_0) - \lambda \theta. \tag{8.3}
\end{aligned}$$

Since $V_\alpha(i_0, \varphi, v_\lambda) \le V^*(i_0, \lambda)$, we deduce from (8.2) and (8.3) that

$$V_\alpha^*(i_0) > V_\alpha(i_0, \varphi) + \lambda\delta.$$

This inequality holds for all $\lambda > 0$, and letting $\lambda \to \infty$ yields a contradiction. Hence, we conclude that $\tilde\lambda < \infty$.   □

We will also need the following preliminary result.

Lemma 8.5. *Suppose that there exists a constant $\lambda \ge 0$ and a policy φ^* in Φ such that*

$$V_\alpha(i_0, \varphi^*, u) = \theta \quad and \quad V_\alpha(i_0, \varphi^*, v_\lambda) = V^*(i_0, \lambda).$$

Then the policy φ^* is discounted constrained optimal.

Proof. Let $\varphi \in \Phi$ be an arbitrary constrained policy, that is, φ satisfies $V_\alpha(i_0, \varphi, u) \le \theta$. We have $V_\alpha(i_0, \varphi^*, v_\lambda) \ge V_\alpha(i_0, \varphi, v_\lambda)$, and thus

$$V_\alpha(i_0, \varphi^*) - \lambda V_\alpha(i_0, \varphi^*, u) \ge V_\alpha(i_0, \varphi) - \lambda V_\alpha(i_0, \varphi, u). \qquad (8.4)$$

Recalling that

$$V_\alpha(i_0, \varphi^*, u) = \theta \ge V_\alpha(i_0, \varphi, u),$$

it follows from (8.4) that

$$V_\alpha(i_0, \varphi^*) \ge V_\alpha(i_0, \varphi),$$

which shows that φ^* is constrained optimal.   □

We now state our first main result in this section.

Theorem 8.6. *We suppose that Assumptions 8.1, 8.2, and 8.3 are satisfied, and that, in addition, the discount rate $\alpha > 0$ is such that $\alpha + c > 0$, with c from Assumption 8.1(a). Then there exists a discounted constrained optimal policy that is a stationary policy that randomizes in, at most, one state.*

Proof. We shall distinguish two cases, $\tilde\lambda = 0$ and $\tilde\lambda > 0$; see (8.1).
   Case $\tilde\lambda = 0$. Choose a sequence $\{\lambda_k\}$ of positive numbers such that $\lambda_k \downarrow 0$ and such that, in addition,

$$f_{\lambda_k}^*(i) \to f^*(i) \quad \forall\, i \in S$$

for some $f^* \in \mathbb{F}$ (recall that, under Assumption 8.2(a), the set $\mathbb{F}$ is compact). Therefore, we have

$$V_\alpha(i_0, f^*_{\lambda_k}, u) = V_u(\lambda_k) \leq \theta,$$

and using a continuity argument (see our Theorem 3.9 or [46, Lemma 3.2]), we can take the limit as $k \to \infty$ to obtain

$$V_\alpha(i_0, f^*, u) \leq \theta,$$

that is, f^* is a "constrained" policy.

Choose now an arbitrary policy $\varphi \in \Phi$ such that $V_\alpha(i_0, \varphi, u) \leq \theta$. We have

$$V_\alpha(i_0, f^*_{\lambda_k}) - \lambda_k V_\alpha(i_0, f^*_{\lambda_k}, u) \geq V_\alpha(i_0, \varphi) - \lambda_k V_\alpha(i_0, \varphi, u).$$

Taking the limit as $k \to \infty$, recalling that $V_\alpha(i_0, f^*_{\lambda_k}, u)$ is bounded in k (see (3.7)), and using Theorem 3.9 we obtain

$$V_\alpha(i_0, f^*) \geq V_\alpha(i_0, \varphi).$$

This shows that $f^* \in \mathbb{F}$ is discounted constrained optimal.

Case $\tilde{\lambda} > 0$. Suppose first that there exists $\lambda > 0$ with $V_u(\lambda) = \theta$. In this case, the policy f^*_λ verifies the conditions of Lemma 8.5 and, therefore, it is constrained optimal.

Henceforth, we will suppose that $V_u(\lambda) \neq \theta$ for all $\lambda > 0$. We choose two monotone sequences $\lambda_k \uparrow \tilde{\lambda}$ and $\delta_k \downarrow \tilde{\lambda}$ such that, in addition,

$$f^*_{\lambda_k} \to f^1 \quad \text{and} \quad f^*_{\delta_k} \to f^2$$

for some $f^1, f^2 \in \mathbb{F}$. By [46, Lemma 3.6], the policies f^1 and f^2 are in $\mathbb{F}^*_{\tilde{\lambda}}$. On the other hand, since

$$V_\alpha(i_0, f^*_{\lambda_k}, u) = V_u(\lambda_k) > \theta \quad \text{and} \quad V_\alpha(i_0, f^*_{\delta_k}, u) = V_u(\delta_k) \leq \theta,$$

we deduce

$$V_\alpha(i_0, f^1, u) \geq \theta \quad \text{and} \quad V_\alpha(i_0, f^2, u) \leq \theta. \tag{8.5}$$

If any of the inequalities in (8.5) is an equality, then from Lemma 8.5 we get that f^1 or f^2 is constrained optimal. Thus, we consider the case

$$V_\alpha(i_0, f^1, u) > \theta \quad \text{and} \quad V_\alpha(i_0, f^2, u) < \theta.$$

Starting from the policies $f^1, f^2 \in \mathbb{F}_{\tilde{\lambda}}^*$, we construct the following sequence $\{f_n\}_{n \geq 0}$ of policies:

$$f_n(i) := \begin{cases} f^2(i), & \text{if } i \geq n, \\ f^1(i), & \text{if } i < n. \end{cases}$$

From the definition of $\mathbb{F}_{\tilde{\lambda}}^*$, it follows that $\{f_n\} \subseteq \mathbb{F}_{\tilde{\lambda}}^*$. Note also that $f_0 = f^2$ and $\lim_n f_n = f^1$, and so

$$V_\alpha(i_0, f_0, u) = V_\alpha(i_0, f^2, u) < \theta < V_\alpha(i_0, f^1, u) = \lim_n V_\alpha(i_0, f_n, u).$$

Again, if there exists $n \geq 0$ such that $V_\alpha(i_0, f_n, u) = \theta$, then the policy f_n is constrained optimal; recall Lemma 8.5. Otherwise, there exists some n^* with

$$V_\alpha(i_0, f_{n^*}, u) < \theta < V_\alpha(i_0, f_{n^*+1}, u). \tag{8.6}$$

Given $0 \leq p \leq 1$, we define $\pi_p \in \Phi_s$ as the randomized stationary policy that randomizes, with probabilities p and $1 - p$, between f_{n^*} and f_{n^*+1}, respectively. (Observe that π_p randomizes only at state $i = n^*$.) Since the set of discount optimal stationary policies for the reward rate $r - \tilde{\lambda}u$ is convex [46, Lemma 3.3], it follows that

$$V_\alpha(i_0, \pi_p, v_{\tilde{\lambda}}) = V^*(i_0, \tilde{\lambda}) \quad \forall \, 0 \leq p \leq 1. \tag{8.7}$$

Moreover, it is easily seen that $p \mapsto V_\alpha(i_0, \pi_p, u)$ is continuous on $[0, 1]$; see Theorem 3.9. Therefore, from the inequalities in (8.6), we derive that there exists $0 < p^* < 1$ such that

$$V_\alpha(i_0, \pi_{p^*}, u) = \theta.$$

This fact, together with (8.7) and Lemma 8.5, shows that π_{p^*} is discounted constrained optimal. $\qquad \square$

We conclude this section by addressing the case when, instead of considering a fixed initial state $i_0 \in S$, we consider an initial distribution μ on S. We will assume that μ satisfies

$$\mu(w) = \sum_{i \in S} w(i)\mu(i) < \infty,$$

where w is the Lyapunov function in Assumption 8.1. Then, the discounted reward $V_\alpha(\mu, \varphi)$ is defined as

$$V_\alpha(\mu, \varphi) := \sum_{i \in S} V_\alpha(i, \varphi)\mu(i),$$

which is finite as a consequence of (3.7) and the fact that $\mu(w) < \infty$. Similarly, we define $V_\alpha(\mu, \varphi, u)$.

In this context, we say that a policy $\varphi^* \in \Phi$ with $V_\alpha(\mu, \varphi^*, u) \leq \theta$ is discounted constrained optimal if, for each policy $\varphi \in \Phi$ that satisfies $V_\alpha(\mu, \varphi, u) \leq \theta$, we have $V_\alpha(\mu, \varphi^*) \geq V_\alpha(\mu, \varphi)$. The corresponding result is the following.

Theorem 8.7. *Suppose that Assumptions* 8.1, 8.2, *and* 8.3 *hold, where Assumption* 8.3(c) *is replaced with:*

(c) *There exists a policy $\varphi \in \Phi$ such that $V_\alpha(\mu, \varphi, u) < \theta$, where the initial distribution μ satisfies $\mu(w) < \infty$.*

If the discount rate $\alpha > 0$ is such that $\alpha + c > 0$, then there exists a discounted constrained optimal policy that is stationary and that randomizes in, at most, one state.

The proof of Theorem 8.7 goes along the same lines as that of Theorem 8.6, and is, therefore, omitted.

8.3. Average reward constrained CMCs

Given the constrained control model $\mathcal{M}$ in Sec. 8.1, our goal in this section is to find policies that maximize the long-run expected average reward (corresponding to the reward rate function r) with the restriction that the long-run expected average cost (for the cost rate u) does not exceed a given constant. We mainly follow [131].

We suppose that the constrained control model verifies the following conditions.

Assumption 8.8. There exists a Lyapunov function w on S such that

(a) For some constants $c > 0$ and $b \geq 0$, and a finite set $C \subset S$

$$\sum_{j \in S} q_{ij}(a) w(j) \leq -cw(i) + b \cdot \mathbf{I}_C(i) \quad \forall\, (i, a) \in K.$$

(b) There exists a constant $M > 0$ such that $|r(i, a)| \leq Mw(i)$ and $|u(i, a)| \leq Mw(i)$ for every $(i, a) \in K$.

Assumption 8.8(a) has already been used throughout the text; see, for instance, Assumptions 2.10(a) or 3.11(a).

Given a control policy $\varphi \in \Phi$, the total expected reward of φ over the time horizon $[0, T]$ was defined in (3.2) as

$$J_T(i, \varphi) := E_i^{\varphi}\left[\int_0^T r(t, x(t), \varphi)dt\right] \quad \forall \, i \in S.$$

Similarly, we define the total expected cost of φ as

$$J_{u,T}(i, \varphi) := E_i^{\varphi}\left[\int_0^T u(t, x(t), \varphi)dt\right] \quad \forall \, i \in S,$$

where $u(t, i, \varphi)$ is defined as in (3.1); see also (3.11). The long-run expected average cost (or average cost, for short) of the policy $\varphi \in \Phi$ is defined as

$$J_u(i, \varphi) := \limsup_{T \to \infty} \frac{1}{T} J_{u,T}(i, \varphi). \tag{8.8}$$

At this point, recall that the average reward $J(i, \varphi)$ of the policy φ was defined in (3.23) as a "lim inf." This is because r is interpreted as a *reward rate*. On the other hand, $J_u(i, \varphi)$ is defined as a "lim sup" because u is interpreted as a *cost rate*.

It is also worth noting that, as a consequence of Assumption 8.8 (see (3.24)), for each $\varphi \in \Phi$ and $i \in S$

$$|J(i, \varphi)| \le bM/c \quad \text{and} \quad |J_u(i, \varphi)| \le bM/c.$$

Now we are ready to define the constrained control problem. Fix an initial state $i \in S$ and a constant $\theta_0 \in \mathbb{R}$. We want to maximize $J(i, \varphi)$ over the set of control policies φ in Φ that verify $J_u(i, \varphi) \le \theta_0$. In short, this control problem will be written as

$$\text{maximize } J(i, \varphi) \quad \text{s.t.} \quad J_u(i, \varphi) \le \theta_0, \tag{8.9}$$

where "s.t." means "subject to." More precisely, given $i \in S$, we say that a control policy $\varphi^* \in \Phi$ with $J_u(i, \varphi^*) \le \theta_0$ is optimal for the constrained control problem (8.9) if $J(i, \varphi) \le J(i, \varphi^*)$ for every $\varphi \in \Phi$ such that $J_u(i, \varphi) \le \theta_0$. The corresponding optimal reward function is denoted by $V^*(i, \theta_0)$, that is,

$$V^*(i, \theta_0) := \sup_{\varphi \in \Phi}\{J(i, \varphi) : J_u(i, \varphi) \le \theta_0\} \tag{8.10}$$

for $i \in S$.

We introduce another assumption.

Assumption 8.9.

(a) There exist constants $c' > 0$ and $b' \geq 0$, and a finite set $C' \subset S$ such that

$$\sum_{j \in S} q_{ij}(a) w^2(j) \leq -c' w^2(i) + b' \cdot \mathbf{I}_{C'}(i) \quad \forall\, (i, a) \in K,$$

where w is the Lyapunov function in Assumption 8.8.

(b) For every $i \in S$, $q(i) \leq w(i)$.

(c) Every stationary policy $\varphi \in \Phi_s$ is irreducible.

In fact, as a consequence of Theorem 2.13, the condition in Assumption 8.8(a) is implied by Assumption 8.9(a), but for clarity of exposition we prefer to state them separately. Note that the condition in Assumption 8.9(b) was also imposed in Assumptions 5.5(a) and 7.2(b).

Assumption 8.8(a), together with Assumption 8.9(c), implies that each $\varphi \in \Phi_s$ has a unique invariant probability measure, denoted by μ_φ, for which $\mu_\varphi(w)$ is finite (by Theorem 2.5). Therefore, as in (3.25) and Remark 3.12, the average reward (or gain) and the average cost of a stationary policy $\varphi \in \Phi_s$ are constant (i.e., they do not depend on the initial state):

$$g(\varphi) := \sum_{j \in S} r(j, \varphi) \mu_\varphi(j) = J(i, \varphi)$$

and

$$g_u(\varphi) := \sum_{j \in S} u(j, \varphi) \mu_\varphi(j) = J_u(i, \varphi)$$

for every $i \in S$.

Observe that $\mu_\varphi(w^2) < \infty$ for every $\varphi \in \Phi_s$, by Assumption 8.9(a) and Theorem 2.5. Moreover, by (2.13) applied to the Lyapunov function w^2, that is,

$$E_i^\varphi[w^2(x(t))] \leq e^{-c't} w^2(i) + b'/c'(1 - e^{-c't}) \quad \forall\, i \in S,\ t \geq 0,$$

we obtain, by integration with respect to μ_φ and taking the limit as $t \to \infty$, that

$$\sup_{\varphi \in \Phi_s} \int_S w^2 d\mu_\varphi \leq b'/c'. \tag{8.11}$$

We also need to introduce the following standard compactness-continuity assumptions (see, e.g., Assumption 3.3).

Assumption 8.10.

(a) For each $i \in S$, the set $A(i)$ is compact.
(b) For every $i, j \in S$, the functions $a \mapsto q_{ij}(a)$, $a \mapsto r(i,a)$, and $a \mapsto u(i,a)$ are continuous on $A(i)$.

The continuity of $a \mapsto \sum_{j \in S} q_{ij}(a)w(j)$ is also a standard assumption (see Assumption 3.3(b)). In our context, however, Assumption 8.9(a) implies that the series $\sum_{j \in S} q_{ij}(a)w(j)$ converges uniformly on $A(i)$ and, hence, it is continuous (Remark 4.6(ii)).

Finally, we note that as a consequence of Theorem 2.11 (see also Remark 2.12), the control model $\mathcal{M}$ is w-exponentially ergodic uniformly on Φ_s.

In the rest of this section, we suppose that Assumptions 8.8, 8.9, and 8.10 are satisfied. It follows from these assumptions that our control model satisfies the hypotheses of Theorem 3.20 for the cost rate (or the reward rate) u under the long-run expected average cost (or reward) optimality criterion. So, let us define

$$\theta_{\min} := \min_{\varphi \in \Phi_s} g_u(\varphi) \quad \text{and} \quad \theta_{\max} := \max_{\varphi \in \Phi_s} g_u(\varphi). \tag{8.12}$$

Therefore, regarding the constant θ_0 in (8.9), to avoid trivial cases we will assume that

$$\theta_{\min} \le \theta_0 \le \theta_{\max}. \tag{8.13}$$

We need to introduce some more notation. Let $\mathcal{B}_w(K)$ be as in (2.9), and let $\mathcal{C}_w(K)$ be the subset of functions $v \in \mathcal{B}_w(K)$ such that $a \mapsto v(i,a)$ is continuous on $A(i)$ for every $i \in S$. In particular, by Assumptions 8.8(b) and 8.10(b), the functions r and u are in $\mathcal{C}_w(K)$.

Finally, let $\mathcal{P}_w(S)$ be the family of probability measures on S for which the integral of w is finite. In particular, the invariant probability measure μ_φ of a stationary control policy $\varphi \in \Phi_s$ is in $\mathcal{P}_w(S)$. Also, let $\mathcal{P}_w(K)$ be the set of probability measures μ on K such that

$$\sum_{i \in S} w(i)\mu(\{i\} \times A(i)) < \infty.$$

For each policy φ in Φ_s, we define $\hat{\mu}_\varphi \in \mathcal{P}_w(K)$ as follows: given $i \in S$ and $B \in \mathbb{B}(A(i))$,

$$\hat{\mu}_\varphi(\{i\} \times B) := \mu_\varphi(i)\varphi(B|i),$$

and let

$$\Gamma := \{\hat{\mu}_\varphi : \varphi \in \Phi_{\mathsf{s}}\} \subseteq \mathcal{P}_w(K). \tag{8.14}$$

In $\mathcal{P}_w(K)$ we will consider the w-weak topology, that is, the smallest topology for which

$$\hat{\mu} \mapsto \int_K v \, d\hat{\mu}$$

is continuous for every $v \in \mathcal{C}_w(K)$. With this topology, $\mathcal{P}_w(K)$ is a Borel space (see [38, Appendix A.5]).

Our next result is a standard characterization of the invariant probability measures of stationary policies (see [92, Sec. 3], for instance). In our context, however, we can prove it under weaker hypotheses because the state space S is denumerable. To obtain this characterization, we recall from Theorem 2.5(ii) that, for each $\varphi \in \Phi_{\mathsf{s}}$,

$$\sum_{i \in S} \mu_\varphi(i) q_{ij}(\varphi) = 0 \quad \forall \, j \in S. \tag{8.15}$$

Lemma 8.11. *A probability measure $\hat{\mu} \in \mathcal{P}_w(K)$ is in Γ if and only if*

$$\int_K (Lv) d\hat{\mu} = 0 \quad \forall \, v \in \mathcal{B}_1(S), \tag{8.16}$$

where $(Lv)(i,a) := \sum_{j \in S} q_{ij}(a) v(j)$ for $(i,a) \in K$.

Proof. (Necessity) First, we prove the *only if* part. To this end, fix $v \in \mathcal{B}_1(S)$ and $\hat{\mu} \in \Gamma$ (and so, $\hat{\mu} = \hat{\mu}_\varphi$ for some $\varphi \in \Phi_{\mathsf{s}}$). Then

$$\int_K (Lv) d\hat{\mu} = \sum_{i \in S} \int_{A(i)} \left[\sum_{j \in S} q_{ij}(a) v(j) \right] \varphi(da|i) \mu_\varphi(i)$$

$$= \sum_{i \in S} \sum_{j \in S} q_{ij}(\varphi) v(j) \mu_\varphi(i) \tag{8.17}$$

$$= \sum_{j \in S} v(j) \sum_{i \in S} q_{ij}(\varphi) \mu_\varphi(i) = 0, \tag{8.18}$$

where the interchange of the sum and the integral in (8.17) follows from the fact that

$$\sum_{j \in S} |q_{ij}(a) v(j)| \leq 2q(i).$$

(recall the notation in (2.1)), while the interchange of the sums in (8.18) follows from

$$\sum_{i \in S} \sum_{j \in S} |q_{ij}(\varphi)| \mu_\varphi(i) = \sum_{j \in S} \sum_{i \in S} |q_{ij}(\varphi)| \mu_\varphi(i)$$

$$= \sum_{j \in S} -2 q_{jj}(\varphi) \mu_\varphi(j) \quad [\text{by } (8.15)]$$

$$\le 2 \sum_{j \in S} q(j) \mu_\varphi(j)$$

$$\le 2 \sum_{j \in S} w(j) \mu_\varphi(j) < \infty \quad [\text{by Assumption 8.9(b)}].$$

(Sufficiency) Let us now prove the *if* part. We suppose that, for some $\hat{\mu} \in \mathcal{P}_w(K)$, the condition (8.16) holds. Hence, by a standard result on the disintegration of measures [68, Proposition D.8], there exists $\varphi \in \Phi_{\mathrm{s}}$ such that, for each $i \in S$ and $B \in \mathbb{B}(A(i))$,

$$\hat{\mu}(\{i\} \times B) = \mu(i) \varphi(B|i),$$

where $\mu \in \mathcal{P}_w(S)$ denotes the marginal of $\hat{\mu}$ on S. Hence, as in the proof of the *only if* part, and letting $v(\cdot) := \mathbf{I}_{\{j_0\}}(\cdot)$ for $j_0 \in S$, we can show that $\sum_{i \in S} \mu(i) q_{ij_0}(\varphi) = 0$ for every $j_0 \in S$. Therefore, by Theorem 2.5(ii), μ is necessarily the invariant probability measure of φ, and, thus, $\hat{\mu} = \hat{\mu}_\varphi$. This completes the proof. □

Before proving our next result, we recall the following.

Facts from convex analysis   Let F be a real-valued function defined on an interval $I \subseteq \mathbb{R}$. We say that F is convex if, for each $x, y \in I$ and every $0 \le \lambda \le 1$,

$$F(\lambda x + (1 - \lambda) y) \le \lambda F(x) + (1 - \lambda) F(y);$$

we say that F is concave if, for each $x, y \in I$ and every $0 \le \lambda \le 1$,

$$F(\lambda x + (1 - \lambda) y) \ge \lambda F(x) + (1 - \lambda) F(y).$$

The hypergraph of the function F is defined as

$$\{(x, y) \in \mathbb{R}^2 : x \in I, \ y \ge F(x)\},$$

while the hypograph of F is

$$\{(x, y) \in \mathbb{R}^2 : x \in I, \ y \leq F(x)\}.$$

We have the following characterization. The function F is convex (respectively, concave) if and only if the hypergraph (respectively, the hypograph) of F is a convex set.

If the function F is convex or concave, then it is continuous on the interior of the interval I.

We say that $d \in \mathbb{R}$ is a subdifferential of F at $x \in I$ if

$$F(y) - F(x) \geq d \cdot (y - x) \quad \forall \ y \in I.$$

Similarly, $d \in \mathbb{R}$ is a superdifferential of F at $x \in I$ if

$$F(y) - F(x) \leq d \cdot (y - x) \quad \forall \ y \in I.$$

If the function F is convex (respectively, concave) and x is an interior point of I, then there exists a subdifferential (respectively, superdifferential) of F at x.

Our next lemma establishes some useful convexity and compactness results.

Lemma 8.12. *The following statements hold:*

(i) *The set Γ in (8.14) is convex and compact with respect to the w-weak topology.*

(ii) *The function $\theta \mapsto V(\theta) := \sup\{\int_K r d\hat{\mu} : \ \hat{\mu} \in \Gamma, \ \int_K u d\hat{\mu} \leq \theta\}$ is concave, nondecreasing, and continuous on $[\theta_{\min}, \theta_{\max}]$.*

Proof. *Proof of (i).* A direct consequence of Lemma 8.11 is that a convex combination of measures in Γ lies in Γ. Hence, Γ is convex.

Regarding the compactness statement, note that for every $n \geq 1$ the set $\{(i, a) \in K : w^2(i) \leq nw(i)\}$ is compact in K (because $\lim_i w(i) = \infty$). Also, by (8.11),

$$\sup_{\hat{\mu} \in \Gamma} \int_K w^2 d\hat{\mu} = \sup_{\varphi \in \Phi_s} \int_S w^2 d\mu_\varphi < \infty.$$

Therefore, from [38, Corollary A.30(c)], the set Γ is compact.

Proof of (ii). It follows directly from the convexity of Γ in part (i) that the function V is concave and nondecreasing on $[\theta_{\min}, \theta_{\max}]$; hence, it is continuous on $(\theta_{\min}, \theta_{\max}]$. To prove the continuity in $\theta_{\min}$, consider a sequence $\theta^{(n)} > \theta_{\min}$ such that $\theta^{(n)} \downarrow \theta_{\min}$. By monotonicity of V, it turns out that

$$\ell := \lim_n V(\theta^{(n)}) \geq V(\theta_{\min}).$$

By standard continuity-compactness arguments, the supremum in the definition of V is indeed attained. Let $\hat{\mu}_n \in \Gamma$, for $n \geq 1$, be such that

$$\int_K r d\hat{\mu}_n = V(\theta^{(n)}) \quad \text{and} \quad \int_K u d\hat{\mu}_n \leq \theta^{(n)}.$$

Since the set Γ is compact, we can suppose without loss of generality that $\hat{\mu}_n \to \hat{\mu}^* \in \Gamma$ in the w-weak topology. Therefore, $\hat{\mu}^* \in \Gamma$ satisfies that

$$\int_K r d\hat{\mu}^* = \ell \geq V(\theta_{\min}) \quad \text{and} \quad \int_K u d\hat{\mu}^* \leq \theta_{\min},$$

and so $\ell = V(\theta_{\min})$ follows from the definition of V. This proves the continuity statement. $\qquad\square$

In the sequel, we will suppose that the constant θ_0 is in the open interval $(\theta_{\min}, \theta_{\max})$. The two extreme cases, namely, $\theta_0 = \theta_{\min}$ and $\theta_0 = \theta_{\max}$, will be treated separately; see Propositions 8.15 and 8.16 below. We state our main result in this section.

Theorem 8.13. *Suppose that Assumptions 8.8, 8.9, and 8.10 are satisfied. For every $i \in S$, consider the constrained problem (8.9), i.e.,*

$$\text{maximize } J(i, \varphi) \quad \text{s.t.} \quad J_u(i, \varphi) \leq \theta_0,$$

where $\theta_{\min} < \theta_0 < \theta_{\max}$. Then:

(i) *The optimal reward function $V^*(i, \theta_0)$ in (8.10) does not depend on $i \in S$, and $V(\theta_0) \equiv V^*(i, \theta_0)$, with V as in Lemma 8.12(ii).*

(ii) *There exist $\lambda_0 \leq 0$ and $h \in \mathcal{B}_w(S)$ such that*

$$V(\theta_0) = \max_{a \in A(i)} \left\{ r(i, a) + \lambda_0(u(i, a) - \theta_0) + \sum_{j \in S} q_{ij}(a)h(j) \right\} \quad \forall\, i \in S.$$

(iii) *There exists a stationary policy, which randomizes in at most one state, that is optimal for (8.9) for every $i \in S$.*

Proof. Proof of (i). The hypograph of the function V is convex (by Lemma 8.12(ii)). Now observe that, for every $\hat{\mu} \in \Gamma$, the pair $(\int_K u d\hat{\mu}, \int_K r d\hat{\mu}) \in \mathbb{R}^2$ belongs to such a hypograph.

Let $-\lambda_0 \geq 0$ be a superdifferential of V at θ_0 (the superdifferential is nonnegative because the function V is nondecreasing). The hypograph of V is, thus, contained in the half-space

$$\{(x,y) \in \mathbb{R}^2 : \lambda_0(x - \theta_0) + (y - V(\theta_0)) \leq 0\}.$$

Therefore, given an arbitrary $\hat{\mu} \in \Gamma$,

$$\int_K (r + \lambda_0(u - \theta_0))d\hat{\mu} \leq V(\theta_0),$$

and so

$$\max_{\hat{\mu} \in \Gamma} \int_K (r + \lambda_0(u - \theta_0))d\hat{\mu} \leq V(\theta_0), \tag{8.19}$$

while

$$V(\theta_0) = \max_{\hat{\mu} \in \Gamma} \left\{ \int_K r d\hat{\mu} : \int_K u d\hat{\mu} \leq \theta_0 \right\}, \tag{8.20}$$

where the two maxima in (8.19) and (8.20) are indeed attained because the functions $\int_K r d\hat{\mu}$ and $\int_K u d\hat{\mu}$ are continuous on the compact set Γ.

We will next show that (8.19) holds with equality. To this end, let $\hat{\mu}^* \in \Gamma$ attain the maximum in (8.20). Recalling that $\lambda_0 \leq 0$, it follows that

$$V(\theta_0) = \int_K r d\hat{\mu}^* \leq \int_K (r + \lambda_0(u - \theta_0))d\hat{\mu}^*$$

$$\leq \max_{\hat{\mu} \in \Gamma} \int_K (r + \lambda_0(u - \theta_0))d\hat{\mu}.$$

This proves that (8.19) holds with equality, i.e.,

$$\max_{\varphi \in \Phi_s} \int_K (r + \lambda_0(u - \theta_0))d\hat{\mu}_\varphi = V(\theta_0)$$

$$= \max_{\varphi \in \Phi_s} \left\{ \int_K r d\hat{\mu}_\varphi : \int_K u d\hat{\mu}_\varphi \leq \theta_0 \right\}.$$

The left-most term of this equality equals the maximal expected long-run average reward of a control problem with reward rate given by $r + \lambda_0(u - \theta_0)$. This reward rate satisfies the hypotheses of Theorem 3.20, and, thus, there exists a solution $(V(\theta_0), h) \in \mathbb{R} \times \mathcal{B}_w(S)$ to the corresponding AROE, that is, for every $i \in S$,

$$V(\theta_0) = \max_{a \in A(i)} \left\{ r(i,a) + \lambda_0(u(i,a) - \theta_0) + \sum_{j \in S} q_{ij}(a)h(j) \right\}. \tag{8.21}$$

In particular,

$$V(\theta_0) = \sup_{\varphi \in \Phi} \tilde{V}(i, \varphi) \tag{8.22}$$

with

$$\tilde{V}(i, \varphi) := \liminf_{T \to \infty} \frac{1}{T} E_i^\varphi \left[\int_0^T [r(t, x(t), \varphi) + \lambda_0(u(t, x(t), \varphi) - \theta_0)]dt \right]$$

for $i \in S$ and $\varphi \in \Phi$.

Proceeding with the proof, fix an initial state $i \in S$ and a policy $\varphi \in \Phi$ that satisfies the constraint $J_u(i, \varphi) \leq \theta_0$. Then, since $\lambda_0 \leq 0$,

$$J(i, \varphi) \leq J(i, \varphi) + \liminf_{T \to \infty} \frac{1}{T} E_i^\varphi \left[\int_0^T \lambda_0[u(t, x(t), \varphi) - \theta_0]dt \right]$$

$$\leq \liminf_{T \to \infty} \frac{1}{T} E_i^\varphi \left[\int_0^T [r(t, x(t), \varphi) + \lambda_0(u(t, x(t), \varphi) - \theta_0)]dt \right]$$

$$= \tilde{V}(i, \varphi) \leq V(\theta_0) \quad [\text{by (8.22)}].$$

Therefore, $V^*(i, \theta_0) \leq V(\theta_0)$ for every $i \in S$.

On the other hand, the inequality $V^*(i, \theta_0) \geq V(\theta_0)$ is obvious because $V^*(i, \theta_0)$ is defined as a supremum over the set of all control policies, whereas $V(\theta_0)$ is a maximum over the set of stationary policies. Hence, we have shown that $V^*(i, \theta_0) = V(\theta_0)$ for every $i \in S$. This completes the proof of (i).

Proof of (ii). This statement directly follows from (8.21).

Proof of (iii). Since $\theta_{\min} < \theta_0 < \theta_{\max}$, the same argument as in the proof of Theorem 8.6 or [46, Theorem 2.1] shows that there are two canonical policies for the AROE (8.21), f_1^* and f_2^* in $\mathbb{F}$, which differ in at most one state, say $i_0 \in S$, and such that

$$g_u(f_1^*) \leq \theta_0 \leq g_u(f_2^*). \tag{8.23}$$

Given $0 \leq \beta \leq 1$, let $\varphi_\beta^* \in \Phi_s$ coincide with f_1^* and f_2^* in the states $i \neq i_0$, and let $\varphi_\beta^*(\cdot|i_0)$ randomize between $f_1^*(i_0)$ and $f_2^*(i_0)$ with probabilities β and $1-\beta$, respectively. As a consequence of Theorem 3.20(iv), the policy φ_β^* is gain optimal for the reward rate $r + \lambda_0(u - \theta_0)$ for every $0 \leq \beta \leq 1$.

Observe now that $\beta \mapsto \varphi_\beta^*$ is continuous, where we consider in Φ_s the topology of the componentwise weak convergence (this topology was introduced in (2.5)). By the continuity of $\varphi \mapsto g_u(\varphi)$ on Φ_s (Corollary 3.18), it follows that

$$\beta \mapsto g_u(\varphi_\beta^*)$$

is continuous. Hence, by (8.23), $g_u(\varphi_{\beta^*}^*) = \theta_0$ for some $\beta^* \in [0,1]$, and, thus, $\varphi_{\beta^*}^*$ is an optimal policy for the constrained problem (8.9). $\qquad \square$

In Theorem 8.13 we have shown that the constrained control problem (8.9) is equivalent to an unconstrained problem that depends on an *unknown* constant $\lambda_0 \leq 0$. In some sense, this fact is not very useful because the constant λ_0 is obtained from the function V, which is precisely the function that we want to determine. To handle this situation we propose our next result, which is based on Ky-Fan's minimax theorem [40, Theorem 8].

Theorem 8.14. *Suppose that Assumptions 8.8, 8.9, and 8.10 are satisfied, and consider the constrained problem (8.9) with $\theta_{\min} < \theta_0 < \theta_{\max}$.*
For each $\lambda \leq 0$, let $(g(\lambda), h_\lambda) \in \mathbb{R} \times \mathcal{B}_w(S)$ be a solution of the AROE

$$g(\lambda) = \max_{a \in A(i)} \left\{ r(i,a) + \lambda(u(i,a) - \theta_0) + \sum_{j \in S} q_{ij}(a)h_\lambda(j) \right\} \quad \forall \ i \in S.$$

Then $V(\theta_0) = \min_{\lambda \leq 0} g(\lambda)$.

Proof. Define the function $H : \Gamma \times (-\infty, 0] \to \mathbb{R}$ as

$$H(\hat{\mu}, \lambda) := \int_K (r + \lambda(u - \theta_0))d\hat{\mu}.$$

Obviously, H is concave (in fact, it is linear) on Γ for fixed λ, and it is convex (linear) on $(-\infty, 0]$ for every $\hat{\mu}$. Moreover, the set Γ is convex and compact (by Lemma 8.12), and the mapping H is continuous (in the w-weak topology) on Γ for fixed λ. Hence, from [40, Theorem 8],

$$\max_{\hat{\mu} \in \Gamma} \inf_{\lambda \leq 0} H(\hat{\mu}, \lambda) = \inf_{\lambda \leq 0} \max_{\hat{\mu} \in \Gamma} H(\hat{\mu}, \lambda).$$

Now, on the one hand, $\inf_{\lambda \leq 0} H(\hat{\mu}, \lambda)$ equals $-\infty$ if $\int_K u d\hat{\mu} > \theta_0$, and equals $\int_K r d\hat{\mu}$ otherwise. Therefore, by Theorem 8.13,

$$V(\theta_0) = \max_{\hat{\mu} \in \Gamma} \inf_{\lambda \leq 0} H(\hat{\mu}, \lambda).$$

On the other hand, by arguments similar to those in Theorem 8.13,

$$\max_{\hat{\mu} \in \Gamma} H(\hat{\mu}, \lambda) = g(\lambda).$$

Therefore, $V(\theta_0) = \inf_{\lambda \leq 0} g(\lambda)$. As a consequence of Theorem 8.13, this infimum is attained at λ_0. This completes the proof. $\qquad\square$

Now we address the case when the constraint constant is $\theta_0 = \theta_{\min}$.

Proposition 8.15. *Suppose that Assumptions 8.8, 8.9, and 8.10 are satisfied. There exists a deterministic stationary policy that is optimal for the constrained problem*

$$\text{maximize } J(i, \varphi) \quad \text{s.t.} \quad J_u(i, \varphi) \leq \theta_{\min}.$$

for every $i \in S$.

Proof. We solve the following AROE-like equation (cf. (3.36)) for an average cost minimization problem

$$\theta_{\min} = \min_{a \in A(i)} \left\{ u(i, a) + \sum_{j \in S} q_{ij}(a)h(j) \right\} \quad \forall \, i \in S,$$

where $h \in \mathcal{B}_w(S)$. Let $A'(i) \subseteq A(i)$ be the nonempty compact set of minima of the above equation. The set of corresponding canonical policies (i.e., the policies $f \in \mathbb{F}$ such that $f(i) \in A'(i)$ for each $i \in S$) is the set of policies f in $\mathbb{F}$ for which $g_u(f) = \theta_{\min}$. Then we consider the following nested AROE

$$V^* = \max_{a \in A'(i)} \left\{ r(i, a) + \sum_{j \in S} q_{ij}(a)h'(j) \right\} \quad \forall \, i \in S,$$

for $h' \in \mathcal{B}_w(S)$. We deduce from the definition of the function V (recall Lemma 8.12) that, necessarily, $V^* = V(\theta_{\min})$, and let $f^* \in \mathbb{F}$ be such that $g_u(f^*) = \theta_{\min}$ and $g(f^*) = V(\theta_{\min})$.

Now let $\varphi \in \Phi$ be a control policy such that $J_u(i, \varphi) \leq \theta_{\min}$. For every $\theta_0 > \theta_{\min}$, we have $J_u(i, \varphi) \leq \theta_0$, and so $J(i, \varphi) \leq V(\theta_0)$, by Theorem 8.13(i). By continuity of the function V (Lemma 8.12(ii)), we conclude that $J(i, \varphi) \leq V(\theta_{\min}) = g(f^*)$, and the optimality of f^* for the constrained problem follows. $\qquad\square$

It is worth mentioning that the proof of this proposition is different from the one in the case $\theta_{\min} < \theta_0 < \theta_{\max}$ analyzed in Theorem 8.13 because the superdifferential of V at $\theta_{\min}$ can be infinite.

Finally, we consider the case $\theta_0 = \theta_{\max}$.

Proposition 8.16. *Suppose that Assumptions 8.8, 8.9, and 8.10 are satisfied. Then any canonical policy for the (unconstrained) control problem* *with reward rate r is optimal for the constrained problem*

$$\max \ J(i, \varphi) \quad \text{s.t.} \quad J_u(i, \varphi) \leq \theta_{\max}.$$

for every $i \in S$.

Proof.   We solve the AROE (which is the same as (3.36))

$$g^* = \max_{a \in A(i)} \left\{ r(i, a) + \sum_{j \in S} q_{ij}(a) h(j) \right\} \quad \forall \ i \in S,$$

where $h \in \mathcal{B}_w(S)$. Here, g^* is the optimal gain of the unconstrained problem. Since for every $f \in \mathbb{F}$, $g_u(f) \leq \theta_{\max}$, it follows that $g^* = V(\theta_{\max})$.

It is clear that any control policy $\varphi \in \Phi$ such that $J_u(i, \varphi) \leq \theta_{\max}$ verifies $J(i, \varphi) \leq g^*$, and thus any canonical policy for the unconstrained control problem is constrained optimal. $\qquad\square$

The above proof is not as obvious as it seems: we are not claiming that the control problem in Proposition 8.16 is unconstrained, although its solutions correspond to those of the unconstrained problem. Indeed, not every policy $\varphi \in \Phi$ verifies $J_u(i, \varphi) \leq \theta_{\max}$ (although every stationary policy does) because $J_u(i, \varphi)$ is defined as a "lim sup," and not as a "lim inf."

8.4.   Pathwise constrained CMCs

The expected average reward optimality criterion studied in Secs 3.4 and 3.5 was generalized in Theorem 3.27 to pathwise average reward optimality; see Sec. 3.6. Here, after [131], we consider CMCs with pathwise constraints in which the *expected* averages in (8.9) are replaced with the *pathwise* averages in (8.24), below.

In the remainder of this section, we suppose that Assumptions 8.8, 8.9, and 8.10 hold.

First of all, we introduce some notation. Given $\varphi \in \Phi$, $i \in S$, and $T > 0$, we define the total pathwise cost (cf. (3.44))

$$J_{u,T}^0(i, \varphi) := \int_0^T u(t, x(t), \varphi) dt$$

and the long-run pathwise average cost (cf. (3.45))

$$J_u^0(i, \varphi) := \limsup_{T \to \infty} \frac{1}{T} J_{u,T}^0(i, \varphi).$$

As in Lemma 3.25, for every stationary $\varphi \in \Phi_s$ and $i \in S$, we have

$$J_u^0(i, \varphi) = \lim_{T \to \infty} \frac{1}{T} J_{u,T}^0(i, \varphi) = g_u(\varphi) \quad P_i^\varphi\text{-a.s.}$$

The pathwise constrained control problem is defined as follows. Given an initial state $i \in S$ and a constant $\theta_0 \in \mathbb{R}$, the problem is

$$\text{maximize } J^0(i, \varphi) \quad \text{s.t.} \quad J_u^0(i, \varphi) \le \theta_0. \tag{8.24}$$

Formally, we say that a stationary policy $\varphi^* \in \Phi_s$ with $g_u(\varphi^*) \le \theta_0$ is optimal for the pathwise constrained CMC (8.24) if for each $\varphi \in \Phi$ such that $J_u^0(i, \varphi) \le \theta_0 \; P_i^\varphi\text{-a.s.}$, we have $J^0(i, \varphi) \le g(\varphi^*)$ with P_i^φ-probability one.

Theorem 8.17. *Suppose that Assumptions 8.8, 8.9, and 8.10 are satisfied. Consider the pathwise constrained problem (8.24) with $\theta_{\min} \le \theta_0 \le \theta_{\max}$, and let $\varphi^* \in \Phi_s$ be:*

(a) *as in Theorem 8.13(iii) if $\theta_{\min} < \theta_0 < \theta_{\max}$,*
(b) *as in Proposition 8.15 if $\theta_0 = \theta_{\min}$, or*
(c) *as in Proposition 8.16 if $\theta_0 = \theta_{\max}$.*

Then φ^* is optimal for the pathwise constrained problem (8.24).

Proof. We only prove case (a), that is, $\theta_{\min} < \theta_0 < \theta_{\max}$. The two remaining cases are proved similarly.

Suppose that $\varphi \in \Phi$ verifies

$$J_u^0(i, \varphi) \le \theta_0 \quad P_i^\varphi\text{-a.s.} \tag{8.25}$$

We use Theorem 3.27, on pathwise average optimality, for the control problem given by the AROE in Theorem 8.13(ii), i.e.,

$$V(\theta_0) = \max_{a \in A(i)} \left\{ r(i, a) + \lambda_0(u(i, a) - \theta_0) + \sum_{j \in S} q_{ij}(a) h(j) \right\} \quad \forall \, i \in S.$$

It then follows that

$$\liminf_{T \to \infty} \frac{1}{T} \int_0^T [r(t, x(t), \varphi) + \lambda_0(u(t, x(t), \varphi) - \theta_0)] dt \le V(\theta_0) \quad P_i^\varphi\text{-a.s.}$$

Since $\lambda_0 \le 0$,

$$J^0(i, \varphi) + \lambda_0(J_u^0(i, \varphi) - \theta_0) \le V(\theta_0) \quad P_i^\varphi\text{-a.s.}$$

Hence, by (8.25), $J^0(i, \varphi) \leq V(\theta_0)$ P_i^{φ}-a.s. Recalling that $g_u(\varphi^*) = \theta_0$ and that $g(\varphi^*)$ equals $V(\theta_0)$, the optimality of φ^* follows. $\qquad\square$

8.5. The vanishing discount approach to constrained CMCs

By analogy with the results in Theorem 6.10, we are interested in studying whether the "limit" of α-discounted θ_α-constrained optimal policies is average θ_0-constrained optimal. We base our approach on [133].

In what follows, we suppose that Assumptions 8.8, 8.9, and 8.10 hold, where the cost rate u *is assumed to be nonnegative*, i.e., $u(i, a) \geq 0$ for each $(i, a) \in K$.

Given an initial state $i_0 \in S$, a discount rate $\alpha > 0$, and a constraint constant θ_α that satisfies the conditions in Assumption 8.3(c), we know that there exists a stationary policy φ^*_α in Φ_s that is discounted constrained optimal, that is, by Theorem 8.6 (using the notation in (3.10)),

$$V_\alpha(i_0, \varphi^*_\alpha, u) \leq \theta_\alpha$$

and, for each $\varphi \in \Phi$ such that $V_\alpha(i_0, \varphi, u) \leq \theta_\alpha$, we have $V_\alpha(i_0, \varphi) \leq V_\alpha(i_0, \varphi^*_\alpha)$. Finally, we suppose that there exists θ_0 such that

$$\lim_{\alpha\downarrow 0} \alpha\theta_\alpha = \theta_0 \in (\theta_{\min}, \theta_{\max}], \qquad (8.26)$$

where $\theta_{\min}$ and $\theta_{\max}$ are the bounds in (8.12) of the range of the average constrained problem.

Our convergence results are summarized in the next proposition.

Proposition 8.18. *Suppose that Assumptions 8.8, 8.9, and 8.10 hold. In addition, we suppose that the cost rate function u is nonnegative, and that $i_0 \in S$ is a fixed initial state for which Assumption 8.3(c) is satisfied for every (small) $\alpha > 0$. Under these conditions, the following statements hold:*

(i) *Any limit policy in Φ_s of $\{\varphi^*_\alpha\}$ as $\alpha \downarrow 0$ is optimal for the average constrained problem (8.9), with θ_0 as in (8.26).*

(ii) *The following limit holds (we use the notation introduced in Lemma 8.12):*

$$\lim_{\alpha\downarrow 0} \alpha V_\alpha(i_0, \varphi^*_\alpha) = V(\theta_0).$$

Proof.   *Proof of* (i). We know that Φ_s is a compact metric space with the topology of componentwise weak convergence; see (2.5). Hence, suppose that, for some sequence of discount rates $\alpha_n \downarrow 0$ and some φ^* in Φ_s, we have $\varphi^*_{\alpha_n} \to \varphi^*$.

By Theorem 6.9,

$$\theta_0 = \lim_{n\to\infty} \alpha_n \theta_{\alpha_n} \geq \lim_{n\to\infty} \alpha_n V_{\alpha_n}(i_0, \varphi^*_{\alpha_n}, u) = g_u(\varphi^*).$$

Therefore, $g_u(\varphi^*) \leq \theta_0$, and so $g(\varphi^*) \leq V(\theta_0)$. We are going to show that the previous inequality holds, in fact, with equality. To this end, we proceed by contradiction and, hence, we suppose that

$$g(\varphi^*) < V(\theta_0).$$

By the continuity of the function V proved in Lemma 8.12(ii), there exists $\varepsilon > 0$ such that $\theta_{\min} \leq \theta_0 - \varepsilon$ and

$$g(\varphi^*) < V(\theta_0 - \varepsilon) \leq V(\theta_0).$$

Let φ^*_ε in Φ_s be an average constrained optimal policy for (8.9) when the constraint constant θ_0 is replaced with $\theta_0 - \varepsilon$. In particular,

$$g(\varphi^*) < g(\varphi^*_\varepsilon) = V(\theta_0 - \varepsilon). \tag{8.27}$$

Also, $g_u(\varphi^*_\varepsilon) \leq \theta_0 - \varepsilon$ and so, by Lemma 6.8,

$$\lim_{n\to\infty} \alpha_n(V_{\alpha_n}(i_0, \varphi^*_\varepsilon, u) - \theta_{\alpha_n}) \leq -\varepsilon < 0.$$

Consequently, for sufficiently large n, we have $V_{\alpha_n}(i_0, \varphi^*_\varepsilon, u) \leq \theta_{\alpha_n}$. Therefore, the policy φ^*_ε satisfies the restriction of the α_n-discounted θ_{α_n}-constrained problem, and, thus, for large n

$$V_{\alpha_n}(i_0, \varphi^*_\varepsilon) \leq V_{\alpha_n}(i_0, \varphi^*_{\alpha_n}).$$

Multiplying both sides of this inequality by α_n and letting $n \to \infty$ yields, by Theorem 6.9, $g(\varphi^*_\varepsilon) \leq g(\varphi^*)$, which contradicts (8.27).

Hence, we have shown that $g_u(\varphi^*) \leq \theta_0$ and $g(\varphi^*) = V(\theta_0)$. Therefore, the policy φ^* is average constrained optimal for (8.9).

   Proof of (ii). By (3.7), $\{\alpha V_\alpha(i_0, \varphi^*_\alpha)\}_{\alpha>0}$ is bounded. Now choose an arbitrary sequence $\alpha_n \downarrow 0$ such that $\alpha_n V_{\alpha_n}(i_0, \varphi^*_{\alpha_n})$ converges. There exists a further subsequence, denoted by $\{n'\}$, such that $\varphi^*_{\alpha_{n'}} \to \varphi^*$ for some φ^*

in Φ_s. Therefore, by Theorem 6.9 and part (i) of this proposition,

$$\lim_n \alpha_n V_{\alpha_n}(i_0, \varphi_{\alpha_n}^*) = \lim_{n'} \alpha_{n'} V_{\alpha_{n'}}(i_0, \varphi_{\alpha_{n'}}^*) = g(\varphi^*) = V(\theta_0).$$

Since the limit of any convergent subsequence of $\alpha V_\alpha(i_0, \varphi_\alpha^*)$ is $V(\theta_0)$, we reach the desired result. $\square$

Proposition 8.18 does not address the case $\lim_{\alpha \downarrow 0} \alpha \theta_\alpha = \theta_{min}$. However, the proof of this proposition shows that, when $\lim_{\alpha \downarrow 0} \alpha \theta_\alpha = \theta_{min}$, the inequality

$$\limsup_{\alpha \downarrow 0} \alpha V_\alpha(i_0, \varphi_\alpha^*) \leq V(\theta_{min})$$

holds. Our next example shows that this inequality can be strict.

Example 8.19. Consider the CMC with state space $S = \{0, 1\}$, action sets

$$A(0) = \{0\}, \quad A(1) = \{1, 2, 3\},$$

and transition rates given by

$$q_{00}(0) = -1, \quad q_{11}(1) = -2, \quad q_{11}(2) = q_{11}(3) = -4.$$

The cost rate function is defined as

$$u(0, 0) = 0, \quad u(1, 1) = 3, \quad u(1, 2) = 5, \quad u(1, 3) = 10,$$

whereas the reward rate function is

$$r(0, 0) = 1, \quad r(1, 1) = r(1, 2) = r(1, 3) = 0.$$

There exist three deterministic stationary policies $\mathbb{F} = \{f_1, f_2, f_3\}$, which are identified with their value at state 1, that is, $f_i(1) = i$ for $i = 1, 2, 3$. The family of randomized stationary policies Φ_s is identified with

$$\Delta := \{(x, y) \in \mathbb{R}^2 \; : \; x \geq 0, \; y \geq 0, \; x + y \leq 1\},$$

where $(x, y) \in \Delta$ corresponds to the policy φ in Φ_s such that

$$\varphi(\{1\}|1) = x, \quad \varphi(\{2\}|1) = y, \quad \varphi(\{3\}|1) = 1 - x - y.$$

Therefore,

$$u(0, (x, y)) = 0, \quad u(1, (x, y)) = 10 - 7x - 5y,$$

and

$$[q_{ij}(x,y)]_{i,j} = \begin{pmatrix} -1 & 1 \\ 4 - 2x & 2x - 4 \end{pmatrix}.$$

Let us now solve the discounted constrained problem for an arbitrary discount rate $\alpha > 0$. We consider the initial state $i_0 = 0$ and the constraint constant

$$\theta_\alpha = \frac{4}{\alpha(\alpha + 4)}.$$

A simple calculation shows that, for each discount rate $\alpha > 0$ and each policy $(x, y) \in \Delta$,

$$V_\alpha(0, (x, y), u) = \frac{10 - 7x - 5y}{\alpha(5 + \alpha - 2x)}$$

and

$$V_\alpha(0, (x, y)) = \frac{4 + \alpha - 2x}{\alpha(5 + \alpha - 2x)}.$$

Moreover, the set of policies that satisfy the constraint $V_\alpha(0, (x, y), u) \leq \theta_\alpha$ is

$$\{(x, y) \in \Delta \; : \; 6\alpha + 20 \leq (7\alpha + 20)x + (5\alpha + 20)y\},$$

that is, the triangle in Δ with vertices

$$(1/2, 1/2), \quad (1, 0), \quad \text{and} \quad \left(\frac{6\alpha + 20}{7\alpha + 20}, 0\right).$$

Among these policies, the one that maximizes $V_\alpha(0, (x, y))$ is $(1/2, 1/2)$.

Therefore, for each $\alpha > 0$, the unique α-discounted θ_α-constrained optimal policy is $\varphi_\alpha^* \equiv (1/2, 1/2)$, that is, φ_α^* randomizes between f_1 and f_2 with probabilities $1/2$ and $1/2$. The corresponding optimal reward is

$$V_\alpha(0, \varphi_\alpha^*) = \frac{3 + \alpha}{\alpha(4 + \alpha)}.$$

Now we turn our attention to the average constrained problem. The invariant probability distribution of the policy $(x, y) \in \Delta$ is given by

$$\left(\frac{4 - 2x}{5 - 2x}, \; \frac{1}{5 - 2x}\right).$$

The corresponding expected average cost and average reward are

$$g_u(x,y) = \frac{10 - 7x - 5y}{5 - 2x} \quad \text{and} \quad g(x,y) = \frac{4 - 2x}{5 - 2x}.$$

The constraint constant is $\theta_0 = \lim_{\alpha \downarrow 0} \alpha \theta_\alpha = 1$. At this point, note that $\theta_{\min} = 1$ and $\theta_{\max} = 2$. So $\lim_{\alpha \downarrow 0} \alpha \theta_\alpha = \theta_{\min}$, which is precisely the case missing in Proposition 8.18.

The set of policies that satisfy the constraint $g_u(x,y) \le 1$ is

$$\{(x, 1 - x) \ : \ 0 \le x \le 1\},$$

that is, the set of policies that randomize between f_1 and f_2, or, equivalently, the edge of Δ with vertices $(1,0)$ and $(0,1)$. The policy that maximizes $g(x, 1 - x)$, for $0 \le x \le 1$, is $(0,1) = f_2$. Hence, the unique average constrained optimal policy is f_2 and the corresponding optimal reward is $V(\theta_0) = V(1) = 4/5$.

It is now obvious that the limit policies of φ_α^* as $\alpha \downarrow 0$ are not average constrained optimal, and also that

$$\frac{3}{4} = \lim_{\alpha \downarrow 0} \alpha V_\alpha(0, \varphi_\alpha^*) < V(\theta_0) = \frac{4}{5}.$$

Therefore, Proposition 8.18 does not necessarily hold when $\theta_0 = \theta_{\min}$.

If $\lim_{\alpha \downarrow 0} \alpha \theta_\alpha = \theta_0 > \theta_{\min}$, then a Slater-like condition as in Assumption 8.3(c) holds for the average constrained problem. Indeed, there exists a policy $f \in \mathbb{F}$ such that $g_u(f) = \theta_{\min} < \theta_0$. If $\theta_0 = \theta_{\min}$, however, the average constrained problem is somehow degenerate because it cannot be solved by means of an average reward optimality equation with a Lagrange multiplier (see Theorem 8.13). Instead, it is solved by means of two nested optimality equations (see Proposition 8.15). This explains why the vanishing discount approach may fail.

8.6. Conclusions

In this chapter we have studied constrained CMCs in the discounted case, and also under the average reward optimality criterion, with both expected and pathwise constraints. Further, we have analyzed the vanishing discount approach for constrained CMCs and we have provided a counterexample for which the typical vanishing discount results do not hold.

We note that for both the discounted and the average constrained problems, studied in Secs 8.2 and 8.3, we follow the *Lagrange multipliers approach*. The convexity issues in the corresponding proofs are, however, different. Indeed, for the discounted constrained problem, we deal with "convexity in the class of stationary policies," whereas for the average constrained problem, we deal with "convexity in the class of invariant state-action measures."

We should also mention that the linear programming approach to CMCs (see, e.g., [66, 68, 70]) is especially well-suited for constrained problems. Indeed, once the control problem is transformed into an equivalent linear programming problem, in order to deal with a constrained control problem one just has to add a supplementary (linear) constraint, so that, technically, the unconstrained and constrained problems are similar.

Interestingly, our main results in this chapter, Theorems 8.6 and 8.13, recover a fairly general result on constrained optimization. Namely, when considering a constrained problem with m constraints, there exists an optimal policy in Φ_s that is a randomization of, at most, $m + 1$ deterministic stationary policies in $\mathbb{F}$; see [42]. As already noted, constrained optimality is the first of the optimality criteria studied so far for which there do not necessarily exist *deterministic* optimal policies. This is also the case for the stochastic game models studied in Chapters 10 and 11.

Chapter 9

Applications

9.1. Introduction

In the previous chapters, we have established our main theoretical results for continuous-time CMCs under various optimality criteria. In this chapter, we illustrate these results by means of examples of practical interest. Namely, we will show applications to queueing systems (Sec. 9.2), birth-and-death processes (Sec. 9.3), controlled population systems with catastrophes (Sec. 9.4), and epidemic processes (Sec. 9.5).

One of our main goals in this chapter is to check whether the assumptions that we have imposed so far on our control model are verifiable in practical models. In particular, one of our immediate concerns is to show that the Lyapunov conditions (on w, w^2, and even w^3), which are somehow our basic requirements for the control model, are not restrictive in practice. We will also compare our assumptions with others that have been proposed in the literature.

As another application, we will analyze the numerical approximation of the optimal reward and an optimal policy of a CMC under the discounted and the average reward optimality criteria. To this end, the controlled birth-and-death process in Sec. 9.3 and the controlled population system in Sec. 9.4 (called an upwardly skip-free process by Anderson [6, Sec. 9.1]) are solved numerically, under the discounted and the average reward optimality criteria. To do so, we use the approximation theorems in Secs 4.3 and 4.4, together with the policy iteration algorithm described in Sec. 4.2. For numerical analysis of the controlled population systems, we also refer the reader to [134–136].

For the interested reader, let us now give some references on queueing systems and population models. Queueing systems have been studied by a huge number of authors, e.g., Kitaev and Rykov [90], and Sennott [147]. For population models, we can cite the works by Albright and Winston [1], Allen [2], Bartholomew [9], and Iosifescu and Tautu [78]. Finally, controlled epidemic processes have been studied by Bailey [8], Lefèvre [97, 98], and Wickwire [164], among others. Our examples here are mainly borrowed from [52, 105, 131].

Next, we present a discussion about the assumptions that we have imposed, in the previous chapters, on our control model

$$\mathcal{M} = \{S, A, (A(i)), (q_{ij}(a)), (r(i, a))\}.$$

These can be roughly classified as follows:

(i) Lyapunov or drift conditions on a Lyapunov function w on S;
(ii) irreducibility of stationary policies;
(iii) continuity-compactness hypotheses;
(iv) uniform w-exponential ergodicity.

Note that the assumptions in (ii) and (iv) are required only for average reward optimality and related criteria. The verification of the conditions in (ii) and (iii) is usually straightforward because these conditions are part of the definition itself of the control model. Hence, the main difficulty relies in checking (i) and (iv).

Regarding the uniform w-exponential ergodicity in (iv) we have at hand, basically, two different sufficient conditions: the *monotonicity conditions* in Assumption 2.7 (see Theorem 2.8), and the *strengthened Lyapunov conditions* in Assumption 2.10 (see Theorem 2.11). Moreover, note that such strengthened Lyapunov conditions (imposed on the functions w^2 or w^3) are also needed when dealing with optimality criteria such as variance minimization (Assumption 5.5(b)) or Blackwell optimality (Assumption 7.2(c)).

Interestingly enough, in the examples presented below we will see that, in fact, these Lyapunov conditions on w^2 or w^3 are not restrictive at all, and they are satisfied by a wide class of models of practical interest. More precisely, we are going to show that, under suitable conditions, our examples here verify the following assumptions, which we state for clarity of exposition.

Assumption 9.1. There exists a Lyapunov function w on S such that:

(a) For each integer $k \geq 1$ (and then for every real number $k \geq 1$; see Theorem 2.13) there exist constants $c > 0$ and $d > 0$, and a finite set

$D \subset S$ (all depending on k) such that

$$\sum_{j \in S} q_{ij}(a)w^k(j) \le -cw^k(i) + d \cdot \mathbf{I}_D(i) \quad \forall \ (i,a) \in K.$$

(b) For every $i \in S$, $q(i) := \sup_{a \in A(i)} \{-q_{ii}(a)\} \le w(i)$.
(c) The reward function r is in $\mathcal{B}_w(K)$.

Our next assumption contains the usual compactness-continuity requirements.

Assumption 9.2.

(a) For all $i \in S$, the action set $A(i)$ is compact.
(b) The functions $q_{ij}(a)$ and $r(i,a)$ are continuous on $A(i)$ for each $i,j \in S$.

Finally, we state our irreducibility condition for average reward CMCs.

Assumption 9.3. Each randomized stationary policy in Φ_s is irreducible.

Under these assumptions, our control model satisfies all the conditions needed for each of the optimality criteria studied in the previous chapters. In particular, the following theorems hold:

– Theorem 3.7 on discount optimality,
– Theorem 3.20 on average optimality,
– Theorem 3.27 on pathwise average optimality,
– Theorems 5.3 and 5.4 on bias and overtaking optimality,
– Theorem 5.10 on variance optimality,
– Theorem 6.13 on n-discount optimality,
– Theorems 6.14 and 6.17 on strong discount optimality, and
– Theorem 7.5 on Blackwell optimality.

Moreover, concerning *constrained* control problems, as in Chapter 8, if the cost rate u verifies Assumptions 9.1(c) and 9.2(b), and the constraint constant satisfies adequate conditions (recall Assumption 8.3 and (8.13)), then the following results hold as well:

– Theorem 8.6 on discounted constrained optimality,
– Theorem 8.13 on average constrained optimality, and
– Theorem 8.17 on pathwise average constrained optimality.

In the examples studied in Secs 9.2 to 9.5, we will prove that Assumptions 9.1, 9.2, and 9.3 hold, so that all the aforementioned theorems hold for these examples.

We conclude this chapter with some important comments in Sec. 9.6.

9.2.   Controlled queueing systems

The following example is a simplified version of the queueing system studied in [45, Sec. VI] and [131, Sec. 5].

We consider a multiple-server queueing system with state space

$$S = \{0, 1, 2, \ldots\},$$

where the state variable is interpreted as the total number of customers being served or waiting for service. The customers arrive at a rate $\lambda > 0$, while the (noncontrolled) service rate is $\mu > 0$ (assumed to be the same for each server).

The action space is the compact set $A = [a_1, a_2] \times [b_1, b_2]$, where the action $a \in [a_1, a_2]$ allows the decision-maker to modify (either increase or decrease) the customers' arrival rate, whereas the action $b \in [b_1, b_2]$ allows the decision-maker to control the service rate.

Therefore, if the system is at state $i \geq 0$, then the transition rate to the state $i + 1$ (that is, a customer arrives) is

$$q_{i,i+1}(a, b) = \lambda + h_1(i, a) \quad \forall \, (i, a, b) \in K, \tag{9.1}$$

where $h_1(i, a)$ is the control on the arrival rate. When $i \geq 1$, the transition rate to the state $i - 1$ (meaning that a customer has been served) is

$$q_{i,i-1}(a, b) = \mu i + h_2(i, b) \quad \forall \, (a, b) \in A, \tag{9.2}$$

where $h_2(i, b)$ is the control on the service rate. Then we have

$$-q_{ii}(a, b) = \lambda + \mu i + h_1(i, a) + h_2(i, b) \quad \forall \, i \geq 1, \, (a, b) \in A, \tag{9.3}$$

and

$$q_{00}(a, b) = -q_{01}(a, b) \quad \forall \, (a, b) \in A. \tag{9.4}$$

We suppose that the transition rates (9.1) and (9.2) are positive for every $(i, a, b) \in K$. Consequently, every stationary policy in Φ_s is irreducible.

The functions $h_1(i, \cdot)$ and $h_2(i, \cdot)$ are assumed to be continuous on A for each $i \in S$. This implies that $q_{ij}(\cdot, \cdot)$ is continuous on A for every $i, j \in S$. Besides, we impose the following growth conditions on h_1 and h_2:

$$\lim_{i \to \infty} \frac{\sup_{a \in [a_1, a_2]} |h_1(i, a)|}{i + 1} = 0 \quad \text{and} \quad \lim_{i \to \infty} \frac{\sup_{b \in [b_1, b_2]} |h_2(i, b)|}{i + 1} = 0. \tag{9.5}$$

When the state of the system is $i \in S$, the controller receives a reward at a rate pi for some $p > 0$. Furthermore, when the action $(a, b) \in A$ is chosen, the controller incurs a cost at a rate $c(i, a, b)$. The function $c(i, \cdot, \cdot)$ is assumed to be continuous on A for each $i \in S$, and it verifies that

$$\sup_{(a,b)\in A} |c(i, a, b)| \le M(i+1) \quad \forall\, i \in S, \tag{9.6}$$

for some constant $M > 0$. To complete the specification of our queueing system, we suppose that the controller's net reward rate is

$$r(i, a, b) = pi - c(i, a, b).$$

For this control system, let us consider the Lyapunov function

$$w(i) = C(i+1) \quad \forall\, i \in S,$$

for some constant $C \ge 1$. By the definition of the reward rate (see also (9.6)), it follows that r is in $\mathcal{B}_w(K)$.

It is clear from (9.3) and (9.4) that we can choose $C \ge 1$ large enough so that

$$q(i) = \sup_{(a,b)\in A} \{-q_{ii}(a, b)\} \le w(i) \quad \forall\, i \in S.$$

This shows that Assumptions 9.1(b), 9.1(c), 9.2, and 9.3 hold. Thus, it only remains to prove Assumption 9.1(a). This is done in the next result.

Lemma 9.4. *The queueing system satisfies Assumption 9.1(a), that is, given an integer $k \ge 1$, there exist constants $c > 0$ and $d > 0$, and a finite set $D \subset S$ such that*

$$\sum_{j\in S} q_{ij}(a, b) w^k(j) \le -cw^k(i) + d \cdot \mathbf{I}_D(i) \quad \forall\, (i, a, b) \in K.$$

Proof. Fix $k \ge 1$ and $i > 0$. We will develop the expressions of $q_{ij}(a, b) w^k(j)$ for $i - 1 \le j \le i + 1$ as a power series of $(i + 1)$. More precisely, we will consider the terms $(i+1)^{k+1}$ and $(i+1)^k$, plus a residual term. We have

$$\begin{aligned}
q_{i,i+1}(a, b) w^k(i+1) &= C^k(\lambda + h_1(i, a)) \cdot (i+2)^k \\
&= C^k(\lambda + h_1(i, a)) \cdot ((i+1)+1)^k \\
&= C^k(\lambda + h_1(i, a)) \cdot (i+1)^k + R_1(i, a) \\
&= (\lambda + h_1(i, a)) \cdot w^k(i) + R_1(i, a), \tag{9.7}
\end{aligned}$$

where the residual term $R_1(i, a)$ is a sum of products of polynomials in i (of degree less than k) multiplied by $h_1(i, a)$. Hence, by (9.5), the residual R_1 verifies

$$\lim_{i\to\infty} \frac{\sup_{a\in[a_1,a_2]} |R_1(i, a)|}{w^k(i)} = 0. \tag{9.8}$$

Similarly, $q_{i,i-1}(a,b)w^k(i-1)$ equals

$$
\begin{aligned}
C^k(\mu i + h_2(i,b)) \cdot i^k \\
&= C^k(\mu(i+1) - \mu + h_2(i,b)) \cdot ((i+1) - 1)^k \\
&= C^k \mu(i+1)^{k+1} + C^k(-\mu(k+1) + h_2(i,b)) \cdot (i+1)^k + R_2(i,b) \\
&= \frac{\mu}{C} \cdot w^{k+1}(i) + (-\mu(k+1) + h_2(i,b)) \cdot w^k(i) + R_2(i,b), \quad (9.9)
\end{aligned}
$$

and, as a consequence of (9.5), the residual term $R_2(i,b)$ verifies that

$$
\lim_{i \to \infty} \frac{\sup_{b \in [b_1,b_2]} |R_2(i,b)|}{w^k(i)} = 0. \quad (9.10)
$$

Finally, we have that $q_{ii}(a,b)w^k(i)$ equals

$$
\begin{aligned}
-C^k(\lambda + \mu i + h_1(i,a) + h_2(i,b)) \cdot (i+1)^k \\
&= -C^k \mu \cdot (i+1)^{k+1} + C^k(\mu - \lambda - h_1(i,a) - h_2(i,b)) \cdot (i+1)^k \\
&= -\frac{\mu}{C} \cdot w^{k+1}(i) + (\mu - \lambda - h_1(i,a) - h_2(i,b)) \cdot w^k(i). \quad (9.11)
\end{aligned}
$$

Hence, summing (9.7), (9.9), and (9.11), we obtain for all $i > 0$ and $(a,b) \in A$,

$$
\begin{aligned}
\sum_{j \in S} q_{ij}(a,b)w^k(j) &= -\mu k \cdot w^k(i) + R_1(i,a) + R_2(i,b) \\
&= w^k(i)\left(-\mu k + \frac{R_1(i,a) + R_2(i,b)}{w^k(i)} \right).
\end{aligned}
$$

By (9.8) and (9.10), we can choose I_0 large enough so that

$$
-\mu k + \frac{R_1(i,a) + R_2(i,b)}{w^k(i)} \leq -\frac{\mu k}{2} \quad \forall\, i > I_0,\ (a,b) \in A.
$$

Putting $c = \mu k / 2$, this yields

$$
\sum_{j \in S} q_{ij}(a,b)w^k(j) \leq -c w^k(i) \quad \forall\, i > I_0,\ (a,b) \in A.
$$

Obviously, choosing $d > 0$ sufficiently large we have

$$
\sum_{j \in S} q_{ij}(a,b)w^k(j) \leq -c w^k(i) + d \quad \forall\, i \leq I_0,\ (a,b) \in A.
$$

Finally, we let $D := \{0, 1, \ldots, I_0\}$ to complete the proof of the lemma. $\quad\square$

Summarizing, our queueing model satisfies all the conditions for the optimality criteria listed in Sec. 9.1, and so, optimal policies for these optimality criteria indeed exist. Furthermore, if the decision-maker wants to maximize his/her return given by the reward rate pi, interpreting the cost rate $c(i, a, b)$ as a constraint, then constrained optimal policies also exist.

9.3. A controlled birth-and-death process

We consider the model studied in [52, Example 7.1].

The state space is $S = \{0, 1, 2, \ldots\}$, and the state variable stands for the population size. The population's natural birth rate is $\lambda > 0$, while the death rate is assumed to be controlled by the decision-maker. More precisely, we consider the compact action space $A = [\mu_1, \mu_2]$, for $0 < \mu_1 < \mu_2$, so that the death rate corresponds to an action $a \in A$ chosen by the controller. In addition, we suppose that, when the population decreases, either one or two individuals die with probabilities $p_1 > 0$ and $p_2 > 0$, respectively, with $p_1 + p_2 = 1$.

Therefore, the transition rates are, for $i = 0$,

$$q_{01}(a) = \lambda = -q_{00}(a) \quad \forall \, a \in A,$$

whereas for $i = 1$,

$$q_{10}(a) = a, \; q_{11}(a) = -(a + \lambda), \; q_{12}(a) = \lambda \quad \forall \, a \in A.$$

For $i \geq 2$, the transition rates of the system are given by

$$q_{ij}(a) = \begin{cases} p_2 a i, & \text{for } j = i - 2, \\ p_1 a i, & \text{for } j = i - 1, \\ -(a + \lambda)i, & \text{for } j = i, \\ \lambda i, & \text{for } j = i + 1, \\ 0, & \text{otherwise,} \end{cases}$$

for each $a \in A$. Our definition of the transition rates implies that every stationary policy in Φ_s is irreducible. Also note that, for each $i, j \in S$, the transition rates $q_{ij}(a)$ are continuous functions of $a \in A$.

We suppose that the controller receives a reward pi per unit time when the population size is $i \in S$, where $p > 0$ is a given constant. Moreover, we suppose that the cost rate when taking the action $a \in A$ in state $i \in S$ is $c(i, a)$. We assume that $c(i, \cdot)$ is continuous on A for each fixed $i \in S$, and also that, for some constant $M > 0$,

$$\sup_{a \in A} |c(i, a)| \leq M(i + 1) \quad \forall \, i \in S.$$

The decision-maker's net reward rate function is

$$r(i, a) = pi - c(i, a) \quad \forall \ (i, a) \in K.$$

For this control model, we choose a Lyapunov function of the form

$$w(i) = C(i + 1), \quad \text{for } i \in S,$$

where the constant $C \geq 1$ verifies that $C > \mu_2 + \lambda$, and so $q(i) \leq w(i)$ for each $i \in S$. It is clear that Assumptions 9.1(b), 9.1(c), 9.2, and 9.3 hold. Concerning Assumption 9.1(a) we have the following.

Lemma 9.5.

(i) *There exist constants $c_1, c_2 \in \mathbb{R}$ and $d_1, d_2 \geq 0$ such that*

$$\sum_{j \in S} q_{ij}(a)w(j) \leq -c_1 w(i) + d_1 \quad \text{and} \quad \sum_{j \in S} q_{ij}(a)w^2(j) \leq -c_2 w^2(i) + d_2$$

for all $(i, a) \in K$.

(ii) *Suppose that the birth-and-death process satisfies $\mu_1(1 + p_2) > \lambda$. Then, for each integer $k \geq 1$, there exist constants $c_k > 0$ and $d_k > 0$, and a finite set $D_k \subset S$ such that*

$$\sum_{j \in S} q_{ij}(a)w^k(j) \leq -c_k w^k(i) + d_k \cdot \mathbf{I}_{D_k}(i) \quad \forall \ (i, a) \in K.$$

Proof. *Proof of (i).* A direct calculation yields, for $i \geq 2$ and $a \in A$,

$$\sum_{j \in S} q_{ij}(a)w(j) = (\lambda - a(1 + p_2))w(i) + C(a(1 + p_2) - \lambda)$$

$$\leq (\lambda - \mu_1(1 + p_2))w(i) + C(\mu_2(1 + p_2) - \lambda).$$

Similarly, we obtain, for $i \geq 2$ and $a \in A$,

$$\sum_{j \in S} q_{ij}(a)w^2(j) = \left(2(\lambda - a(1 + p_2)) + \frac{a(3 + 5p_2) - \lambda}{i + 1}\right)w^2(i)$$

$$- C^2(\lambda + a(1 + 3p_2)),$$

and part (i) easily follows.

Proof of (ii). The proof of this part is similar to that of Lemma 9.4. The idea is to write $q_{ij}(a)w^k(j)$ as a power series of $(i + 1)$, in which we keep the terms of degree $k + 1$ and k. Proceeding this way, for $i \geq 2$ and $a \in A$ we obtain

$$q_{i,i-2}(a)w^k(i - 2) = \frac{p_2 a}{C} \cdot w^{k+1}(i) - p_2 a(1 + 2k) \cdot w^k(i) + R_{-2}(i, a),$$

$$q_{i,i-1}(a)w^k(i - 1) = \frac{p_1 a}{C} \cdot w^{k+1}(i) - p_1 a(1 + k) \cdot w^k(i) + R_{-1}(i, a),$$

$$q_{ii}(a)w^k(i) = -\frac{a+\lambda}{C} \cdot w^{k+1}(i) + (a+\lambda) \cdot w^k(i),$$

$$q_{i,i+1}(a)w^k(i) = \frac{\lambda}{C} \cdot w^{k+1}(i) + \lambda(k-1) \cdot w^k(i) + R_1(i,a),$$

where the residual terms verify that

$$\lim_{i\to\infty} \frac{\sup_{a\in A} |R_\ell(i,a)|}{w^k(i)} = 0 \quad \text{for } \ell = -2, -1, 1. \tag{9.12}$$

Summing the above equations yields

$$\sum_{j\in S} q_{ij}(a)w^k(j) = -k(a(1+p_2) - \lambda) \cdot w^k(i) + \sum_\ell R_\ell(i,a)$$

$$= w^k(i)\left(-k(a(1+p_2) - \lambda) + \frac{\sum R_\ell(i,a)}{w^k(i)} \right)$$

$$\leq w^k(i)\left(-k(\mu_1(1+p_2) - \lambda) + \frac{\sum R_\ell(i,a)}{w^k(i)} \right).$$

By (9.12), and since $\mu_1(1+p_2) > \lambda$, there exists I_0 such that, for $i > I_0$ and $a \in A$,

$$\sum_{j\in S} q_{ij}(a)w^k(j) \leq -\frac{k(\mu_1(1+p_2) - \lambda)}{2} \cdot w^k(i),$$

and there exists $d_k > 0$ such that, for $i \leq I_0$ and $a \in A$,

$$\sum_{j\in S} q_{ij}(a)w^k(j) \leq -\frac{k(\mu_1(1+p_2) - \lambda)}{2} \cdot w^k(i) + d_k.$$

This completes the proof of the lemma. $\qquad\square$

From Lemma 9.5(i) we obtain that for a discount rate $\alpha > 0$ such that

$$\alpha + \mu_1(1+p_2) - \lambda > 0 \tag{9.13}$$

there exist discount optimal policies.

On the other hand, under the conditions of Lemma 9.5(ii) we conclude that our controlled birth-and-death process satisfies Assumptions 9.1–9.3, and so there exist optimal policies for *all* the optimality criteria given in Sec. 9.1.

Our condition in Lemma 9.5(ii), namely $\mu_1(1+p_2) > \lambda$, is in fact the usual ergodicity requirement for birth-and-death processes. Indeed, $p_1 + 2p_2 = 1 + p_2$ is the expected decrease of the population when a death occurs, and so $\mu_1(1+p_2) > \lambda$ means that the minimal death rate is larger than the birth rate.

Note that, in [52, Example 7.1], the conditions imposed on the control model are: (i) $\mu_1 > \lambda$, and (ii) $p_2 \leq \mu_1/2\mu_2$. Our condition here is weaker because (i) implies that $\mu_1(1 + p_2) > \lambda$, and then the condition (ii) is not needed (in particular, we can have $\mu_1 < \lambda$). Let us make some general comments on this fact.

Remark 9.6. The reason why our condition $\mu_1(1 + p_2) > \lambda$ is weaker than those imposed in [52, Example 7.1] is that, here, we use different sufficient conditions for w-exponential ergodicity. More precisely, in [52] the authors use the monotonicity conditions in Assumption 2.7. In particular, the controlled process must be stochastically monotone, and the following Lyapunov condition must be satisfied:

$$\sum_{j \in S} q_{ij}(a)w(j) \leq -cw(i) + d \cdot \mathbf{I}_{\{0\}}(i) \quad \forall \, (i, a) \in K. \tag{9.14}$$

With our Lyapunov conditions on the functions w^k proved in Lemma 9.5(ii), however, we do not need monotonicity conditions and, furthermore, the finite set in the Lyapunov condition (9.14) does not need to be necessarily $\{0\}$; it can be any finite set $D_k \subset S$ (recall Assumption 2.10(a) and Theorem 2.11). Actually, a key feature in the proof of Lemmas 9.4 and 9.5(ii) is to choose the set $D_k = \{0, 1, \ldots, I_0\}$ for "large" I_0.

To conclude, in practice, the assumptions for exponential ergodicity in Theorem 2.11 are easier to verify than the hypotheses of Theorem 2.8. Hence, our conditions here on the parameters of the control model are less restrictive than those in [52, Example 7.1].

Finite state approximations   The controlled birth-and-death process in this section will be denoted by

$$\mathcal{M} := \{S, A, (q_{ij}(a)), (r(i, a))\}.$$

Now we analyze the finite state approximations studied in Secs 4.3 and 4.4. To this end, let us define the following *finite state and finite action* control model $\mathcal{M}_n$ for each given $n \geq 1$:

- The state space is $S_n := \{0, 1, \ldots, n\}$.
- Fix an arbitrary integer $m \geq 1$. For given $n \geq 1$, consider the partition of A with diameter $(\mu_2 - \mu_1)/mn$:

$$\mathcal{P}_n := \left\{ \mu_1 + \frac{\ell}{mn}(\mu_2 - \mu_1) : 0 \leq \ell \leq mn \right\}. \tag{9.15}$$

The action set of the truncated model $\mathcal{M}_n$ is $\mathcal{P}_n$ for each $i \in S_n$.

- We define the transition rates

$$q^n_{ij}(a) := q_{ij}(a) \quad \forall \, 0 \le i < n, \ 0 \le j \le n, \ a \in \mathcal{P}_n,$$

and

$$q^n_{n,n-2}(a) := p_2 a n, \quad q^n_{n,n-1}(a) := p_1 a n, \quad q^n_{nn}(a) := -an \quad \forall \, a \in \mathcal{P}_n. \tag{9.16}$$

(It is easy to check that these transition rates verify that $q^n_{ij}(a) \ge 0$ if $i \ne j$, and $\sum_{j \in S_n} q^n_{ij}(a) = 0$ for every $a \in \mathcal{P}_n$ and $i, j \in S_n$.)

- The reward rate is given by

$$r_n(i, a) := r(i, a) \quad \forall \, i \in S_n, \ a \in \mathcal{P}_n. \tag{9.17}$$

The set of stationary policies is denoted by $\mathbb{F}_n$; hence, $f : S_n \to A$ is in $\mathbb{F}_n$ if $f(i) \in \mathcal{P}_n$ for each $i \in S_n$. It is easily seen that each $f \in \mathbb{F}_n$ is irreducible, and we denote by μ^n_f its invariant probability measure on S_n.

Discounted reward approximations Choose a discount rate $\alpha > 0$ such that (9.13) holds; recall Lemma 9.5(i). The policy iteration algorithm for discounted reward CMCs studied in Sec. 4.2.1 converges in a *finite number of steps* to a discount optimal policy of

$$\mathcal{M}_n = \{S_n, A, (\mathcal{P}_n), (q^n_{ij}(a)), (r_n(i, a))\}$$

because the state and action spaces are both finite. Hence, to approximate the optimal discounted reward and a discount optimal policy of $\mathcal{M}$ we will use the optimal discounted reward V^{*n}_α and a discount optimal policy $f^*_n \in \mathbb{F}_n$ of $\mathcal{M}_n$, obtained by means of the policy iteration algorithm.

As a direct consequence of Theorem 4.11 and Proposition 4.12 — in particular, observe that the asymptotically dense requirement in Definition 4.8(ii) is satisfied by $\mathcal{P}_n$ — we obtain our next result.

Theorem 9.7. *Consider the controlled birth-and-death process $\mathcal{M}$ and let the discount rate $\alpha > 0$ satisfy $\alpha + \mu_1(1 + p_2) - \lambda > 0$. If V^{*n}_α and $f^*_n \in \mathbb{F}_n$, for $n \ge 1$, are the optimal discounted reward and a discount optimal policy, respectively, of the truncated control model $\mathcal{M}_n$, then, for all $i \in S$,*

$$\lim_{n \to \infty} V^{*n}_\alpha(i) = V^*_\alpha(i),$$

the optimal discounted reward of $\mathcal{M}$, and, in addition, any limit policy of $\{f^*_n\}_{n \ge 1}$ in $\mathbb{F}$ is discount optimal for $\mathcal{M}$.

For the numerical application, we consider the following values of the parameters:

$$\lambda = 5.7, \quad \mu_1 = 4, \quad \mu_2 = 7, \quad p_1 = 0.6, \quad p_2 = 0.4.$$

We choose $\alpha = 0.15$ and note that (9.13) is satisfied. The reward rate function is

$$r(i, a) = (p - (a - \mu_2)^2 \log(1 + \mu_2 - a)) \cdot i \quad \forall \ (i, a) \in S \times A,$$

with $p = 10$. The interpretation of the cost rate function

$$c(i, a) = (a - \mu_2)^2 \log(1 + \mu_2 - a) \cdot i$$

is that, if μ_2 is the natural death rate of the population, then the controller can decrease this death rate by using an adequate medical policy. Clearly, the cost of the medical policy must be a decreasing function of the controlled death rate. Finally, let the integer m in the definition of the action sets (9.15) be $m = 2$.

For each $1 \leq n \leq 60$, we determine the optimal 0.15-discounted reward V_α^{*n} and a discount optimal policy f_n^* of the truncated control model $\mathcal{M}_n$ by using the policy iteration algorithm in Sec. 4.2.1.

Given the initial states $i = 5, 10, 15$, we show the optimal discounted rewards $V_\alpha^{*n}(i)$ — see Figure 9.1 — and the optimal actions $f_n^*(i)$ — see Figure 9.2 — as functions of n. Empirically, we observe that the optimal rewards and actions quickly converge, and that they become stable for relatively small truncation sizes n.

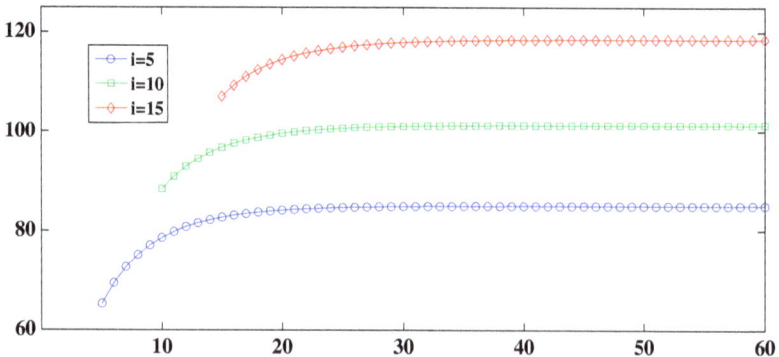

Fig. 9.1   The optimal rewards $V_\alpha^{*n}(i)$ for $i = 5, 10, 15$.

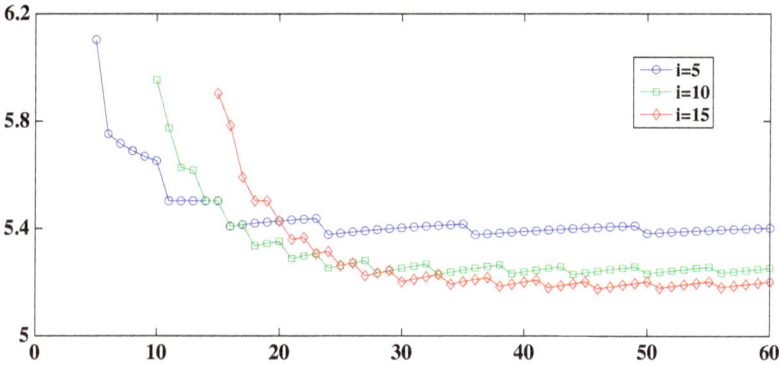

Fig. 9.2   The optimal actions $f_n^*(i)$ for $i = 5$, 10, 15.

Average reward approximations   Consider the controlled birth-and-death model $\mathcal{M}$ and the finite state and action truncations $\mathcal{M}_n$, for $n \geq 1$, defined above. Note that, at this point, we cannot use the approximation results in Sec. 4.4 because the action sets of the control models $\mathcal{M}$ and $\mathcal{M}_n$ are not the same.

Next, we describe the procedure that we will use to find approximations of the optimal gain g^* and a gain optimal policy for $\mathcal{M}$. Its convergence will be proved in Theorem 9.8.

For the truncated control models $\mathcal{M}_n$, the policy iteration algorithm defined in Sec. 4.2.2 converges in a *finite number of steps* to a gain optimal policy of $\mathcal{M}_n$ because the state and action spaces are both finite. Consider now the following approximation procedure.

Step I. Set $n = 1$.

Step II. Choose a stationary policy $f_0 \in \mathbb{F}_n$. Set $s = 0$.

Step III. Determine the gain g and the bias h of the policy f_s under $\mathcal{M}_n$ by solving the Poisson equation for f_s:
$$g = r_n(i, f_s) + \sum_{j \in S_n} q_{ij}^n(f_s)h(j) \quad \forall \, i \in S_n,$$
subject to $\mu_{f_s}^n(h) = 0$.

Step IV. Find $f_{s+1} \in \mathbb{F}_n$ such that, for each $i \in S_n$,
$$r_n(i, f_{s+1}) + \sum_{j \in S_n} q_{ij}^n(f_{s+1})h(j)$$
$$= \max_{a \in \mathcal{P}_n} \left\{ r_n(i, a) + \sum_{j \in S_n} q_{ij}^n(a)h(j) \right\}.$$

If $f_{s+1} = f_s$, then set $g_n := g$ and $f_n^* := f_s$, increase n by one and return to Step II.

If $f_{s+1} \neq f_s$, then replace s with $s + 1$ and return to Step III.

This procedure yields a real-valued sequence $\{g_n\}_{n \geq 1}$ and a sequence $f_n^* \in \mathbb{F}_n$ of policies (which are the optimal gain and an average optimal policy of the control model $\mathcal{M}_n$).

Before proving our main result, we note the following fact, which follows from the proof of Proposition 4.12: under the conditions of Lemma 9.5(ii), given arbitrary $n \geq 1$, $i \in S_n$, $a \in \mathcal{P}_n$, and $k \geq 1$,

$$\sum_{j \in S_n} q_{ij}^n(a) w^k(j) \leq -c_k w^k(i) + d_k \cdot \mathbf{I}_{D_k}(i), \tag{9.18}$$

where c_k, d_k, and D_k are from Lemma 9.5(ii).

Theorem 9.8. *Consider the controlled birth-and-death process $\mathcal{M}$, with $\mu_1(1 + p_2) > \lambda$, and let $\{g_n\}$ and $\{f_n^*\}$ be the sequences obtained by application of the iterative procedure Steps I–IV defined above.*

Then $\lim_n g_n = g^*$, the optimal gain of $\mathcal{M}$. Furthermore, any limit policy in $\mathbb{F}$ of $\{f_n^*\}_{n \geq 1}$ is gain optimal for $\mathcal{M}$.

Proof.   The control models $\mathcal{M}_n$ with state space S_n and action sets $\mathcal{P}_n$ do not satisfy Assumption 4.13 because their action sets are not the same as $\mathcal{M}$'s. Therefore, let us define $\overline{\mathcal{M}}_n$ as follows.

We fix $n \geq 1$. The state space of $\overline{\mathcal{M}}_n$ is S_n and the action sets are $A = [\mu_1, \mu_2]$ for each $i \in S_n$. The transition and reward rates are defined as $q_{ij}^n(a)$ and $r_n(i, a)$, for $i, j \in S_n$ and $a \in \mathcal{P}_n$, respectively; see (9.15)–(9.17). In $A = [\mu_1, \mu_2]$, we define the transition rates $q_{ij}^n(a)$ and the reward rates $r_n(i, a)$ as the piecewise linear functions that coincide with $q_{ij}^n(a)$ and $r_n(i, a)$ on $\mathcal{P}_n$, respectively.

It is clear that the so-defined $q_{ij}^n(a)$ are indeed transition rates, i.e., $q_{ij}^n(a) \geq 0$ if $i \neq j$, and

$$\sum_{j \in S_n} q_{ij}^n(a) = 0 \quad \forall \, i \in S_n, \, a \in A.$$

Let us check that the control models $\overline{\mathcal{M}}_n$, for $n \geq 1$, satisfy Assumption 4.13, and also the condition in Theorem 4.21(b).

Regarding Assumptions 4.13(a) and (c), and the condition in Theorem 4.21(b), note that, for all $n \geq 1$ and each integer $k \geq 1$, given $i \in S_n$ and $a \in A$,

$$\sum_{j \in S_n} q_{ij}^n(a) w^k(j) \leq \max_{a \in \mathcal{P}_n} \left\{ \sum_{j \in S_n} q_{ij}^n(a) w^k(j) \right\} \leq -c_k w^k(i) + d_k \cdot \mathbf{I}_{D_k}(i).$$

The leftmost inequality holds because the function $a \mapsto \sum_{j \in S_n} q_{ij}^n(a) w^k(j)$ is piecewise linear, whereas the rightmost inequality is derived from (9.18). This yields the desired result.

By this same argument (because the involved functions are piecewise linear in $a \in A$), we have

$$q_n(i) := \sup_{a \in A} \{-q_{ii}^n(a)\} = \max_{a \in \mathcal{P}_n} \{-q_{ii}^n(a)\} \leq w(i) \quad \forall \, i \in S_n,$$

and Assumption 4.13(b) follows. Similarly, we can prove that Assumption 4.13(e) is satisfied. Moreover, Assumptions 4.13(d) and (f) trivially hold.

Finally, it is clear that $\{\overline{\mathcal{M}}_n\}_{n \geq 1}$ converges to $\mathcal{M}$ because the functions $q_{ij}^n(\cdot)$ and $r_n(i, \cdot)$ converge uniformly on A to $q_{ij}(\cdot)$ and $r(i, \cdot)$, respectively, for each fixed $i, j \in S$ (this is because these functions are piecewise linear interpolations, with diameter converging to zero, of continuous functions on a compact interval).

Therefore, as a consequence of Theorem 4.21, the optimal gains $\{g_n^*\}$ of $\overline{\mathcal{M}}_n$ converge to the optimal gain g^* of $\mathcal{M}$, and any limit policy of optimal policies of $\overline{\mathcal{M}}_n$ is optimal for $\mathcal{M}$.

Hence, to complete the proof, it remains to show that Steps II–IV above (for a control model with *finite action space*) in fact solve the average reward control model $\overline{\mathcal{M}}_n$ (with *continuous action space* $A = [\mu_1, \mu_2]$). Indeed, in the improvement Step IV, since

$$r_n(i, a) + \sum_{j \in S_n} q_{ij}^n(a) h(j)$$

is a piecewise linear function of $a \in A$, we have

$$\max_{a \in A} \left\{ r_n(i, a) + \sum_{j \in S_n} q_{ij}^n(a) h(j) \right\} = \max_{a \in \mathcal{P}_n} \left\{ r_n(i, a) + \sum_{j \in S_n} q_{ij}^n(a) h(j) \right\}.$$

So, the algorithm described in Steps II–IV corresponds, precisely, to the policy iteration algorithm for $\overline{\mathcal{M}}_n$ when the initial policy is in $\mathbb{F}_n$ (equivalently, if we start the policy iteration algorithm for $\overline{\mathcal{M}}_n$ with f_0 in $\mathbb{F}_n$, then the subsequent policies f_s are also in $\mathbb{F}_n$). Therefore, the optimal gain g_n of $\mathcal{M}_n$ equals g_n^*, the optimal gain of $\overline{\mathcal{M}}_n$, and $f_n^* \in \mathbb{F}_n$ is a gain optimal policy for $\overline{\mathcal{M}}_n$. By Theorem 4.21, the proof is now complete. $\square$

For our numerical application, we fix the following values of the parameters:

$$\lambda = 5, \quad \mu_1 = 4.5, \quad \mu_2 = 7, \quad p_1 = 0.75, \quad p_2 = 0.25,$$

and so, $\mu_1(1 + p_2) > \lambda$ holds. The reward rate function is

$$r(i, a) = 10i - (a - \mu_2)^2 \log(1 + \mu_2 - a) \cdot i \quad \forall \, (i, a) \in K.$$

Finally, the constant m in the definition of the partitions $\mathcal{P}_n$ is $m = 2$.

For each $1 \leq n \leq 60$ we solve the truncated control model $\mathcal{M}_n$, under the average reward optimality criterion, by using the policy iteration procedure (see Steps II–IV above).

In Figure 9.3 we show the optimal gain g_n^* of $\overline{\mathcal{M}}_n$ as a function of n. Observe that, as a consequence of Remark 4.24 and Lemma 9.5(ii), we have $|g_n^* - g^*| = \mathrm{O}(n^{-\beta})$ for all $\beta > 0$ (recall the O notation introduced in Sec. 1.4).

To study the convergence of the optimal policies, Figure 9.4 shows the optimal actions $f_n^*(i)$, for $i = 5$, 10, 15, of $\overline{\mathcal{M}}_n$ as a function of n.

Empirically, it is clear from Figures 9.3 and 9.4 that the sequences $\{g_n^*\}$ and $\{f_n^*(i)\}$ converge. We deduce the approximate values $g^* = 10.66$ and

$$f^*(5) = 5.3125, \quad f^*(10) = 5.2000, \quad \text{and} \quad f^*(15) = 5.1625$$

for a gain optimal policy for $\mathcal{M}$.

Observe that the graph in Figure 9.4 is somehow saw-shaped. This is because the optimal actions $f_n^*(i)$ are in $\mathcal{P}_n$, which is a partition with "large" diameter for small values of n. The graph in Figure 9.3 is smoother because g_n^* is obtained by averaging the $r_n(i, a)$.

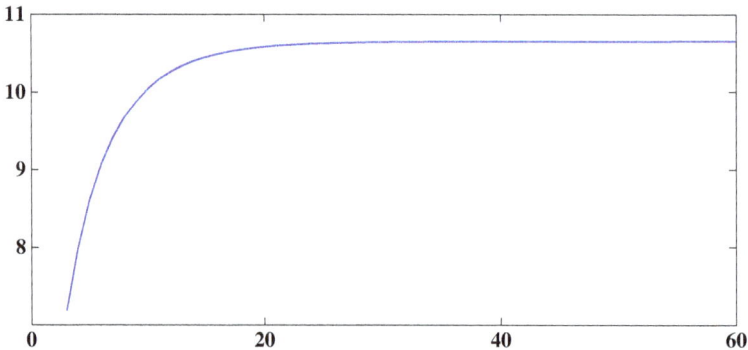

Fig. 9.3   The optimal gain g_n^*.

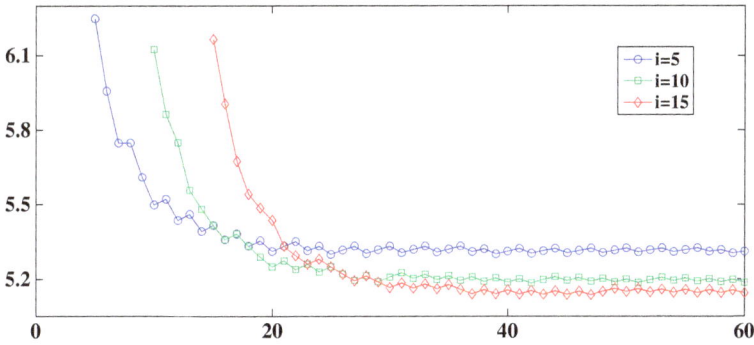

Fig. 9.4   The optimal actions $f_n^*(i)$ for $i = 5$, 10, 15.

9.4.   A population system with catastrophes

We consider a generalization of [52, Example 7.2]. This is the example introduced in Sec. 1.1.1; see also [136, Sec. IV].

The state space is $S = \{0, 1, 2, \ldots\}$, and the state variable stands for the population size. The natural birth and death rates are, respectively, $\lambda > 0$ and $\mu > 0$. The birth rate is controlled by actions $a \in [0, a_2]$ of the decision-maker (this can be interpreted as controlling the immigration rate).

When the population size is $i \in S$, we suppose that a catastrophe occurs at a rate $d(i, b) \geq 0$, which is controlled by actions $b \in [b_1, b_2]$ of the decision-maker. In [52, Example 7.2] it is assumed that when such an event occurs, the population decreases by one or two. Here, we suppose that the catastrophe can have an *arbitrary size*; that is, for each $i \geq 1$, we consider a probability distribution $\{\gamma_i(j)\}_{1 \leq j \leq i}$, where $\gamma_i(j)$ is the probability that j individuals perish in the catastrophe. We denote by E_{γ_i} the expectation operator associated with the distribution $\{\gamma_i(j)\}_{1 \leq j \leq i}$, and by X_i the random number of perished individuals; this notation will be used in Lemma 9.9 below.

Hence, our population model is as follows. The action space is $A = [0, a_2] \times [b_1, b_2]$, where $a \in [0, a_2]$ is the controlled immigration rate, and $b \in [b_1, b_2]$ is the control corresponding to the catastrophes. The action set in state $i = 0$ is

$$A(0) = [a_1, a_2], \quad \text{for } 0 < a_1 < a_2,$$

and the corresponding transition rates are

$$q_{01}(a) = a = -q_{00}(a) \quad \forall \, a \in A(0).$$

For each $i \geq 1$, the action set is $A(i) = A$, and the transition rates are

$$
q_{ij}(a,b) = \begin{cases} 0, & \text{for } j > i+1, \\ \lambda i + a, & \text{for } j = i+1, \\ -(\lambda + \mu)i - a - d(i,b), & \text{for } j = i, \\ \mu i + d(i,b)\gamma_i(1), & \text{for } j = i-1, \\ d(i,b)\gamma_i(i-j), & \text{for } 0 \leq j < i-1, \end{cases}
$$

for every $(a,b) \in A$. We make the hypothesis that the function $b \mapsto d(i,b)$ is continuous on $[b_1, b_2]$ for each fixed $i \in S$, and that there exists $M > 0$ such that

$$
\sup_{b \in [b_1, b_2]} |d(i,b)| \leq M(i+1) \quad \forall\, i \in S. \tag{9.19}
$$

It follows that each policy in Φ_s is irreducible (note that the particular definition of $A(0)$ means that the state 0 is not absorbing under any policy).

When the population size is $i \in S$, the controller receives a reward at a rate pi, for some $p > 0$. The cost rate for controlling the immigration and the catastrophe rates is $c(i,a,b)$. Hence, we consider the following net reward rate

$$
r(i,a,b) = pi - c(i,a,b) \quad \forall\, (i,a,b) \in K.
$$

We suppose that the function $c(i,\cdot,\cdot)$ is continuous on $A(i)$ for each $i \in S$ and that, furthermore, for some constant $M' > 0$,

$$
\sup_{(a,b) \in A(i)} |c(i,a,b)| \leq M'(i+1) \quad \forall\, i \in S.
$$

We consider the Lyapunov function $w(i) = C(i+1)$ for $i \in S$, where the constant $C \geq 1$ satisfies

$$
C > \lambda + \mu + a_2 + M,
$$

so that $q(i) \leq w(i)$ for every $i \in S$.

Assumptions 9.1(b), 9.1(c), 9.2, and 9.3 hold. To conclude, we study Assumption 9.1(a).

Lemma 9.9.

(i) *There exist constants $c_1, c_2 \in \mathbb{R}$ and $d_1, d_2 \geq 0$ such that, for all $(i,a,b) \in K$,*

$$
\sum_{j \in S} q_{ij}(a,b)w(j) \leq -c_1 w(i) + d_1
$$

and

$$\sum_{j\in S} q_{ij}(a,b)w^2(j) \leq -c_2 w^2(i) + d_2.$$

(ii) *If $\mu > \lambda$, then Assumption 9.1(a) holds.*

(iii) *Suppose now that $\mu \leq \lambda$, and that there exist $\varepsilon > 0$ and $\eta_0 > 2$ such that, for sufficiently large i,*

$$\lambda - \mu + \varepsilon \leq \inf_{b\in[b_1,b_2]} d(i,b) \cdot \left(E_{\gamma_i}[X_i/(i+1)] \right.$$
$$\left. - \frac{\eta_0 - 1}{2} E_{\gamma_i}[(X_i/(i+1))^2] \right).$$

Then there exist constants $c_{\eta_0} > 0$ and $d_{\eta_0} > 0$, and a finite set $D_{\eta_0} \subset S$ such that

$$\sum_{j\in S} q_{ij}(a,b)w^{\eta_0}(j) \leq -c_{\eta_0} w^{\eta_0}(i) + d_{\eta_0} \cdot \mathbf{I}_{D_{\eta_0}}(i) \quad \forall\, (i,a,b) \in K.$$

$$(9.20)$$

In particular, the condition in Assumption 9.1(a) holds for every real number $1 \leq k \leq \eta_0$.

Proof. *Proof of (i).* A direct calculation shows that, for states $i > 0$ and actions $(a,b) \in A$,

$$\sum_{j\in S} q_{ij}(a,b)w(j) = (\lambda - \mu - d(i,b)E_{\gamma_i}[X_i/(i+1)])\, w(i) + R(a - \lambda + \mu)$$
$$\leq (\lambda - \mu)w(i) + R(a_2 - \lambda + \mu), \qquad (9.21)$$

and

$$\sum_{j\in S} q_{ij}(a,b)w^2(j) \leq \left(2(\lambda - \mu) + \frac{|2a_2 - \lambda + 3\mu|}{2} \right)$$
$$\cdot w^2(i) + R^2(a_2 - \lambda + \mu).$$

Statement (i) readily follows.

Proof of (ii). Fix a state $i \geq 1$ and a real number $\eta \geq 1$. We have

$$q_{i,i+1}(a,b)w^{\eta}(i+1) = \frac{\lambda}{C} \cdot w^{\eta+1}(i) + (a + \lambda(\eta - 1)) \cdot w^{\eta}(i)$$
$$+ R_0(i,a,b),$$

where the residual term verifies

$$\lim_{i\to\infty} \sup_{(a,b)\in A} \frac{|R_0(i,a,b)|}{w^{\eta}(i)} = 0.$$

Similarly,

$$q_{ii}(a,b)w^\eta(i) = -\frac{\lambda+\mu}{C} \cdot w^{\eta+1}(i) + (\lambda+\mu-a-d(i,b)) \cdot w^\eta(i).$$

Finally,

$$\sum_{j=0}^{i-1} q_{i,j}(a,b)w^\eta(j) = \frac{\mu}{C} \cdot w^{\eta+1}(i) - \mu(\eta+1)w^\eta(i)$$

$$+ d(i,b)E_{\gamma_i}[w^\eta(i-X_i)] + R_1(i,a,b),$$

and the residual term is such that

$$\lim_{i\to\infty} \sup_{(a,b)\in A} \frac{|R_1(i,a,b)|}{w^\eta(i)} = 0.$$

Summing up these expressions, we obtain that, for all $i \geq 1$ and $(a,b) \in A$, $\sum_{j\in S} q_{ij}(a,b)w^\eta(j)$ equals

$$w^\eta(i) \cdot \left(-\eta(\mu-\lambda) + d(i,b)\left(E_{\gamma_i}[Z_i^\eta] - 1\right) + \frac{\sum_\ell R_\ell(i,a,b)}{w^\eta(i)} \right), \qquad (9.22)$$

with $Z_i := 1 - \frac{X_i}{i+1}$. Now, recalling that $\mu > \lambda$, and since $0 \leq Z_i \leq 1$, by letting $c_\eta = \eta(\mu-\lambda)/2$ we derive that

$$\sum_{j\in S} q_{ij}(a,b)w^\eta(j) \leq -c_\eta w^\eta(i) + d_\eta \cdot \mathbf{I}_{D_\eta}(i) \quad \forall\ (i,a,b) \in K,$$

with $D_\eta = \{0,1,\ldots,I_\eta\}$ for sufficiently large I_η, and large d_η. Hence, Assumption 9.1(a) holds.

Proof of (iii). Starting from (9.22) and using the inequality

$$(1-h)^\eta \leq 1 - \eta h + \frac{\eta(\eta-1)}{2}h^2$$

(which holds for $\eta \geq 2$ and $0 \leq h \leq 1$) for $\eta = \eta_0$ and $h = 1-Z_i = X_i/(i+1)$, the proof for expression (9.20) follows by arguments similar to those used in (ii), but this time for $c_{\eta_0} = \varepsilon\eta_0/2$. Once (9.20) is established, Assumption 9.1(a) for $1 \leq k \leq \eta_0$ follows directly from Theorem 2.13. $\square$

We make some observations about the hypotheses of Lemma 9.9.

Remark 9.10.

(i) Regarding the discounted reward optimality criterion, it follows from (9.21) that there exist optimal policies for discount rates $\alpha > 0$ such that $\alpha > \lambda - \mu$. In particular, if $\mu \geq \lambda$, then we can choose any $\alpha > 0$. Suppose now, for instance, that there exists $D > 0$ such that

$$\inf_{b \in [b_1, b_2]} d(i, b) \geq Di \quad \forall \, i > 0.$$

From the inequality in (9.21), it follows that the constant c_1 in Lemma 9.9(i) can be any $c_1 < \mu - \lambda + D$. In this case, the discount rate $\alpha > 0$ must be such that $\alpha > \lambda - \mu - D$. So, if $\mu + D \geq \lambda$ (and, in particular, this allows the birth rate λ to be larger than the death rate μ) then we can choose an arbitrary discount rate $\alpha > 0$.

(ii) If the standard ergodicity requirement $\mu > \lambda$ holds, we can obtain Assumption 9.1(a) without imposing any restriction on the size of the catastrophe, and without making any special assumption about the catastrophe size probabilities $\gamma_i(j)$. Hence, our conditions here are, by far, weaker than those in $\mathbf{E_1}$ and $\mathbf{E_2}$ in [52, Example 7.2].

(iii) Furthermore, if the condition in Lemma 9.9(iii) is satisfied, then we can have $\lambda > \mu$ provided that the first two moments of the size of the catastrophe satisfy some suitable condition. As an illustration, it is easily seen that if $d(i, b)$ is of the form $d(i, b) = Dib$, for $D > 0$, and

$$E_{\gamma_i}[X_i] > 2(\lambda - \mu)/Db_1 \quad \text{for large } i, \tag{9.23}$$

then the condition in Lemma 9.9(iii) holds. Indeed, to see this, use the fact that

$$E_{\gamma_i}[(X_i/(i+1))^2] \leq E_{\gamma_i}[(X_i/(i+1))].$$

The condition (9.23) is particularly mild because, as the population i grows, the expected size of the catastrophe is expected to grow as well.

Therefore, the strengthened Lyapunov condition in Assumption 9.1(a) is a powerful tool, which, despite its restrictive appearance, is verified under very weak assumptions on the parameters of the control model.

Summarizing, if $\mu > \lambda$ then our population system satisfies all of our assumptions, and so, there exist optimal policies for all the optimality criteria that we have studied in the previous chapters. If, on the other hand, the condition in Lemma 9.9(iii) holds, then there exist optimal policies for all the criteria given in Sec. 9.1 (except perhaps for variance and Blackwell optimality in case that $\eta_0 < 3$).

Finite state approximations We denote by

$$\mathcal{M} := \{S, A, (A(i)), (q_{ij}(a, b)), (r(i, a, b))\}$$

the controlled population system defined above.

Next, we define an approximating sequence of *finite state and finite action* CMCs. For given $n \geq 1$, define the control model $\mathcal{M}_n$ as follows:

- The state space is $S_n := \{0, 1, \ldots, n\}$.
- Regarding the action sets, fix an integer $m \geq 1$. We consider the following partitions

$$\mathcal{P}_n(0) := \left\{ a_1 + \frac{\ell_1}{mn}(a_2 - a_1) : 0 \leq \ell_1 \leq mn \right\}$$

of the action set $A(0)$, and

$$\mathcal{P}_n(i) := \left\{ \left(\frac{\ell_1 a_2}{mn}, \ b_1 + \frac{\ell_2}{mn}(b_2 - b_1) \right) : 0 \leq \ell_1, \ell_2 \leq mn \right\}$$

of the action set $A(i)$, for $0 < i \leq n$.

The action sets of the control model $\mathcal{M}_n$ are the $\mathcal{P}_n(i)$, for $i \in S_n$.

- For $i, j \in S_n$ and $(a, b) \in \mathcal{P}_n(i)$ such that $(i, j) \neq (n, n)$, we define

$$q_{ij}^n(a, b) := q_{ij}(a, b), \quad \text{and} \quad q_{nn}^n(a, b) := -\mu n - d(n, b). \tag{9.24}$$

- The reward rate is $r_n(i, a, b) := r(i, a, b)$ for $i \in S_n$ and $(a, b) \in \mathcal{P}_n$.

Note that every $f \in \mathbb{F}_n$ is irreducible.

Discounted reward approximations   Choose a discount rate $\alpha > 0$ according to Remark 9.10(i). The policy iteration algorithm for the discounted reward optimality criterion (Sec. 4.2.1) converges in a *finite number of steps* to a discount optimal policy of

$$\mathcal{M}_n = \{S_n, A, (\mathcal{P}_n(i)), (q_{ij}^n(a)), (r_n(i, a))\}$$

because $\mathcal{M}_n$ has finite state space and action sets.

Therefore, we use the optimal discounted reward V_α^{*n} and a discount optimal policy $f_n^* \in \mathbb{F}_n$ of $\mathcal{M}_n$ as approximations of the optimal discounted reward and a discount optimal policy of $\mathcal{M}$.

We note that the condition in Definition 4.8(ii) for the convergence of $\mathcal{M}_n$ to $\mathcal{M}$ is verified. Hence, as a direct consequence of Theorem 4.11 and Proposition 4.12, we obtain our next result.

Theorem 9.11. *Consider the controlled population system with catastrophes $\mathcal{M}$ and let the discount rate $\alpha > 0$ satisfy the conditions in Remark 9.10(i). Then, for all $i \in S$,*

$$\lim_{n \to \infty} V_\alpha^{*n}(i) = V_\alpha^*(i),$$

the optimal discounted reward of $\mathcal{M}$*, and, furthermore, any limit policy of* $\{f_n^*\}$ *in* $\mathbb{F}$ *is discount optimal for* $\mathcal{M}$*.*

For the numerical computations, we choose the values of the parameters

$$\lambda = 2, \quad \mu = 3, \quad a_1 = 1, \quad a_2 = 5, \quad b_1 = 5, \quad b_2 = 8.$$

The catastrophe rate is given by $d(i, b) = ib/10$ for $i > 0$ and $b \in [5, 8]$. The distribution $\{\gamma_i(j)\}$ of the catastrophe size is a truncated geometric distribution with parameter $\gamma = 0.8$; more precisely, given $i > 0$,

$$\gamma_i(j) = \frac{\gamma^{j-1}(1 - \gamma)}{1 - \gamma^i} \quad \text{for } 1 \le j \le i.$$

Finally, the net reward rate is

$$r(i, a, b) = (10 - (a - 2)^2 - 0.5(b - 8)^2)i.$$

The interpretation of the term $(a - 2)^2$ is that we suppose that there is a natural immigration rate (which equals 2), and that augmenting or diminishing this natural immigration rate implies a cost for the controller. Similarly, the term $(b - 8)^2$ means that there is a natural catastrophe rate (which equals 8), and the controller incurs a cost when decreasing it. The discount rate is $\alpha = 0.1$ and the constant m in the definition of the partitions $\mathcal{P}_n(i)$ is $m = 2$.

For every $1 \le n \le 60$, we solved the discounted control problem for $\mathcal{M}_n$. Given the initial states $i = 5, 10, 15$, the optimal discounted rewards $V_\alpha^{*n}(i)$ and the optimal actions $a_n^*(i)$ and $b_n^*(i)$ are displayed in Figures 9.5 and 9.6, respectively, as functions of n. In these figures, we empirically observe convergence as well.

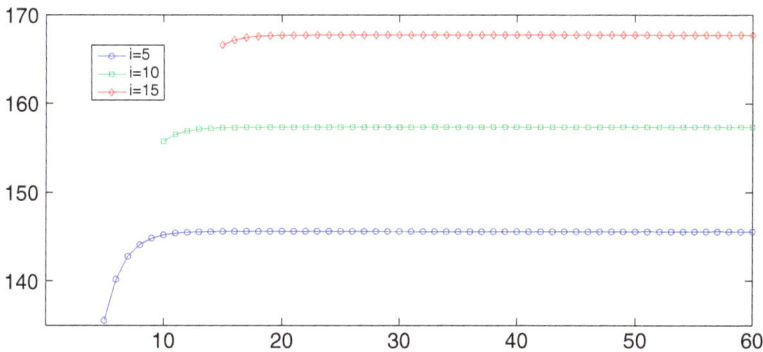

Fig. 9.5 The optimal rewards $V_\alpha^{*n}(i)$ for $i = 5, 10, 15$.

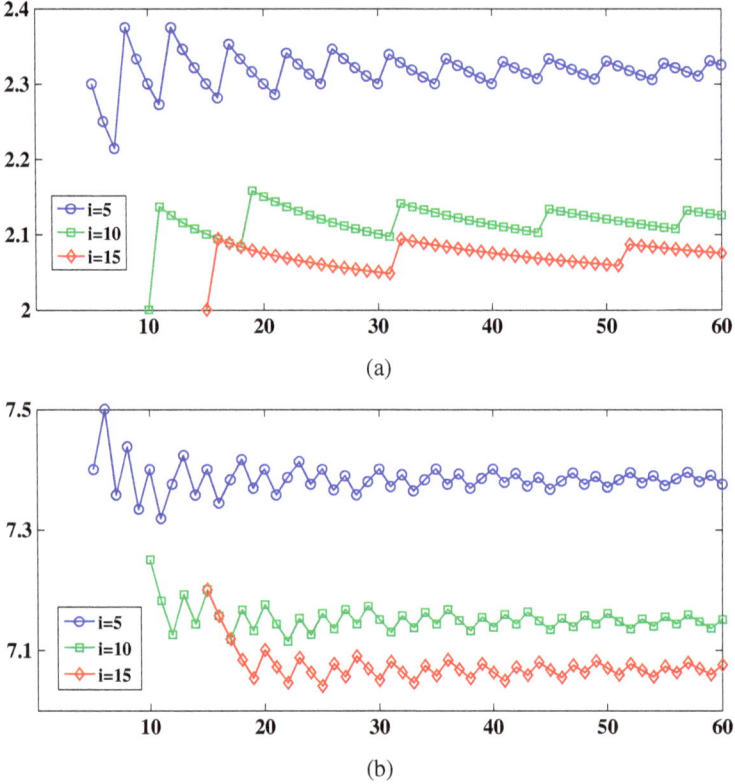

Fig. 9.6   The optimal policies $f_n^*(i)$ for $i = 5, 10, 15$. (a) The optimal actions $a_n^*(i)$. (b) The optimal actions $b_n^*(i)$.

Average reward approximations   Consider the sequence of finite state and action control models $\mathcal{M}_n$, for $n \geq 1$, defined above. Each $\mathcal{M}_n$ can be solved, under the average reward criterion, by applying the policy iteration procedure used for the controlled birth-and-death process (see the Steps II–IV of the algorithm introduced in Sec. 9.3). This yields a sequence of optimal gains $\{g_n\}_{n\geq 1}$, and a sequence of optimal policies $\{f_n^*\}_{n\geq 1}$.

As in Sec. 9.3, we cannot directly use Theorem 4.21 because $\mathcal{M}$ and the $\mathcal{M}_n$ do not have the same action sets. The proof of convergence in Theorem 9.12 is similar to that of Theorem 9.8, and we skip some details.

Theorem 9.12. *Consider the controlled population process $\mathcal{M}$, which satisfies either condition* (ii) *or condition* (iii) *in Lemma 9.9. Moreover, let $\{g_n\}$*

and $\{f_n^*\}$ be the sequences obtained by application of the policy iteration algorithm to the control models $\mathcal{M}_n$.

Then $\lim_n g_n = g^*$, the optimal gain of $\mathcal{M}$. Moreover, any limit policy of $\{f_n^*\}_{n \geq 1}$ is gain optimal for $\mathcal{M}$.

Proof. As in the proof of Theorem 9.8, the control models with state space S_n and action sets $\mathcal{P}_n(i)$ do not satisfy Assumption 4.13 because their action sets do not coincide with those of $\mathcal{M}$. Hence, given $n \geq 1$, let us define the control model $\overline{\mathcal{M}}_n$ as follows.

The state space of $\overline{\mathcal{M}}_n$ is S_n and its action sets are $A(i)$ for each $i \in S_n$. Since $A(i) = A$ for each $i \geq 1$, to simplify the exposition we will deal with states $i \geq 1$; the arguments for $i = 0$ and $A(0)$ are similar.

Inside a "square" of the partition $\mathcal{P}_n(i)$ given by the indices

$$(\ell_1, \ell_2), \ (\ell_1 + 1, \ell_2), \ (\ell_1, \ell_2 + 1), \ (\ell_1 + 1, \ell_2 + 1), \tag{9.25}$$

define the transition rates $q_{ij}^n(a, b)$, for given $i, j \in S_n$, as a function of the form

$$\alpha + \beta a + \gamma b + \delta a b, \tag{9.26}$$

which coincides with the above defined $q_{ij}^n(a, b)$ (see (9.24)) on the vertices of the square (note that such a function exists and, besides, it is unique). An important remark is that, on such a function, the maximum and the minimum on the square are necessarily attained at one of its vertices (this result can be proved by elementary calculus). We proceed similarly to define $r_n(i, a, b)$. Clearly, the functions $q_{ij}^n(a, b)$ and $r_n(i, a, b)$ are continuous on A (this is because the functions coincide on the edges of the squares (9.25)).

It is easy to check that the so-defined $q_{ij}^n(a, b)$ are indeed transition rates. For instance, to prove that $\sum_{j \in S_n} q_{ij}^n(a, b) = 0$ we note that, inside the square (9.25), this function is a sum of functions of the form (9.26) and, hence, it is also of the form (9.26). Since its value at the four vertices is zero, we deduce that the function is zero as well inside the square.

The control models $\mathcal{M}$ and $\overline{\mathcal{M}}_n$, for $n \geq 1$, satisfy Assumption 4.13 and the condition in Theorem 4.21(b); the proof is similar to that of Theorem 9.8, and it is omitted. In addition, we can prove that $\overline{\mathcal{M}}_n \to \mathcal{M}$. Consequently, by Theorem 4.21, the optimal gains and the optimal policies of $\overline{\mathcal{M}}_n$ converge to those of $\mathcal{M}$.

Finally, as in Theorem 9.8, the policy iteration algorithm for $\mathcal{M}_n$ in Sec. 9.3 (Steps II–IV) coincides with the policy iteration algorithm for the "continuous-action" control model $\overline{\mathcal{M}}_n$. Thus, the optimal gain g_n of $\mathcal{M}_n$ equals the optimal gain g_n^* of $\overline{\mathcal{M}}_n$, and the policy f_n^* is indeed optimal for $\overline{\mathcal{M}}_n$. This completes the proof of the theorem. $\square$

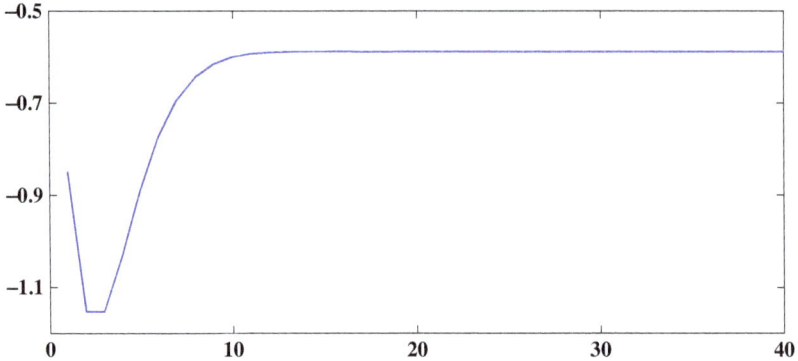

Fig. 9.7   The optimal gain g_n^*. Case $\mu > \lambda$.

For the numerical experiment, we fix the values of the parameters

$$\lambda = 2, \quad \mu = 3, \quad a_1 = 1, \quad a_2 = 5, \quad b_1 = 5, \quad b_2 = 8.$$

The distribution of the catastrophe size is a truncated geometric distribution with parameter $\gamma = 0.8$. The catastrophe rate is

$$d(i, b) = ib/10 \quad \forall\, i \geq 1, \; b \in [5, 8].$$

We consider the reward rate

$$r(i, a, b) = (2 - 0.1(a - 2)^2 - 0.05(b - 8)^2)i.$$

Finally, the constant m in the partitions $\mathcal{P}_n(i)$ of the action set is $m = 2$.

For every $1 \leq n \leq 40$ we solve the control model $\mathcal{M}_n$ by using the policy iteration procedure.

Figure 9.7 displays the optimal gain g_n^* of the truncated model $\mathcal{M}_n$ as a function of n. Figure 9.8 shows the optimal actions $f_n^*(i) = (a_n^*(i), b_n^*(i))$, for $i = 2, 5, 8$, of the control models $\mathcal{M}_n$. It is clear from Figures 9.7 and 9.8 that the optimal gains and the optimal actions quickly converge to those of the original control model $\mathcal{M}$. Note that, as a consequence of Remark 4.24 and Lemma 9.9(ii) (indeed, we have $\mu > \lambda$), we obtain $|g_n^* - g^*| = \mathrm{O}(n^{-\beta})$ for every $\beta > 0$. We derive the approximate value $g^* = -0.5883$.

As noted, the previous example satisfies $\mu > \lambda$. Next, we propose another numerical example for which this condition fails. Fix the values of the parameters $\lambda = 2$ and $\mu = 1.5$, while the values of the rest of the parameters remain the same. The usual ergodicity condition $\mu > \lambda$ does not hold, although (9.23) does because $E_{\gamma_i}[X_i] \to 5 > 2(\lambda - \mu)/Db_1$. Therefore, Lemma 9.9(iii) is verified for $\eta_0 = 5/2$.

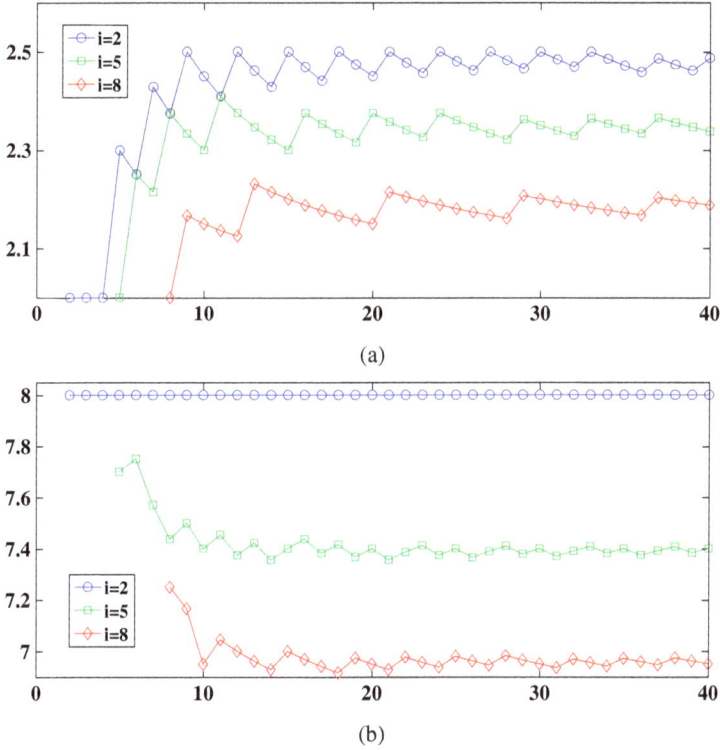

Fig. 9.8   The optimal policies $f_n^*(i)$ for $i = 2, 5, 8$. Case $\mu > \lambda$. (a) The optimal actions $a_n^*(i)$. (b) The optimal actions $b_n^*(i)$.

For every $1 \leq n \leq 40$ we solved the truncated control model $\mathcal{M}_n$. In Figures 9.9 and 9.10 we display the optimal gain g_n^* of $\mathcal{M}_n$ as a function of n, and the optimal actions $f_n^*(i) = (a_n^*(i), b_n^*(i))$, for $i = 2, 5, 8$, of the control model $\mathcal{M}_n$.

As a consequence of Remark 4.24 and Lemma 9.9, we have $|g_n^* - g^*| = O(n^{-1/3})$. We obtain the approximate value $g^* = 0.3767$. As was to be expected from the convergence rate orders, the convergence $g_n^* \to g^*$ is faster in Figure 9.7 than in Figure 9.9.

9.5.   Controlled epidemic processes

The state space in the examples studied in Secs 9.2 to 9.4 is the set of nonnegative integers. In our next example, we deal with a tridimensional state space.

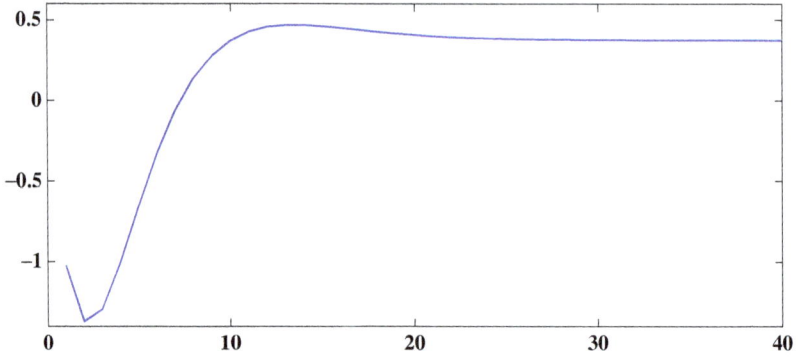

Fig. 9.9   The optimal gain g_n^*. Case $\mu < \lambda$.

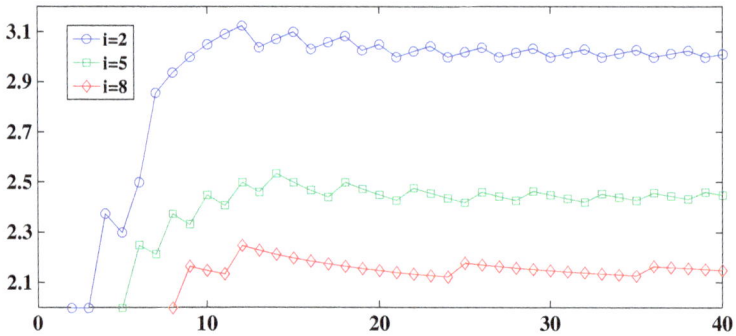

(a)

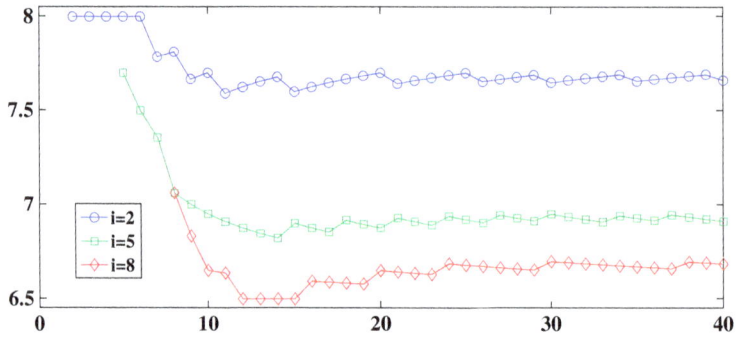

(b)

Fig. 9.10   The optimal policies $f_n^*(i)$ for $i = 2, 5, 8$. Case $\mu < \lambda$. (a) The optimal actions $a_n^*(i)$. (b) The optimal actions $b_n^*(i)$.

We consider a birth-and-death epidemic process based on the model studied in [98]; see also [105]. The population is divided into three disjoint classes:

- the infectives: that is, individuals who are infected and can transmit the disease,
- the susceptibles: individuals who are not infected by the disease, but who are exposed to contagion, and
- the immunized: individuals who cannot be infected.

The state space is

$$S = \{0, 1, 2, \ldots\} \times \{0, 1, 2, \ldots\} \times \{0, 1, 2, \ldots\},$$

where the state variable $(i, j, k) \in S$ denotes the number of infectives, susceptibles, and immunized, respectively.

We suppose that all the classes of individuals are subject to natural birth and death rates. More precisely, let $\lambda_I > 0$ and $\mu_I > 0$ denote the birth and death rates of the infectives, respectively. By $\lambda_S > 0$ and $\mu_S > 0$ we denote the respective birth and death rates of the susceptibles. Finally, let $\mu_K > 0$ be the death rate of the immunized (we suppose that all new-born individuals are either infected or susceptible).

We assume that infection of a susceptible can happen either by propagation of the disease (due to the presence of infectives), or by an external source of contagion. We will also suppose that an infected individual who recovers from the disease becomes immunized.

The action space is $A = [a_1, a_2] \times [b_1, b_2]$. Here, an action $a \in [a_1, a_2]$ stands for the level of a quarantine program, applied to the infective individuals and protecting the susceptible individuals. The action $b \in [b_1, b_2]$ is interpreted as the level of a medical treatment applied to the infected and susceptible individuals.

From the state $(i, j, k) \in S$ transitions to the following states can happen:

(i) $(i+1, j-1, k)$: a susceptible has been infected, with rate $\lambda_1(a)\sqrt{ij} + \lambda_2(b)j$. Here, $\sqrt{ij}$ corresponds to the internal propagation of the infection (this expression was proposed in [107]), $\lambda_1(a)$ is the effect of the quarantine program, and $\lambda_2(b)$ corresponds to contagion from an external source (corrected according to the level of the medical treatment).

(ii) $(i, j-1, k)$: death of a susceptible, at a rate $\mu_S j$.

(iii) $(i, j+1, k)$: birth of a susceptible, with rate $\lambda_S(i+j+k+1)$.

(iv) $(i, j - 1, k + 1)$: a susceptible is immunized, with corresponding rate $\mu(b)j$. This transition rate depends on the level of the medical treatment.

(v) $(i-1, j, k+1)$: an infected is immunized, at a rate $\mu(b)i$. This recovery rate also depends on the medical treatment applied to the population.

(vi) $(i - 1, j, k)$: death of an infected, with rate $\mu_I i$.

(vii) $(i + 1, j, k)$: birth of an infected, at a rate $\lambda_I(i + j + k + 1)$.

(viii) $(i, j, k - 1)$: death of an immunized, with corresponding rate $\mu_K k$.

We assume that the functions λ_1, λ_2, and μ are all positive and continuous. Clearly, the functions λ_1 and λ_2 are decreasing functions of a and b, respectively, while μ is an increasing function of b.

Note that the transition rates in (iii) and (vii) are proportional to the total size of the population plus one. This means that the state $(0, 0, 0) \in S$ is not absorbing and, hence, every stationary policy is irreducible.

The cost rate, which is to be minimized by the decision-maker according to a suitable optimality criterion, depends on the number of infectives, the cost of the quarantine program, and the level of the medical treatment applied to the population. More precisely, we will consider (see [98, 105]):

- A social cost $g(i)$ (per time unit) due to the presence of i infected individuals, which is an increasing function of i such that

$$0 \le g(i) \le M(i + 1) \quad \forall\, i \ge 0,$$

for some constant $M > 0$.

- A continuous and increasing function $h : [a_1, a_2] \to \mathbb{R}$, which denotes the cost rate of the quarantine program per susceptible individual.

- A cost rate function

$$\hat{h} : \{0, 1, 2, \ldots\} \times \{0, 1, 2, \ldots\} \times [b_1, b_2] \to \mathbb{R},$$

where $\hat{h}(i, j, b)$ is the cost rate of the medical treatment applied to the infectives and the susceptibles. This function is increasing in its three arguments, and it is such that $\hat{h}(i, j, \cdot)$ is continuous on $[b_1, b_2]$ for each fixed $i, j \ge 0$. Moreover, we suppose that

$$0 \le \hat{h}(i, j, b) \le M(i + j + 1) \quad \forall\, i, j \ge 0, \; b \in [b_1, b_2].$$

Hence, the decision-maker considers the cost rate

$$c(i, j, k, a, b) = g(i) + h(a)j + \hat{h}(i, j, b) \quad \forall\, (i, j, k, a, b) \in K.$$

We choose a Lyapunov function of the form $w(i, j, k) = C(i+j+k+1)$, for some constant $C \geq 1$. Observe that this function is nondecreasing on S provided that the state space S is "enumerated" in the order

$$(0,0,0), \ (1,0,0), \ (0,1,0), \ (0,0,1), \ (2,0,0), \ (1,1,0),$$
$$(1,0,1), \ (0,2,0), \ (0,1,1), \ (0,0,2), \ (3,0,0), \ (2,1,0), \dots.$$

It is clear that Assumptions 9.1(c), 9.2, and 9.3 hold. Thus, we must prove Assumptions 9.1(a)–(b).

Given $I = (i, j, k) \in S$ and $(a, b) \in A$,

$$-q_{II}(a, b) = \lambda_1(a)\sqrt{ij} + \lambda_2(b)j + \mu(b)(i + j)$$
$$+ \mu_I i + \mu_S j + \mu_K k + (\lambda_S + \lambda_I)(i + j + k + 1).$$

Taking into account that $\sqrt{ij} \leq \frac{1}{2}(i + j)$ and, besides, the continuity and compactness conditions, it is clear that we can choose $C \geq 1$ in the definition of w such that

$$q(I) = \sup_{(a,b) \in A} \{-q_{II}(a, b)\} \leq w(I) \quad \forall \ I \in S.$$

Now, we verify Assumption 9.1(a).

Lemma 9.13. *Suppose that the controlled epidemic process satisfies*

$$\lambda_I + \lambda_S < \min\{\mu_I, \mu_S, \mu_K\}.$$

Then, for every integer $r \geq 1$, there exist constants $c > 0$ and $d > 0$, and a finite set $D \subset S$ such that

$$\sum_{J \in S} q_{IJ}(a, b)w^r(J) \leq -cw^r(I) + d \cdot \mathbf{I}_D(I) \quad \forall \ (I, a, b) \in K.$$

Proof. Observe that the function w is proportional to the population size plus one. Hence, according to the value of w, and starting from the state $I = (i, j, k) \in S$, we can make the following classification of the previously enumerated transitions (i)–(viii) to the state $J \in S$.

The total population increases by one, that is, a new individual is born: transitions (iii) and (vii). The corresponding sum of the $q_{IJ}(a, b)w^r(J)$ equals

$$\sum_J q_{IJ}(a, b)w^r(J) = C^r(i + j + k + 2)^r(\lambda_I + \lambda_S)(i + j + k + 1)$$

$$= \frac{\lambda_I + \lambda_S}{C} \cdot w^{r+1}(I) + r(\lambda_I + \lambda_S) \cdot w^r(I) + R_1(I).$$

$$(9.27)$$

Here, the residual term verifies that

$$\lim_{I\to\infty} \frac{R_1(I)}{w^r(I)} = 0,$$

when $I \to \infty$ in the enumeration of the state space S.

The total population decreases by one, meaning that an individual dies: transitions (ii), (vi), and (viii). For these states J we have

$$\sum_J q_{IJ}(a,b)w^r(J) = C^r(i+j+k)^r(\mu_I i + \mu_S j + \mu_K k)$$

$$= (\mu_I i + \mu_S j + \mu_K k) \cdot w^r(I)$$
$$- rC(\mu_I i + \mu_S j + \mu_K k) \cdot w^{r-1}(I) + R_2(I), \quad (9.28)$$

where the residual term is such that

$$\lim_{I\to\infty} \frac{R_2(I)}{w^r(I)} = 0.$$

The total population remains the same, that is, an individual is immunized or a susceptible is infected: transitions (i), (iv), and (v). For such states J,

$$q_{II}(a,b)w^r(I) + \sum_J q_{IJ}(a,b)w^r(I)$$

equals

$$-\frac{\lambda_I + \lambda_S}{C} \cdot w^{r+1}(I) - (\mu_I i + \mu_S j + \mu_K k) \cdot w^r(I). \quad (9.29)$$

It should be noted that (9.27)–(9.29) do not depend on the action $(a,b) \in A$ because the actions only affect the transitions in which the total size of the population remains the same. Summing (9.27)–(9.29), we obtain that $\sum_{J\in S} q_{IJ}(a,b)w^r(J)$ equals

$$rw^r(I)\left(\lambda_I + \lambda_S - \frac{\mu_I i + \mu_S j + \mu_K k}{i+j+k+1} + \frac{R_1(I) + R_2(I)}{w^r(I)}\right).$$

Observe that, as $I \to \infty$ (in the enumeration of the state space S), the lim inf of

$$\frac{\mu_I i + \mu_S j + \mu_K k}{i+j+k+1}$$

is $\mu := \min\{\mu_I, \mu_S, \mu_K\}$. Hence, we write

$$\sum_{J\in S} q_{IJ}(a,b)w^r(J) \le rw^r(I)\left(\lambda_I + \lambda_S - \mu + \frac{C\mu}{w(I)} + \frac{R_1(I) + R_2(I)}{w^r(I)}\right).$$

Letting $c = \frac{r}{2}(\mu - (\lambda_I + \lambda_S)) > 0$ and choosing I_0 sufficiently large (in the order of S), yields

$$\sum_{J \in S} q_{IJ}(a,b) w^r(J) \leq -c w^r(I) \quad \forall\, I > I_0, \ (a,b) \in A.$$

Obviously, for sufficiently large $d > 0$ we have

$$\sum_{J \in S} q_{IJ}(a,b) w^r(J) \leq -c w^r(I) + d \quad \forall\, I \leq I_0, \ (a,b) \in A.$$

This completes the proof of the lemma. □

Consequently, for this epidemic control model, we can find optimal policies for all the optimality criteria that we have introduced in the previous chapters (recall the list given in Sec. 9.1).

The usual ergodicity condition (that is, the birth rate is smaller than the death rate) becomes, in this controlled epidemic process,

$$\lambda_I + \lambda_S < \min\{\mu_I, \mu_S, \mu_K\}.$$

The reason for this is that the population can decrease along any of the three axes i, j, k of S and, hence, the total birth rate must be smaller than each of these three death rates.

9.6. Conclusions

In this chapter we have studied four examples of CMCs with a real-world motivation: a queueing system, a controlled birth-and-death process, a controlled population system with catastrophes, and a controlled epidemic process.

It follows from our analysis of these examples that the conditions imposed, in the previous chapters, on the control models are not restrictive in practice. In particular, our sufficient condition for uniform ergodicity in Theorem 2.11 is more easily verifiable than the monotonicity conditions in Theorem 2.8 (see Lemmas 9.5 and 9.9, and Remarks 9.6 and 9.10).

This shows that the hypotheses of the Lyapunov condition type (on the Lyapunov function w or its powers) are useful and well suited to analyze practical CMCs models. Indeed, as seen in our examples, for these Lyapunov conditions to hold we only need the standard ergodicity conditions for non controlled processes (roughly speaking, that the death rate is larger than the birth rate); recall Lemmas 9.5, 9.9, and 9.13.

Moreover, as can be seen in the related literature, there were no results (combined with numerical procedures) to approximate the solutions of non-finite state space continuous-time control models. To this end, the convergence Theorems 9.7, 9.8, 9.11, and 9.12 (derived from the approximation theorems in Secs 4.3 and 4.4) turn out to be powerful and efficient tools. This can be seen from our numerical experiments in Secs 9.3 and 9.4.

Chapter 10

Zero-Sum Markov Games

10.1. Introduction

In the previous chapters, we have analyzed continuous-time control models in which a controller or decision-maker tries to maximize his/her reward according to a suitably defined optimality criterion. In this chapter we deal with a continuous-time zero-sum two-person Markov game model, that is, we assume that *two players* (also known as controllers, or decision-makers) simultaneously try to maximize and minimize, respectively, a given payoff function.

The topic of zero-sum stochastic dynamic games has been widely studied. We have already given some references in Chapter 1. Let us now give some additional ones. Discrete-time stochastic games have been analyzed by, e.g., Hernández-Lerma and Lasserre [71], Küenle [91], Jaśkiewicz and Nowak [81, 119], and Sennott [146]. For differential games, the reader is referred to the works by Ardanuy and Alcalá [7], Hamadène [60], and Ramachandran [139]. Regarding stochastic continuous-time games, semi-Markov games have been studied by Lal and Sinha [94], and Polowczuk [124]. In these models, the players choose their actions at certain discrete random times, so that they can, in fact, be reduced to discrete-time stochastic games. For semi-Markov games, the reader can also consult the papers by Jaśkiewicz [80] and Vega-Amaya [159].

Continuous-time stochastic games in which the players choose their actions *continuously* in time have been studied by Lai and Tanaka [93], Tanaka and Homma [154], and Tanaka and Wakuta [155, 156]. These references, however, deal with the bounded transition and payoff rates case.

Continuous-time Markov games with denumerable state space and unbounded transition and payoff rates (in which the players choose their actions continuously in time) have been analyzed by Guo and Hernández-Lerma for the average payoff criterion in zero-sum games [47], and for the discounted payoff criterion for nonzero-sum games [50]. Guo and Hernández-Lerma have studied, as well, the case of nonhomogeneous zero-sum continuous-time games [49], and extensions to jump Markov games on Borel spaces [51]; see also the paper by Guo and Rieder [55].

In this chapter, we place ourselves in the theoretical setting of [47, 50, 128]. Let us also mention that the hypotheses that we impose on the game model are basically the same (e.g., Lyapunov conditions, continuity and compactness assumptions) as for the control models in the previous chapters, with the obvious modifications to account for a game model.

The zero-sum two-person Markov game model is defined in Sec. 10.2, mainly following [47]. We are interested in the standard optimality criteria, such as discounted and average optimality (Secs 10.3 and 10.4), for which we base our approach on [47, 50]. Furthermore, some properties of average optimal strategies are explored in Sec. 10.5. Our conclusions are stated in Sec. 10.6.

10.2.   The zero-sum Markov game model

The game model   We deal with the following game model (compare with the control model $\mathcal{M}$ in Sec. 2.2):

$$\mathcal{GM} := \{S, A, B, K, Q, r\}.$$

The elements of $\mathcal{GM}$ are the following:

- The denumerable state space is S, which we assume to be the set of nonnegative integers, i.e., $S := \{0, 1, 2, \ldots\}$.
- The Polish spaces A and B are the action spaces for player 1 and player 2, respectively. Given $i \in S$, the Borel subsets $A(i) \subseteq A$ and $B(i) \subseteq B$ stand for the available actions for player 1 and player 2, respectively, at state i. We define the triplets of feasible state-actions as

$$K := \{(i, a, b) \in S \times A \times B : a \in A(i), b \in B(i)\}.$$

- The transition rate matrix is $Q = [q_{ij}(a, b)]$. Given $(i, a, b) \in K$ and $j \in S$, $q_{ij}(a, b)$ denotes the system's transition rate from i to j (see Sec. 2.2). In particular, we have $q_{ij}(a, b) \geq 0$ whenever $j \neq i$.
  The function $(a, b) \mapsto q_{ij}(a, b)$ is assumed to be measurable on $A(i) \times B(i)$ for each fixed $i, j \in S$. Moreover, we suppose that the transition rates are

conservative, that is,

$$\sum_{j \in S} q_{ij}(a, b) = 0 \quad \forall \, (i, a, b) \in K,$$

and *stable*, i.e.,

$$q(i) := \sup_{(a,b) \in A(i) \times B(i)} \{-q_{ii}(a, b)\} < \infty \quad \forall \, i \in S.$$

- The measurable function $r : K \to \mathbb{R}$ is the reward/cost rate function; that is, r is the reward rate function for player 1, and it is interpreted as the cost rate function for player 2.

The game is played as follows. Both players observe the state $i \in S$ of the system at time $t \geq 0$, and then they independently choose some actions $a \in A(i)$ and $b \in B(i)$. On a "small" time interval $[t, t + dt]$, player 1 receives a reward $r(i, a, b)dt$, while player 2 incurs a cost $r(i, a, b)dt$. Then the system makes a transition to the state $j \neq i$ with probability (recall the Landau notation introduced in Sec. 1.4) $q_{ij}(a, b)dt + \mathrm{o}(dt)$ and remains at state i with probability $1 + q_{ii}(a, b)dt + \mathrm{o}(dt)$.

Strategies   Let us define the set Π^1 of *randomized Markov strategies* for player 1. We say that

$$\pi^1 := \{\pi_t^1(C|i) : i \in S, \ C \in \mathbb{B}(A(i)), \ t \geq 0\}$$

is in Π^1 if, for each $t \geq 0$ and $i \in S$, $\pi_t^1(\cdot|i)$ is a probability measure on $A(i)$ and, furthermore, the function

$$t \mapsto \pi_t^1(C|i)$$

is measurable on $[0, \infty)$ for every $i \in S$ and $C \in \mathbb{B}(A(i))$. The family Π^2 of randomized Markov strategies for player 2 is defined similarly.

The interpretation of a Markov strategy is as follows. If player 1 observes the state $i \in S$ at time $t \geq 0$, then player 1 chooses his/her action in $A(i)$ according to the probability measure $\pi_t^1(\cdot|i)$.

Given $\pi^1 \in \Pi^1$ and an arbitrary $i \in S$, if $\pi_t^1(\cdot|i)$ is the same for all $t \geq 0$, then we say that π^1 is a *randomized stationary strategy*. Thus, a stationary strategy can be identified with

$$\pi^1 := \{\pi^1(C|i) : i \in S, \ C \in \mathbb{B}(A(i))\},$$

where $\pi^1(\cdot|i)$ is a probability measure on $A(i)$ for each $i \in S$. We denote the set of stationary strategies for player 1 by Π_{s}^1. The class Π_{s}^2 of stationary strategies for player 2 is given the obvious definition. Clearly, the following inclusions hold:

$$\Pi_s^1 \subseteq \Pi^1 \quad \text{and} \quad \Pi_s^2 \subseteq \Pi^2.$$

Remark 10.1. When dealing with control models, in the previous chapters, we used the terminology *policy* when referring to the actions taken by the controller or decision-maker. Here, for a game model, we will refer to the *strategies* of the players.

The state and actions processes   Given states $i, j \in S$, a pair of strategies $\pi^1 \in \Pi^1$ and $\pi^2 \in \Pi^2$, and $t \geq 0$, let

$$q_{ij}(t, \pi^1, \pi^2) := \int_{B(i)} \int_{A(i)} q_{ij}(a, b) \pi_t^1(da|i) \pi_t^2(db|i), \qquad (10.1)$$

which is well defined and finite as a consequence of the conservativeness and stability of the transition rates. In fact, the transition rates in (10.1) are also conservative and stable because

$$-q_{ii}(t, \pi^1, \pi^2) = \sum_{j \neq i} q_{ij}(t, \pi^1, \pi^2) \leq q(i)$$

for all $i \in S$, $t \geq 0$, $\pi^1 \in \Pi^1$, and $\pi^2 \in \Pi^2$. If π^1 and π^2 are stationary, (10.1) will be simply written as $q_{ij}(\pi^1, \pi^2)$.

Given a pair of randomized Markov strategies $(\pi^1, \pi^2) \in \Pi^1 \times \Pi^2$, the family of matrices $Q^{\pi^1, \pi^2}(t) = [q_{ij}(t, \pi^1, \pi^2)]_{i,j \in S}$, for $t \geq 0$, is a nonhomogeneous and conservative Q-matrix (recall Sec. 2.3). We are, thus, interested in the existence of a nonhomogeneous transition function $P_{ij}^{\pi^1, \pi^2}(s, t)$, for $i, j \in S$ and $t \geq s \geq 0$, such that

$$\lim_{t \downarrow s} \frac{P_{ij}^{\pi^1, \pi^2}(s, t) - \delta_{ij}}{t - s} = q_{ij}(s, \pi^1, \pi^2)$$

for all $s \geq 0$, and $i, j \in S$. The existence of such a transition function is given by [167, Theorem 3]. To ensure its regularity, we impose the following assumption, which, in addition, contains a bounding condition on the reward/cost function. (We recall that $w : S \to [1, \infty)$ is said to be a Lyapunov function on S if w is monotone nondecreasing and $\lim_{i \to \infty} w(i) = \infty$.)

Assumption 10.2. There exists a Lyapunov function w on S such that:

(a) For some constants $c \in \mathbb{R}$ and $d \geq 0$,

$$\sum_{j \in S} q_{ij}(a, b) w(j) \leq -cw(i) + d \quad \forall \, (i, a, b) \in K.$$

(b) There exists a constant $M > 0$ such that

$$|r(i, a, b)| \leq M w(i) \quad \forall \, (i, a, b) \in K.$$

By [44, Theorem 3.1], under Assumption 10.2(a), the transition function $\{P_{ij}^{\pi^1, \pi^2}(s, t)\}$ is regular. In particular, for every initial state $i \in S$ at time $s \geq 0$, and each pair of strategies $(\pi^1, \pi^2) \in \Pi^1 \times \Pi^2$, there exists a right-continuous regular Markov process $\{x^{\pi^1, \pi^2}(t)\}_{t \geq s}$, such that $x^{\pi^1, \pi^2}(s) = i$, with transition rates given by (10.1). We will refer to this Markov process as the *state process* (corresponding to the strategies π^1 and π^2). The state process will be simply denoted by $\{x(t)\}$ when there is no risk of confusion.

We denote by $P_{s,i}^{\pi^1, \pi^2}$ and $E_{s,i}^{\pi^1, \pi^2}$ the associated probability measure and expectation operator, respectively, when the initial state at time $s \geq 0$ is $i \in S$. If $s = 0$, we will write $P_i^{\pi^1, \pi^2}$ and $E_i^{\pi^1, \pi^2}$.

If the strategies π^1 and π^2 are stationary, then the corresponding transition function $P_{ij}^{\pi^1, \pi^2}(s, t)$ is stationary as well, and we will write

$$P_{ij}^{\pi^1, \pi^2}(t) := P_{ij}^{\pi^1, \pi^2}(s, s + t)$$

for all $i, j \in S$ and $t \geq 0$.

As for control models — see (2.13) — Assumption 10.2(a) gives, for each $i \in S$, $s \geq 0$, and $(\pi^1, \pi^2) \in \Pi^1 \times \Pi^2$,

$$E_{s,i}^{\pi^1, \pi^2}[w(x(t))] \leq e^{-c(t-s)} w(i) + \frac{d}{c}(1 - e^{-c(t-s)}) \quad \forall \, t \geq s \qquad (10.2)$$

if the constant c in Assumption 10.2(a) is $c \neq 0$, while if $c = 0$ we have

$$E_{s,i}^{\pi^1, \pi^2}[w(x(t))] \leq w(i) + d(t - s) \quad \forall \, t \geq s.$$

It is worth mentioning that the existence of the regular state process $\{x^{\pi^1, \pi^2}(t)\}$, as well as the bound (10.2), are derived exactly as in the control case studied in Chapter 2. This is because these results concern properties of nonhomogeneous Q-matrices (regardless of whether they come from a control or a game model).

Given strategies $(\pi^1, \pi^2) \in \Pi^1 \times \Pi^2$ and an initial state $i \in S$ at time $s = 0$, we have so far constructed the state process $\{x(t)\}_{t \geq 0}$. In fact, we can also construct the state-actions process. To this end, consider the product space Ω that consists of all the functions from $[0, \infty)$ to K. Namely, $\Omega := K^{[0, \infty)} = \{(x(t), a(t), b(t))\}_{t \geq 0}$, endowed with the product σ-field $\mathcal{F}$.

Proposition 10.3. *Suppose that Assumption 10.2(a) holds. Fix an initial state $i \in S$ at time $s = 0$, and a pair of randomized Markov strategies*

$\pi^1 \in \Pi^1$ and $\pi^2 \in \Pi^2$. *Then there exists a unique probability measure* $\tilde{P}_i^{\pi^1,\pi^2}$ *on* $(\Omega, \mathcal{F})$ *such that*

$$\tilde{P}_i^{\pi^1,\pi^2}\{x(t) = j,\ a(t) \in C_A,\ b(t) \in C_B\} = P_{ij}^{\pi^1,\pi^2}(0,t)\pi_t^1(C_A|j)\pi_t^2(C_B|j)$$

for all $t \geq 0$, $j \in S$, $C_A \in \mathbb{B}(A(j))$, *and* $C_B \in \mathbb{B}(B(j))$.

For a proof of Proposition 10.3, see [47, Proposition 3.1]. The expectation operator corresponding to $\tilde{P}_i^{\pi^1,\pi^2}$ will be denoted by $\tilde{E}_i^{\pi^1,\pi^2}$. Hence, we will refer to the coordinate processes $x(t)$, $a(t)$, and $b(t)$ as the state process and the action processes for players 1 and 2, respectively.

If Assumption 10.2(b) holds, given a state $i \in S$, strategies $\pi^1 \in \Pi^1$ and $\pi^2 \in \Pi^2$, and $t \geq 0$, we define

$$r(t, i, \pi^1, \pi^2) := \int_{B(i)} \int_{A(i)} r(i, a, b)\pi_t^1(da|i)\pi_t^2(db|i),$$

which satisfies a condition similar to Assumption 10.2(b): $|r(t, i, \pi^1, \pi^2)| \leq Mw(i)$. If π^1 and π^2 are stationary strategies, we will write

$$r(i, \pi^1, \pi^2) := \int_{B(i)} \int_{A(i)} r(i, a, b)\pi^1(da|i)\pi^2(db|i).$$

Suppose that i is the initial state, and that the players use the strategies $\pi^1 \in \Pi^1$ and $\pi^2 \in \Pi^2$, respectively. Then their reward/cost rate at time t is $r(x(t), a(t), b(t))$, whose expectation is, thus

$$\tilde{E}_i^{\pi^1,\pi^2}[r(x(t), a(t), b(t))] = E_i^{\pi^1,\pi^2}[r(t, x(t), \pi^1, \pi^2)] \qquad (10.3)$$

(see [47, Equation (3.10)]).

Remark 10.4. For future reference, observe that (10.3) shows that when dealing with expectations such as, e.g., $\tilde{E}_i^{\pi^1,\pi^2}[r(x(t), a(t), b(t))]$, it suffices to integrate $r(t, \cdot, \pi^1, \pi^2)$ with respect to the state process $x(t)$.

Finally, we introduce some notation. Given $i \in S$, let $\overline{A}(i)$ and $\overline{B}(i)$ respectively denote the family of probability measures on $A(i)$ and $B(i)$. Given $\phi \in \overline{A}(i)$ and $\psi \in \overline{B}(i)$, we define, for $j \in S$,

$$q_{ij}(\phi, \psi) := \int_{B(i)} \int_{A(i)} q_{ij}(a, b)\phi(da)\psi(db)$$

and

$$r(i, \phi, \psi) := \int_{B(i)} \int_{A(i)} r(i, a, b)\phi(da)\psi(db).$$

If $\pi^1 \in \Pi_s^1$ and $\pi^2 \in \Pi_s^2$ are stationary strategies, then

$$q_{ij}(\phi, \pi^2) := q_{ij}(\phi, \pi^2(\cdot|i)),$$

and similarly for $q_{ij}(\pi^1, \psi)$, $r(i, \phi, \pi^2)$, and $r(i, \pi^1, \psi)$.

Remark 10.5 (A remark on zero-sum games). In a general two-person game, each player has its own objective function, say

$$v_1(\pi^1, \pi^2) \quad \text{and} \quad v_2(\pi^1, \pi^2),$$

that he/she wishes to maximize. The game is called a *zero-sum game* if

$$v_1(\pi^1, \pi^2) + v_2(\pi^1, \pi^2) = 0.$$

In this case, letting $v(\pi^1, \pi^2) := v_1(\pi^1, \pi^2) = -v_2(\pi^1, \pi^2)$, it follows from the change of sign $v_1 = v$ and $v_2 = -v$ that the players have *opposite aims*, in the sense that player *1* maximizes $\pi^1 \mapsto v(\pi^1, \pi^2)$, while player *2* minimizes $\pi^2 \mapsto v(\pi^1, \pi^2)$.

10.3. Discount optimality

Throughout this section, we suppose that Assumption 10.2 holds. We fix a discount rate $\alpha > 0$ such that $\alpha + c > 0$, where c is the constant in Assumption 10.2(a).

Given an initial state $i \in S$, and a pair of strategies $\pi^1 \in \Pi^1$ and $\pi^2 \in \Pi^2$, we define

$$V_\alpha(i, \pi^1, \pi^2) := \tilde{E}_i^{\pi^1, \pi^2}\left[\int_0^\infty e^{-\alpha t} r(x(t), a(t), b(t)) dt\right].$$

As a consequence of Remark 10.4, we can write

$$V_\alpha(i, \pi^1, \pi^2) = E_i^{\pi^1, \pi^2}\left[\int_0^\infty e^{-\alpha t} r(t, x(t), \pi^1, \pi^2) dt\right].$$

Therefore, $V_\alpha(i, \pi^1, \pi^2)$ is the total expected discounted *reward* for player 1 (when player 2 plays the strategy π^2), and the total expected discounted *cost* for player 2 (when player 1 plays the strategy π^1). Hence (as in Remark 10.5), player 1 wishes to maximize $\pi^1 \mapsto V_\alpha(i, \pi^1, \pi^2)$, while player 2 wishes to minimize $\pi^2 \mapsto V_\alpha(i, \pi^1, \pi^2)$.

By Assumption 10.2(b) and (10.2), we have (cf. (3.7))

$$|V_\alpha(i, \pi^1, \pi^2)| \le \frac{Mw(i)}{\alpha + c} + \frac{dM}{\alpha(\alpha + c)} \quad \forall\, i \in S,\ (\pi^1, \pi^2) \in \Pi^1 \times \Pi^2. \quad (10.4)$$

In particular

$$V_\alpha(\cdot, \pi^1, \pi^2) \in \mathcal{B}_w(S),$$

where $\mathcal{B}_w(S)$ is the space introduced in Sec. 2.3.

Next, we define the lower and upper value functions of the discounted game. Given $i \in S$,

$$L_\alpha(i) := \sup_{\pi^1 \in \Pi^1} \inf_{\pi^2 \in \Pi^2} V_\alpha(i, \pi^1, \pi^2)$$

is called the *lower value* of the discounted game, while

$$U_\alpha(i) := \inf_{\pi^2 \in \Pi^2} \sup_{\pi^1 \in \Pi^1} V_\alpha(i, \pi^1, \pi^2)$$

is the *upper value* of the discounted game. As a direct consequence of (10.4), the lower and upper value functions, L_α and U_α, are in $\mathcal{B}_w(S)$.

It is easily seen that $L_\alpha(i) \leq U_\alpha(i)$ for every $i \in S$. If the lower and upper value functions coincide, that is, $L_\alpha(i) = U_\alpha(i)$ for every $i \in S$, then the function $V_\alpha(i) := L_\alpha(i) = U_\alpha(i)$ is called the *value* of the discounted game. In this case, we say that $(\pi^{*1}, \pi^{*2}) \in \Pi^1 \times \Pi^2$ is a *pair of optimal strategies* (for the α-discounted game) if

$$V_\alpha(i) = \inf_{\pi^2 \in \Pi^2} V_\alpha(i, \pi^{*1}, \pi^2) = \sup_{\pi^1 \in \Pi^1} V_\alpha(i, \pi^1, \pi^{*2}) \quad \forall\, i \in S. \tag{10.5}$$

This is equivalent to saying that the pair $(\pi^{*1}, \pi^{*2}) \in \Pi^1 \times \Pi^2$ is a *saddle point*, in the sense that

$$V_\alpha(i, \pi^1, \pi^{*2}) \leq V_\alpha(i, \pi^{*1}, \pi^{*2}) \leq V_\alpha(i, \pi^{*1}, \pi^2)$$

for every $i \in S$ and $(\pi^1, \pi^2) \in \Pi^1 \times \Pi^2$. In this case, it is easily verified that

$$V_\alpha(\cdot) = V_\alpha(\cdot, \pi^{*1}, \pi^{*2})$$

is the value of the game.

Remark 10.6. A pair of optimal strategies is sometimes called a *Nash equilibrium* or a *noncooperative equilibrium* for the zero-sum game.

Minimax vs. maximin   Let us note that the concept of a noncooperative equilibrium corresponds to "maximin" and "minimax" strategies for players 1 and 2, respectively. Indeed, if player 1 chooses the strategy $\pi^1 \in \Pi^1$, then the worst scenario for player 1 is when player 2 chooses a strategy attaining the infimum $\inf_{\pi^2 \in \Pi^2} V_\alpha(i, \pi^1, \pi^2)$. Consequently, player 1 chooses a *maximin strategy*, i.e., a strategy $\pi^{*1} \in \Pi^1$ attaining the supremum

$$\sup_{\pi^1 \in \Pi^1} \inf_{\pi^2 \in \Pi^2} V_\alpha(i, \pi^1, \pi^2) = \inf_{\pi^2 \in \Pi^2} V_\alpha(i, \pi^{*1}, \pi^2) = L_\alpha(i). \tag{10.6}$$

Similarly, player 2 uses a *minimax strategy*, and tries to find a strategy $\pi^{*2} \in \Pi^2$ attaining the infimum

$$\inf_{\pi^2 \in \Pi^2} \sup_{\pi^1 \in \Pi^1} V_\alpha(i, \pi^1, \pi^2) = \sup_{\pi^1 \in \Pi^1} V_\alpha(i, \pi^1, \pi^{*2}) = U_\alpha(i). \tag{10.7}$$

Observe now that, by (10.6) and (10.7), the definition of a noncooperative equilibrium in (10.5) is equivalent to the existence of the value $V_\alpha(i) = L_\alpha(i) = U_\alpha(i)$.

Before stating our main result on discounted games, we need to impose another assumption.

Assumption 10.7.

(a) For each $i \in S$, the action sets $A(i)$ and $B(i)$ are compact.

(b) For fixed states $i, j \in S$, the functions $r(i, a, b)$, $q_{ij}(a, b)$, and $\sum_{j \in S} q_{ij}(a, b) w(j)$ are continuous on $A(i) \times B(i)$, where the function w is as in Assumption 10.2.

(c) There exists a nonnegative function w' on S, and constants $M' > 0$, $c' \in \mathbb{R}$, and $d' \geq 0$ such that

$$q(i)w(i) \leq M'w(i) \quad \text{and} \quad \sum_{j \in S} q_{ij}(a, b)w'(j) \leq -c'w'(i) + d'$$

for all $(i, a, b) \in K$.

The conditions in Assumption 10.7(a) and (b) are the usual continuity-compactness requirements, while the condition in (c) is needed to use the Dynkin formula (compare with Assumption 2.2).

Our main result for discounted games is the following, where we use the notation $\overline{A}(i)$ and $\overline{B}(i)$ introduced after Remark 10.4.

Theorem 10.8. *Suppose that Assumptions 10.2 and 10.7 hold, and consider a discount rate $\alpha > 0$ such that $\alpha + c > 0$ (with c as in Assumption 10.2(a)). Then the following statements hold:*

(i) *There exists a unique function $u_\alpha \in \mathcal{B}_w(S)$ and a pair of stationary strategies $\pi^{*1} \in \Pi_s^1$ and $\pi^{*2} \in \Pi_s^2$ such that, for every $i \in S$,*

$$\alpha u_\alpha(i) = r(i, \pi^{*1}, \pi^{*2}) + \sum_{j \in S} q_{ij}(\pi^{*1}, \pi^{*2}) u_\alpha(j)$$

$$= \sup_{\phi \in \overline{A}(i)} \left\{ r(i, \phi, \pi^{*2}) + \sum_{j \in S} q_{ij}(\phi, \pi^{*2}) u_\alpha(j) \right\}$$

$$= \inf_{\psi \in \overline{B}(i)} \left\{ r(i, \pi^{*1}, \psi) + \sum_{j \in S} q_{ij}(\pi^{*1}, \psi) u_\alpha(j) \right\}.$$

(ii) *The function u_α in (i) is the value function of the discounted game, and*

$$(\pi^{*1}, \pi^{*2}) \in \Pi_s^1 \times \Pi_s^2$$

is a pair of optimal strategies.

For a proof of statement (i), the reader is referred to Theorem 5.1 in [50]. Once (i) is proved, part (ii) directly follows from the usual "verification" results; see, e.g., Lemma 3.5.

We introduce some terminology. The optimality equations in Theorem 10.8(i) are referred to as the *discounted payoff optimality equations* or the *Shapley equations* in the discounted case. (Actually, Shapley [150] introduced these equations for the average payoff case only; see (10.9) and (10.11) below.) For some continuous-time games, they are also known as the *Bellman–Isaacs* equations or the *Hamilton–Jacobi–Bellman–Isaacs* equations.

10.4.   Average optimality

To study the average payoff optimality criterion, we must strengthen Assumption 10.2. Concerning Assumption 10.9(b) below, we say that a pair of stationary strategies $(\pi^1, \pi^2) \in \Pi_s^1 \times \Pi_s^2$ is *irreducible* if the corresponding state process is irreducible; that is, for every two distinct states $i, j \in S$ there exists $t > 0$ such that $P_{ij}^{\pi^1, \pi^2}(t) > 0$.

Assumption 10.9. Part (a) of Assumption 10.2 is replaced with:

(a) For some constants $c > 0$ and $d \geq 0$, and a finite set $D \subset S$,

$$\sum_{j \in S} q_{ij}(a, b)w(j) \leq -cw(i) + d \cdot \mathbf{I}_D(i) \quad \forall \, (i, a, b) \in K.$$

Moreover:

(b) Each pair of stationary strategies $(\pi^1, \pi^2) \in \Pi_s^1 \times \Pi_s^2$ is irreducible.

Given a pair of arbitrary strategies $(\pi^1, \pi^2) \in \Pi^1 \times \Pi^2$, we define the total expected payoff over the time interval $[0, T]$, when the initial state is $i \in S$, as

$$J_T(i, \pi^1, \pi^2) := \tilde{E}_i^{\pi^1, \pi^2} \left[\int_0^T r(x(t), a(t), b(t))dt \right]$$

$$= E_i^{\pi^1, \pi^2} \left[\int_0^T r(t, x(t), \pi^1, \pi^2)dt \right],$$

where the last equality is due to (10.3). The corresponding long-run *expected average payoff* (or *average payoff*, for short) is defined as

$$J(i, \pi^1, \pi^2) := \limsup_{T \to \infty} \frac{1}{T} J_T(i, \pi^1, \pi^2).$$

Recalling (10.2), Assumption 10.9(a) yields

$$|J(i, \pi^1, \pi^2)| \leq \frac{dM}{c} \quad \forall\, i \in S,\ (\pi^1, \pi^2) \in \Pi^1 \times \Pi^2.$$

In the average payoff game, player 1 tries to maximize his/her average reward $\pi^1 \mapsto J(i, \pi^1, \pi^2)$, while player 2 tries to minimize his/her average cost $\pi^2 \mapsto J(i, \pi^1, \pi^2)$.

As for the discounted case, the *lower value* of the average payoff game is defined as

$$L(i) := \sup_{\pi^1 \in \Pi^1} \inf_{\pi^2 \in \Pi^2} J(i, \pi^1, \pi^2) \quad \forall\, i \in S,$$

and the *upper value* is

$$U(i) := \inf_{\pi^2 \in \Pi^2} \sup_{\pi^1 \in \Pi^1} J(i, \pi^1, \pi^2) \quad \forall\, i \in S.$$

Clearly, $L(i) \leq U(i)$ for all $i \in S$, and when

$$L(i) = U(i) =: V(i) \quad \forall\, i \in S,$$

we say that $V(i)$ is the *value of the average payoff game* when the initial state is $i \in S$.

If the game has a value, then $(\pi^{*1}, \pi^{*2}) \in \Pi^1 \times \Pi^2$ is said to be a *pair of average optimal strategies* if

$$V(i) = \inf_{\pi^2 \in \Pi^2} J(i, \pi^{*1}, \pi^2) = \sup_{\pi^1 \in \Pi^1} J(i, \pi^1, \pi^{*2}) \quad \forall\, i \in S.$$

Equivalently, (π^{*1}, π^{*2}) is a *saddle point* of the average payoff function, that is,

$$J(i, \pi^1, \pi^{*2}) \leq J(i, \pi^{*1}, \pi^{*2}) \leq J(i, \pi^{*1}, \pi^2) \quad \forall\, i \in S,\ (\pi^1, \pi^2) \in \Pi^1 \times \Pi^2,$$

and, in this case, $V(i) = J(i, \pi^{*1}, \pi^{*2})$ for every $i \in S$. Remark 10.6 as well as the comments on maximin and minimax strategies are also valid for average payoff games.

As a consequence of Theorem 2.5, if Assumption 10.9 holds then, for every pair (π^1, π^2) of stationary strategies, the state process $\{x^{\pi^1, \pi^2}(t)\}_{t \geq 0}$ has a unique invariant probability measure, denoted by μ_{π^1, π^2}. This invariant probability measure verifies (by irreducibility in Assumption 10.9(b)) that

$$\mu_{\pi^1, \pi^2}(i) > 0 \quad \forall\, i \in S,$$

and also that

$$\mu_{\pi^1, \pi^2}(w) := \sum_{i \in S} w(i)\mu_{\pi^1, \pi^2}(i) < \infty.$$

In the following, we will also use the notation (similar to (2.21))

$$\mu_{\pi^1,\pi^2}(u) := \sum_{i \in S} u(i)\mu_{\pi^1,\pi^2}(i)$$

for $u \in \mathcal{B}_w(S)$.

Finally, we introduce an exponential ergodicity assumption in which w is the function in Assumption 10.9(a).

Assumption 10.10. The Markov game model is w-exponentially ergodic uniformly on $\Pi_s^1 \times \Pi_s^2$, that is, there exist constants $\delta > 0$ and $R > 0$ such that

$$|E_i^{\pi^1,\pi^2}[u(x(t))] - \mu_{\pi^1,\pi^2}(u)| \leq Re^{-\delta t}||u||_w w(i)$$

for every $(\pi^1, \pi^2) \in \Pi_s^1 \times \Pi_s^2$, $u \in \mathcal{B}_w(S)$, $i \in S$, and $t \geq 0$.

Sufficient conditions for Assumption 10.10 are obtained by reformulating the corresponding sufficient conditions for control models. In particular, Assumption 10.10 holds if the monotonicity conditions in Theorem 2.8 and Remark 2.9 hold, or if the strengthened Lyapunov conditions in Theorem 2.11 and Remark 2.12 are satisfied.

In the rest of this chapter, we assume that Assumptions 10.2, 10.7, 10.9, and 10.10 hold.

The bias of stationary strategies We fix a pair of stationary strategies $(\pi^1, \pi^2) \in \Pi_s^1 \times \Pi_s^2$. Its average payoff $J(i, \pi^1, \pi^2) \equiv g(\pi^1, \pi^2)$ is constant and it verifies that (see (3.25) and Remark 3.12)

$$g(\pi^1, \pi^2) = \lim_{T \to \infty} \frac{1}{T} J_T(i, \pi^1, \pi^2) = \mu_{\pi^1,\pi^2}(r(\cdot, \pi^1, \pi^2)) \quad \forall i \in S.$$

Moreover, as in (3.26), we can also define the *bias* of the pair (π^1, π^2) as

$$h(i, \pi^1, \pi^2) := \int_0^\infty [E_i^{\pi^1,\pi^2}[r(x(t), \pi^1, \pi^2)] - g(\pi^1, \pi^2)]dt \quad \forall i \in S.$$

By Assumptions 10.2(b) and 10.10, the bias $h(\cdot, \pi^1, \pi^2)$ is in $\mathcal{B}_w(S)$ and it satisfies

$$\sup_{(\pi^1,\pi^2)\in\Pi_s^1\times\Pi_s^2} ||h(\cdot, \pi^1, \pi^2)||_w \leq RM/\delta,$$

where M is the constant in Assumption 10.2(b), and R and δ are as in Assumption 10.10.

Under our standing assumptions, the average payoff and the bias of (π^1, π^2) can be characterized by means of the corresponding Poisson equation, as in Proposition 3.14 and Remark 3.15.

Also, as in (5.2), we can interpret the bias of (π^1, π^2) as the "ordinate at the origin" of the total expected payoff $J_T(i, \pi^1, \pi^2)$ as $T \to \infty$. More precisely, for each $i \in S$, $(\pi^1, \pi^2) \in \Pi_s^1 \times \Pi_s^2$, and $T \geq 0$, we have

$$J_T(i, \pi^1, \pi^2) = g(\pi^1, \pi^2)T + h(i, \pi^1, \pi^2) + U(i, \pi^1, \pi^2, T), \qquad (10.8)$$

with

$$\sup_{(\pi^1, \pi^2) \in \Pi_s^1 \times \Pi_s^2} ||U(\cdot, \pi^1, \pi^2, T)||_w \leq R^2 M e^{-\delta T}/\delta.$$

The average payoff optimality equations We say that a constant $g^* \in \mathbb{R}$, a function $h \in \mathcal{B}_w(S)$, and a pair of stationary strategies $(\pi^{*1}, \pi^{*2}) \in \Pi_s^1 \times \Pi_s^2$ form a solution of the *average payoff optimality equations* (also referred to as the *Shapley equations* in the average payoff case) if

$$g^* = r(i, \pi^{*1}, \pi^{*2}) + \sum_{j \in S} q_{ij}(\pi^{*1}, \pi^{*2})h(j) \qquad (10.9)$$

$$= \sup_{\phi \in \overline{A}(i)} \left\{ r(i, \phi, \pi^{*2}) + \sum_{j \in S} q_{ij}(\phi, \pi^{*2})h(j) \right\} \qquad (10.10)$$

$$= \inf_{\psi \in \overline{B}(i)} \left\{ r(i, \pi^{*1}, \psi) + \sum_{j \in S} q_{ij}(\pi^{*1}, \psi)h(j) \right\} \qquad (10.11)$$

for each $i \in S$. If the average optimality equations (10.9)–(10.11) hold, then we say that the pair of strategies $(\pi^{*1}, \pi^{*2}) \in \Pi_s^1 \times \Pi_s^2$ is *canonical* (compare with the definition in Remark 3.22(a) of a canonical policy for a control model).

Our next theorem establishes that the average payoff game has a value, and it characterizes the optimal strategies as solutions of the average optimality equations.

Theorem 10.11. *If Assumptions* 10.2, 10.7, 10.9, *and* 10.10 *are satisfied, then the following statements hold:*

(i) *There exists a solution* $(g^*, h, \pi^{*1}, \pi^{*2})$ *to the average payoff optimality equations* (10.9)–(10.11).

(ii) *The constant* g^* *in* (10.9) *is the value of the Markov game, that is,* $V(\cdot) \equiv g^*$ *(in particular,* g^* *is unique). The function* h *in* (10.9)–(10.11) *is unique up to additive constants.*

(iii) *A pair of stationary strategies is average optimal if and only if it is canonical. Hence, the pair* $(\pi^{*1}, \pi^{*2}) \in \Pi_s^1 \times \Pi_s^2$ *in* (10.9)–(10.11) *is a pair of average optimal strategies.*

Proof.   *Proof of* (i). The proof of statement (i) uses the same technique as that of Theorem 3.20(i). More precisely, starting from Theorem 10.8, we use the vanishing discount method to derive the existence of solutions to the average optimality equations. For details, see Theorem 5.1 in [47].

   Proof of (ii). To prove part (ii), we use the same arguments as in the proof of Theorem 3.20 in order to deduce that

$$J(i, \pi^1, \pi^{*2}) \le J(i, \pi^{*1}, \pi^{*2}) \equiv g(\pi^{*1}, \pi^{*2}) =: g^* \le J(i, \pi^{*1}, \pi^2)$$

for each $i \in S$ and $(\pi^1, \pi^2) \in \Pi^1 \times \Pi^2$. This shows that g^* is the value of the game and, moreover, that the pair of strategies $(\pi^{*1}, \pi^{*2}) \in \Pi_s^1 \times \Pi_s^2$ is optimal.

   Let us now prove the second statement in (ii), that is, we will show that h is unique up to additive constants. To this end, suppose that

$$(g^*, h, \pi^{*1}, \pi^{*2}) \quad \text{and} \quad (g^*, \bar{h}, \bar{\pi}^1, \bar{\pi}^2)$$

are two solutions of the average optimality equations. For fixed $i \in S$, we have

$$g^* = \sup_{\pi^1 \in \Pi^1} J(i, \pi^1, \pi^{*2}) \ge J(i, \bar{\pi}^1, \pi^{*2}).$$

On the other hand,

$$g^* = \inf_{\pi^2 \in \Pi^2} J(i, \bar{\pi}^1, \pi^2) \le J(i, \bar{\pi}^1, \pi^{*2}),$$

and it follows that

$$J(i, \bar{\pi}^1, \pi^{*2}) = g(\bar{\pi}^1, \pi^{*2}) = g^*. \tag{10.12}$$

Also, we derive from (10.10) that

$$g^* \ge r(i, \bar{\pi}^1, \pi^{*2}) + \sum_{j \in S} q_{ij}(\bar{\pi}^1, \pi^{*2}) h(j) \quad \forall\, i \in S.$$

It follows from Corollary 3.16 and (10.12) that the equality necessarily holds for every $i \in S$, i.e.,

$$g^* = r(i, \bar{\pi}^1, \pi^{*2}) + \sum_{j \in S} q_{ij}(\bar{\pi}^1, \pi^{*2}) h(j) \quad \forall\, i \in S. \tag{10.13}$$

Now, from (10.11) for the solution $(g^*, \bar{h}, \bar{\pi}^1, \bar{\pi}^2)$, we get

$$g^* \le r(i, \bar{\pi}^1, \pi^{*2}) + \sum_{j \in S} q_{ij}(\bar{\pi}^1, \pi^{*2}) \bar{h}(j) \quad \forall\, i \in S$$

and, once again from Corollary 3.16 and (10.12), we obtain

$$g^* = r(i, \bar{\pi}^1, \pi^{*2}) + \sum_{j \in S} q_{ij}(\bar{\pi}^1, \pi^{*2})\bar{h}(j) \quad \forall\, i \in S. \tag{10.14}$$

It follows from (10.13) and (10.14) that (g^*, h) and $(g^*, \bar{h})$ are two solutions of the Poisson equation for $(\bar{\pi}^1, \pi^{*2})$ and, therefore, by Proposition 3.14 and Remark 3.15, the functions h and $\bar{h}$ differ by a constant. This completes the proof of statement (ii).

Proof of (iii). It has already been proved that if $(g^*, h, \pi^{*1}, \pi^{*2})$ is a solution of the average optimality equations, then (π^{*1}, π^{*2}) is a pair of average optimal strategies.

Conversely, let $(\pi^{*1}, \pi^{*2}) \in \Pi_s^1 \times \Pi_s^2$ be a pair of average optimal strategies. Then, the value of the game equals its average payoff $g^* = g(\pi^{*1}, \pi^{*2})$, and let $h \in \mathcal{B}_w(S)$ be the bias of (π^{*1}, π^{*2}). We are going to show that $(g^*, h, \pi^{*1}, \pi^{*2})$ is a solution of the average optimality equations.

By the Poisson equation, (10.9) holds. On the other hand, for all $i \in S$, we have

$$g^* \leq \sup_{\phi \in \overline{A}(i)} \left\{ r(i, \phi, \pi^{*2}) + \sum_{j \in S} q_{ij}(\phi, \pi^{*2})h(j) \right\}. \tag{10.15}$$

Using a continuity argument [47, Lemma 7.2], we deduce that the above suprema are indeed attained. Hence, if the inequality in (10.15) is strict for some $i_0 \in S$, then there exists $\pi^1 \in \Pi_s^1$ such that

$$g^* \leq r(i, \pi^1, \pi^{*2}) + \sum_{j \in S} q_{ij}(\pi^1, \pi^{*2})h(j) \quad \forall\, i \in S,$$

with strict inequality at $i = i_0$. By Corollary 3.16, we deduce that

$$g(\pi^{*1}, \pi^{*2}) = g^* < g(\pi^1, \pi^{*2}),$$

which contradicts the fact that (π^{*1}, π^{*2}) is optimal. Hence, (10.15) holds with equality for each $i \in S$, and so (10.10) is satisfied.

A similar argument shows that (10.11) holds. The proof of the theorem is now complete. $\qquad \square$

Theorem 10.11(iii) establishes that the set of canonical strategies coincides with the class of average optimal strategies. As is well known, however, such a result does not necessarily hold in a general setting (recall Remark 3.22(b)). Under our irreducibility condition, this result is indeed true (in particular, Corollary 3.16 is a crucial result to prove that both classes coincide).

As a direct consequence of Theorem 10.11, g^* and h verify the following form of the Shapley equation:

$$g^* = \sup_{\phi \in \overline{A}(i)} \inf_{\psi \in \overline{B}(i)} \left\{ r(i, \phi, \psi) + \sum_{j \in S} q_{ij}(\phi, \psi) h(j) \right\} \qquad (10.16)$$

$$= \inf_{\psi \in \overline{B}(i)} \sup_{\phi \in \overline{A}(i)} \left\{ r(i, \phi, \psi) + \sum_{j \in S} q_{ij}(\phi, \psi) h(j) \right\}$$

for every $i \in S$.

10.5.   The family of average optimal strategies

We conclude our analysis on average games by exploring some properties of the family of average optimal stationary strategies.

Assumptions 10.2, 10.7, 10.9, and 10.10 are assumed to hold throughout this section.

Let $g^* \in \mathbb{R}$ and $h \in \mathcal{B}_w(S)$ be a solution of the average optimality equations (10.9)–(10.11). We know that g^* is unique and that h is unique up to additive constants. Given $i \in S$, let $\overline{A}_0(i) \subseteq \overline{A}(i)$ be the set of probability measures $\phi \in \overline{A}(i)$ such that

$$g^* = \inf_{\psi \in \overline{B}(i)} \left\{ r(i, \phi, \psi) + \sum_{j \in S} q_{ij}(\phi, \psi) h(j) \right\}.$$

Similarly, let $\overline{B}_0(i) \subseteq \overline{B}(i)$ be the family of $\psi \in \overline{B}(i)$ such that

$$g^* = \sup_{\phi \in \overline{A}(i)} \left\{ r(i, \phi, \psi) + \sum_{j \in S} q_{ij}(\phi, \psi) h(j) \right\}.$$

Observe that the sets $\overline{A}_0(i)$ and $\overline{B}_0(i)$ do not depend on the particular solution h of (10.9)–(10.11) because h is unique up to additive constants (see Theorem 10.11(ii)).

Proposition 10.12. *Suppose that Assumptions* 10.2, 10.7, 10.9, *and* 10.10 *are satisfied. Then:*

(i) *For each $i \in S$, the sets $\overline{A}_0(i)$ and $\overline{B}_0(i)$ are nonempty convex compact Borel spaces, with the topology of weak convergence.*

(ii) *A pair of stationary strategies $(\pi^1, \pi^2) \in \Pi_s^1 \times \Pi_s^2$ is average optimal if and only if $\pi^1(\cdot|i) \in \overline{A}_0(i)$ and $\pi^2(\cdot|i) \in \overline{B}_0(i)$ for every $i \in S$.*

Proof. Throughout this proof we assume that $h \in \mathcal{B}_w(S)$ is taken from a solution of the average optimality equations (10.9)–(10.11).

Proof of (i). Let (π^{*1}, π^{*2}) be a pair of average optimal stationary strategies, whose existence is ensured by Theorem 10.11. For each $i \in S$, let $\phi_0 := \pi^{*1}(\cdot|i)$ and $\psi_0 := \pi^{*2}(\cdot|i)$. Then

$$\phi_0 \in \overline{A}_0(i) \quad \text{and} \quad \psi_0 \in \overline{B}_0(i),$$

so that the sets $\overline{A}_0(i)$ and $\overline{B}_0(i)$ are nonempty.

In $\overline{A}(i)$ and $\overline{B}(i)$ we consider the topology of weak convergence (recall the definition given in Sec. 1.4). It follows from Prohorov's theorem [38, Appendix A.5] that $\overline{A}(i)$ and $\overline{B}(i)$ are compact Borel spaces with this topology, because so are $A(i)$ and $B(i)$ (by Assumption 10.7(a)). Therefore, to prove the compactness statement, it suffices to show that the sets $\overline{A}_0(i)$ and $\overline{B}_0(i)$ are closed. To this end, since the function

$$(\phi, \psi) \mapsto r(i, \phi, \psi) + \sum_{j \in S} q_{ij}(\phi, \psi) h(j)$$

is continuous on $\overline{A}(i) \times \overline{B}(i)$ for each $i \in S$ (see [47, Lemma 7.2]), it follows from a standard calculus result that $\overline{A}_0(i)$ and $\overline{B}_0(i)$ are closed.

Finally, it remains to prove that $\overline{A}_0(i)$ and $\overline{B}_0(i)$ are convex. For each fixed $i \in S$ and $\psi \in \overline{B}(i)$, the function

$$\phi \mapsto H(\phi) := r(i, \phi, \psi) + \sum_{j \in S} q_{ij}(\phi, \psi) h(j)$$

is linear on $\overline{A}(i)$ because

$$H(\lambda \phi^1 + (1 - \lambda)\phi^2) = \lambda H(\phi^1) + (1 - \lambda)H(\phi^2)$$

for every $\phi^1, \phi^2 \in \overline{A}(i)$ and $0 \leq \lambda \leq 1$. Therefore,

$$\phi \mapsto \inf_{\psi \in \overline{B}(i)} \left\{ r(i, \phi, \psi) + \sum_{j \in S} q_{ij}(\phi, \psi) h(j) \right\}$$

is concave because it is the infimum of a family of linear functions. Besides, by (10.16),

$$g^* = \sup_{\phi \in \overline{A}(i)} \inf_{\psi \in \overline{B}(i)} \left\{ r(i, \phi, \psi) + \sum_{j \in S} q_{ij}(\phi, \psi) h(j) \right\}.$$

Thus $\overline{A}_0(i)$ is the set of maxima of a concave function, and so, it is convex. A similar argument proves that $\overline{B}_0(i)$ is convex for every $i \in S$.

Proof of (*ii*). If the strategies $\pi^1 \in \Pi_s^1$ and $\pi^2 \in \Pi_s^2$ are such that $\pi^1(\cdot|i) \in \overline{A}_0(i)$ and $\pi^2(\cdot|i) \in \overline{B}_0(i)$ for every $i \in S$, then they are canonical; thus, they are average optimal.

Conversely, if the pair $(\pi^1, \pi^2) \in \Pi_s^1 \times \Pi_s^2$ is average optimal and h^* is the bias of (π^1, π^2), then (g^*, h^*, π^1, π^2) is a solution of the average optimality equations; see the proof of Theorem 10.11(iii). It follows that h^* and h differ by a constant, and so $\pi^1(\cdot|i)$ is in $\overline{A}_0(i)$ and $\pi^2(\cdot|i)$ is in $\overline{B}_0(i)$ for every $i \in S$. □

An important consequence of Proposition 10.12(ii) is that the set of average optimal stationary strategies is a *rectangle* in $\Pi_s^1 \times \Pi_s^2$, which will be denoted by $\Pi_s^{*1} \times \Pi_s^{*2}$. That is, (π^{*1}, π^{*2}) is in $\Pi_s^{*1} \times \Pi_s^{*2}$ if and only if $\pi^{*1}(\cdot|i) \in \overline{A}_0(i)$ and $\pi^{*2}(\cdot|i) \in \overline{B}_0(i)$ for every $i \in S$.

10.6. Conclusions

In this chapter we have analyzed continuous-time two-person zero-sum Markov games under the discounted payoff and the average payoff optimality criteria. We reach results that are similar to the control model counterpart, in particular, concerning the existence of optimal strategies and the corresponding optimality equations.

It should be noted that when dealing with stochastic games one has to consider *randomized* stationary strategies because, in general, there might not exist optimal strategies in the class of deterministic stationary strategies. For control models, however, we noted that, in general, we can find optimal strategies in $\mathbb{F}$, except for constrained control problems; see Chapter 8.

The characterization of average optimal strategies in Sec. 10.5 is very interesting because it shows that the set of optimal stationary strategies is a rectangle. This means that, when we are restricted to the class of average payoff optimal strategies, we are faced again with a two-person stochastic game, in which the actions or strategies of each player are well defined. In particular, this means that the game model for the bias optimality criterion, which will be studied in Chapter 11, is also well defined.

Chapter 11

Bias and Overtaking Equilibria for Markov Games

11.1. Introduction

In this chapter, we consider the game model described in Chapter 10. As a refinement of the average payoff optimality criterion, we introduce (as we did for control models in Chapter 5) the notion of bias and overtaking equilibria.

Overtaking equilibria for games were introduced by Rubinstein [143] for repeated games, and by Carlson, Haurie, and Leizarowitz [20] for a class of differential games. For discrete-time zero-sum Markov games, overtaking optimality has been studied by Nowak [117, 118]. Interestingly, in [117], the notion of bias optimality for zero-sum Markov games is implicitly introduced. Bias and overtaking equilibria for continuous-time denumerable state space zero-sum Markov games were studied by Prieto-Rumeau and Hernández-Lerma [128].

We noted in Chapter 10 that the results on discounted and average payoff optimality for Markov games were somehow similar to the corresponding results for control models (in the sense that the optimal value function is characterized by a dynamic programming optimality equation, and that optimal policies or strategies exist). As we shall see here, this is no longer true for bias and overtaking optimality. This exhibits a (perhaps unexpected) difference between control and game models, which shows that the above mentioned parallelism cannot be carried out for some of the advanced optimality criteria. More insight on this issue is given in the conclusions in Sec. 11.5.

Our plan for this chapter is as follows. Based on [128], we will study in Sec. 11.2 the bias optimality criterion for Markov games. Overtaking equilibria are discussed in Sec. 11.3. Finally, in Sec. 11.4, we propose a counterexample for bias and overtaking optimality, which shows that there might not exist overtaking optimal strategies (as opposed to control models, for which there indeed exist overtaking optimal policies). The counterexample is taken from [128]. We state our conclusions in Sec. 11.5.

11.2. Bias optimality

We consider the game model $\mathcal{GM}$ described in Sec. 10.2. We are interested in refinements of the average payoff optimality criterion. Hence, we suppose that the corresponding hypotheses, Assumptions 10.2, 10.7, 10.9, and 10.10, are satisfied. In what follows, we will use the definition of the bias of a pair of stationary strategies given in Sec. 10.4. Also, we will restrict ourselves to the class of stationary strategies $\Pi_s^1 \times \Pi_s^2$.

Suppose that the two players use the average payoff criterion, and that they choose a pair of optimal strategies $(\pi^1, \pi^2) \in \Pi_s^{*1} \times \Pi_s^{*2}$. Hence, by the interpretation of the bias $h(\cdot, \pi^1, \pi^2)$ in (10.8), the corresponding total expected payoff over $[0, T]$ is (using the notation O in Sec. 1.4)

$$J_T(i, \pi^1, \pi^2) = g^* T + h(i, \pi^1, \pi^2) + \mathrm{O}(e^{-\delta T}), \qquad (11.1)$$

as $T \to \infty$, where $i \in S$ is the initial state, and g^* is the value of the game.

Therefore, if player 1 wants to maximize his/her gain and, *in addition*, the asymptotic growth of his/her finite horizon total expected returns, then player 1 will choose an average payoff optimal strategy $\pi^1 \in \Pi_s^{*1}$ with the largest bias. Similarly, player 2 will select an average optimal strategy $\pi^2 \in \Pi_s^{*2}$ with the smallest bias. Hence, the players will consider a zero-sum game in which the payoff function is given by the bias. This leads to the following definition of a *bias equilibrium*.

Definition 11.1. We say that a pair of average optimal stationary strategies $(\pi^{*1}, \pi^{*2}) \in \Pi_s^{*1} \times \Pi_s^{*2}$ is *bias optimal* if it is a saddle point of the bias function, that is,

$$h(i, \pi^1, \pi^{*2}) \leq h(i, \pi^{*1}, \pi^{*2}) \leq h(i, \pi^{*1}, \pi^2)$$

for every $i \in S$ and every $(\pi^1, \pi^2) \in \Pi_s^{*1} \times \Pi_s^{*2}$. In this case, $H^*(i) := h(i, \pi^{*1}, \pi^{*2})$ is called the *optimal bias function*.

In what follows, we will prove the existence of bias optimal strategies, and we will show how to derive them by means of a pair of nested average optimality equations.

Let $g^* \in \mathbb{R}$ be the value of the average payoff game, and fix a function $h \in \mathcal{B}_w(S)$ that is a solution of the average optimality equations (10.9)–(10.11). Then the bias $h(i, \pi^1, \pi^2)$ of an average optimal pair of strategies $(\pi^1, \pi^2) \in \Pi_s^{*1} \times \Pi_s^{*2}$ and h differ by a constant (recall Theorem 10.11, Remark 3.15, and Corollary 3.23). Hence, for some $z \in \mathbb{R}$,

$$h(i, \pi^1, \pi^2) = h(i) + z \quad \forall i \in S. \tag{11.2}$$

Recalling that $\mu_{\pi^1, \pi^2}(h(\cdot, \pi^1, \pi^2)) = 0$ (see Proposition 3.14 and Remark 3.15), and taking the μ_{π^1, π^2}-expectation in (11.2), yields

$$z = \mu_{\pi^1, \pi^2}(-h). \tag{11.3}$$

Therefore, player 1 (respectively, player 2) will try to maximize (respectively, minimize) the value $\mu_{\pi^1, \pi^2}(-h)$ or, equivalently, the average payoff of the function $-h(i)$, for $i \in S$. This yields the following interpretation: to find bias optimal strategies, the two players must consider a zero-sum average payoff game with reward/cost rate function $-h$, for which the set of feasible strategies is $\Pi_s^{*1} \times \Pi_s^{*2}$.

To formalize this idea, first of all we give the definition of the bias optimality equations. We say that $g \in \mathbb{R}$, $h^0, h^1 \in \mathcal{B}_w(S)$, and $(\pi^1, \pi^2) \in \Pi_s^1 \times \Pi_s^2$ verify the *bias optimality equations* if, for every $i \in S$,

$$g = r(i, \pi^1, \pi^2) + \sum_{j \in S} q_{ij}(\pi^1, \pi^2) h^0(j) \tag{11.4}$$

$$= \sup_{\phi \in \overline{A}(i)} \left\{ r(i, \phi, \pi^2) + \sum_{j \in S} q_{ij}(\phi, \pi^2) h^0(j) \right\} \tag{11.5}$$

$$= \inf_{\psi \in \overline{B}(i)} \left\{ r(i, \pi^1, \psi) + \sum_{j \in S} q_{ij}(\pi^1, \psi) h^0(j) \right\}, \tag{11.6}$$

and, moreover,

$$h^0(i) = \sum_{j \in S} q_{ij}(\pi^1, \pi^2) h^1(j) \tag{11.7}$$

$$= \sup_{\phi \in \overline{A}_0(i)} \left\{ \sum_{j \in S} q_{ij}(\phi, \pi^2) h^1(j) \right\} \tag{11.8}$$

$$= \inf_{\psi \in \overline{B}_0(i)} \left\{ \sum_{j \in S} q_{ij}(\pi^1, \psi) h^1(j) \right\}, \tag{11.9}$$

with $\overline{A}_0(i)$ and $\overline{B}_0(i)$ as in Sec. 10.5.

Note that the equations (11.4)–(11.6) correspond to the average optimality equations (10.9)–(10.11). Therefore, g equals g^*, the value of the average payoff game. Also note that, as a consequence of Theorem 10.11, the strategies $(\pi^1, \pi^2) \in \Pi_s^1 \times \Pi_s^2$ satisfy the equations (11.4)–(11.6) if and only if they are average optimal or, equivalently, (π^1, π^2) is in $\Pi_s^{*1} \times \Pi_s^{*2}$.

Theorem 11.2. *Suppose that Assumptions* 10.2, 10.7, 10.9, *and* 10.10 *are satisfied. Then the following statements hold*:

(i) *There exists a solution* $(g, h^0, h^1, \pi^1, \pi^2)$ *to the bias optimality equations* (11.4)–(11.9). *Besides,* $g = g^*$, *the value of the average game, and* $h^0 = H^*$, *the optimal bias function.*

(ii) *A pair of stationary strategies* $(\pi^1, \pi^2) \in \Pi_s^1 \times \Pi_s^2$ *is bias optimal if and only if* π^1 *and* π^2 *verify the bias optimality equations* (11.4)–(11.9).

Proof. First of all, let $(g^*, h) \in \mathbb{R} \times \mathcal{B}_w(S)$ be a solution of the average optimality equations (11.4)–(11.6).

Now, consider the zero-sum game model $\mathcal{G}\mathcal{M}_0$ with state space S, action sets $\overline{A}_0(i)$ and $\overline{B}_0(i)$, for $i \in S$, transition rates given by $q_{ij}(\phi, \psi)$ for

$$(\phi, \psi) \in \overline{A}_0(i) \times \overline{B}_0(i),$$

and reward/cost rate function $-h(i)$, for $i \in S$.

It is obvious that the game $\mathcal{G}\mathcal{M}_0$ satisfies Assumptions 10.2, 10.9(a), and 10.7(c). Assumption 10.7(a) is derived from Proposition 10.12, while the continuity on $\overline{A}_0(i) \times \overline{B}_0(i)$ of the functions

$$r(i, \phi, \psi), \quad q_{ij}(\phi, \psi), \quad \text{and} \quad \sum_{j \in S} q_{ij}(\phi, \psi) w(j),$$

which is proved in [47, Lemma 7.2], yields Assumption 10.7(b). Finally, it remains to check Assumptions 10.9(b) and 10.10.

The family of probability measures on $\overline{A}_0(i)$ will be denoted by $\mathcal{P}(\overline{A}_0(i))$. Thus, in the game $\mathcal{G}\mathcal{M}_0$, a randomized stationary strategy $\bar{\pi}^1$ for player 1 is determined by $\bar{\pi}^1(\cdot|i) \in \mathcal{P}(\overline{A}_0(i))$ for each $i \in S$. We give a similar definition for $\mathcal{P}(\overline{B}_0(i))$, for $i \in S$.

Given $i \in S$, now define the following projection operator

$$\mathbf{p} : \mathcal{P}(\overline{A}_0(i)) \to \overline{A}_0(i)$$

where, for $\overline{\phi} \in \mathcal{P}(\overline{A}_0(i))$, we define

$$(\mathbf{p}\overline{\phi})(F) := \int_{\overline{A}_0(i)} \phi(F)\overline{\phi}(d\phi)$$

for each measurable set $F \in \mathbb{B}(A(i))$. Note that $\mathbf{p}\bar{\phi}$ is in $\overline{A}_0(i)$ because $\overline{A}_0(i)$ is a convex set of probability measures; see Proposition 10.12. Similarly, we may define the projection $\mathbf{p} : \mathcal{P}(\overline{B}_0(i)) \to \overline{B}_0(i)$ and, for simplicity, we will use the same notation.

Given a pair of randomized stationary strategies $(\bar{\pi}^1, \bar{\pi}^2)$ for $\mathcal{G}\mathcal{M}_0$, we define $\pi^1 := \mathbf{p}\bar{\pi}^1 \in \Pi_s^{*1}$ by means of

$$\pi^1(\cdot|i) := \mathbf{p}(\bar{\pi}^1(\cdot|i)) \quad \forall i \in S.$$

We give a similar definition for $\mathbf{p}\bar{\pi}^2$. From the definition of $\mathbf{p}$ we deduce that the reward/cost rate verifies

$$r(i, \bar{\pi}^1, \bar{\pi}^2) = r(i, \mathbf{p}\bar{\pi}^1, \mathbf{p}\bar{\pi}^2),$$

while the transition rates verify

$$q_{ij}(\bar{\pi}^1, \bar{\pi}^2) = q_{ij}(\mathbf{p}\bar{\pi}^1, \mathbf{p}\bar{\pi}^2) \tag{11.10}$$

for every $i, j \in S$. In other words, given a pair of randomized strategies $(\bar{\pi}^1, \bar{\pi}^2)$ for $\mathcal{G}\mathcal{M}_0$, there exists a pair of nonrandomized strategies $(\mathbf{p}\bar{\pi}^1, \mathbf{p}\bar{\pi}^2)$ for $\mathcal{G}\mathcal{M}_0$ with the same reward rate function and the same transition rates. In particular, since $(\mathbf{p}\bar{\pi}^1, \mathbf{p}\bar{\pi}^2)$ is irreducible (Assumption 10.9(b)), so is $(\bar{\pi}^1, \bar{\pi}^2)$. Similarly, the uniform w-exponential ergodicity in Assumption 10.10 holds on $\mathcal{G}\mathcal{M}_0$. This proves that the game model $\mathcal{G}\mathcal{M}_0$ satisfies Assumptions 10.2, 10.7, 10.9, and 10.10.

Let $\tilde{g} \in \mathbb{R}$ be the value of the average game $\mathcal{G}\mathcal{M}_0$. The corresponding average optimality equations are

$$\tilde{g} = -h(i) + \sum_{j \in S} q_{ij}(\pi^1, \pi^2)h^1(j)$$

$$= \sup_{\phi \in \overline{A}_0(i)} \left\{ -h(i) + \sum_{j \in S} q_{ij}(\phi, \pi^2)h^1(j) \right\}$$

$$= \inf_{\psi \in \overline{B}_0(i)} \left\{ -h(i) + \sum_{j \in S} q_{ij}(\pi^1, \psi)h^1(j) \right\}$$

for some $h^1 \in \mathcal{B}_w(S)$ and $(\pi^1, \pi^2) \in \Pi_s^{*1} \times \Pi_s^{*2}$, for all $i \in S$. Note that, in principle, the strategies π^1 and π^2 should be randomized strategies for the game model $\mathcal{G}\mathcal{M}_0$ and, also, we should take the supremum and infimum over $\mathcal{P}(\overline{A}_0(i))$ and $\mathcal{P}(\overline{B}_0(i))$, respectively. As a consequence of (11.10), however, we can restrict ourselves to the class of nonrandomized strategies for $\mathcal{G}\mathcal{M}_0$ (that is, $\Pi_s^{*1} \times \Pi_s^{*2}$), and we can take the supremum over $\overline{A}_0(i)$ and $\overline{B}_0(i)$, respectively.

If we let $h^0(i) := h(i) + \tilde{g}$ for all $i \in S$, it follows that $(g^*, h^0, h^1, \pi^1, \pi^2)$ indeed solves the bias optimality equations. This proves the existence of solutions to the bias optimality equations.

Suppose now that we have a solution $(g^*, h^0, h^1, \pi^1, \pi^2)$ of the bias optimality equations. Then (π^1, π^2) is average optimal and, in addition, we deduce from (11.7) and Remark 6.6 that

$$h^0(i) = h(i, \pi^1, \pi^2) \quad \forall i \in S.$$

Also, given arbitrary $\bar{\pi}^1 \in \Pi_s^{*1}$, we get from (11.8) that

$$0 \geq -h^0(i) + \sum_{j \in S} q_{ij}(\bar{\pi}^1, \pi^2) h^1(j) \quad \forall i \in S,$$

which, by Corollary 3.16, yields that $z = \mu_{\bar{\pi}^1, \pi^2}(-h^0) \leq 0$ (here, z is taken from (11.2)). Now, recalling (11.2) and (11.3), we have

$$h(i, \bar{\pi}^1, \pi^2) \leq h(i, \pi^1, \pi^2) \quad \forall \bar{\pi}^1 \in \Pi_s^{*1}, \ i \in S.$$

By a similar reasoning, we have

$$h(i, \pi^1, \pi^2) \leq h(i, \pi^1, \bar{\pi}^2) \quad \forall \bar{\pi}^2 \in \Pi_s^{*2}, \ i \in S.$$

This shows that (π^1, π^2) is a pair of bias optimal strategies and, besides, that $h^0 = h(\cdot, \pi^1, \pi^2)$ is the optimal bias function. This completes the proof of statement (i).

Regarding part (ii), we have already proved that if a pair of strategies verifies the bias optimality equations, then it is bias optimal.

The converse follows from Theorem 10.11(iii), which shows that the classes of average optimal and canonical strategies coincide. Applying this result to the game model $\mathcal{GM}_0$ with the reward/cost rate function $-H^*$ yields the desired result.   □

In the next section, we introduce the concept of overtaking optimality, and explore its relations with bias optimality.

11.3. Overtaking optimality

As already seen, the bias optimality criterion arises when the players want to optimize their total expected returns

$$J_T(i, \pi^1, \pi^2)$$

asymptotically as $T \to \infty$. Another approach to this situation is to give a direct definition of such an equilibrium, instead of considering two "nested" games, the average game and the bias game, as we did before. This leads to the following definition of an *overtaking equilibrium* on a subset $\Gamma^1 \times \Gamma^2$ of stationary strategies.

Definition 11.3. A pair of stationary strategies $(\pi^{*1}, \pi^{*2}) \in \Gamma^1 \times \Gamma^2 \subseteq \Pi_s^1 \times \Pi_s^2$ is overtaking optimal in $\Gamma^1 \times \Gamma^2$ if, for each $(\pi^1, \pi^2) \in \Gamma^1 \times \Gamma^2$ and every $i \in S$,

$$\liminf_{T \to \infty} \left[J_T(i, \pi^{*1}, \pi^{*2}) - J_T(i, \pi^1, \pi^{*2}) \right] \geq 0$$

and

$$\limsup_{T \to \infty} \left[J_T(i, \pi^{*1}, \pi^{*2}) - J_T(i, \pi^{*1}, \pi^2) \right] \leq 0.$$

In the context of CMCs (which we may call single-player Markov games), bias optimality is equivalent to overtaking optimality in the class of deterministic stationary policies $\mathbb{F}$; see Theorem 5.3. Moreover, the proof of this result is a straightforward consequence of the interpretation of the bias given in (5.2). As we will show in the following, the relations between bias and overtaking equilibria for Markov games, however, are not as strong as for CMCs.

Our main results on overtaking optimality are stated in the next theorem. We note that Theorem 11.4 below does not address the issue of the existence of a pair of strategies that is overtaking optimal in $\Pi_s^1 \times \Pi_s^2$. In fact, there might not exist overtaking optimal strategies in $\Pi_s^1 \times \Pi_s^2$, as shown in the example given in Sec. 11.4.

Theorem 11.4. *We suppose that Assumptions* 10.2, 10.7, 10.9, *and* 10.10 *are satisfied. Then*:

(i) *If* $(\pi^{*1}, \pi^{*2}) \in \Pi_s^{*1} \times \Pi_s^{*2}$ *is a pair of bias optimal strategies, then it is overtaking optimal in* $\Pi_s^{*1} \times \Pi_s^{*2}$.

(ii) *If* $(\pi^{*1}, \pi^{*2}) \in \Pi_s^1 \times \Pi_s^2$ *is overtaking optimal in* $\Pi_s^1 \times \Pi_s^2$, *then it is bias optimal.*

Proof. *Proof of* (i). Let (π^{*1}, π^{*2}) be as in (i). Moreover, fix an arbitrary pair of average optimal strategies $(\pi^1, \pi^2) \in \Pi_s^{*1} \times \Pi_s^{*2}$. Then, the pair (π^{*1}, π^2) is average optimal as well (see Proposition 10.12(ii)), and so, by (10.8) and (11.1),

$$J_T(i, \pi^{*1}, \pi^2) = g^* T + h(i, \pi^{*1}, \pi^2) + \mathrm{O}(e^{-\delta T}) \qquad (11.11)$$

for a given $i \in S$. Similarly, we have

$$J_T(i, \pi^{*1}, \pi^{*2}) = g^* T + h(i, \pi^{*1}, \pi^{*2}) + \mathrm{O}(e^{-\delta T}). \qquad (11.12)$$

On the other hand, since (π^{*1}, π^{*2}) is bias optimal, by Definition 11.1 we have $h(i, \pi^{*1}, \pi^{*2}) \le h(i, \pi^{*1}, \pi^2)$. Now, subtracting (11.11) from (11.12) yields

$$\lim_{T \to \infty} [J_T(i, \pi^{*1}, \pi^{*2}) - J_T(i, \pi^{*1}, \pi^2)] = h(i, \pi^{*1}, \pi^{*2}) - h(i, \pi^{*1}, \pi^2) \le 0$$

for each $i \in S$. Similarly, we can prove that, for all $i \in S$,

$$\lim_{T \to \infty} [J_T(i, \pi^{*1}, \pi^{*2}) - J_T(i, \pi^1, \pi^{*2})] = h(i, \pi^{*1}, \pi^{*2}) - h(i, \pi^1, \pi^{*2}) \ge 0.$$

This shows that a pair of bias optimal strategies is overtaking optimal in $\Pi_s^{*1} \times \Pi_s^{*2}$.

Proof of (ii). Suppose now that $(\pi^{*1}, \pi^{*2}) \in \Pi_s^1 \times \Pi_s^2$ is overtaking optimal in $\Pi_s^1 \times \Pi_s^2$. Given arbitrary $(\pi^1, \pi^2) \in \Pi_s^1 \times \Pi_s^2$ and a state $i \in S$, we have

$$\liminf_{T \to \infty} [J_T(i, \pi^{*1}, \pi^{*2}) - J_T(i, \pi^1, \pi^{*2})] \ge 0.$$

Recalling the interpretation of the bias given in (10.8), this is equivalent to

$$\lim_{T \to \infty} [(g(\pi^{*1}, \pi^{*2}) - g(\pi^1, \pi^{*2}))T + h(i, \pi^{*1}, \pi^{*2}) - h(i, \pi^1, \pi^{*2})] \ge 0.$$

$$(11.13)$$

Proceeding similarly, we also get

$$\lim_{T \to \infty} [(g(\pi^{*1}, \pi^{*2}) - g(\pi^{*1}, \pi^2))T + h(i, \pi^{*1}, \pi^{*2}) - h(i, \pi^{*1}, \pi^2)] \le 0.$$

$$(11.14)$$

We deduce from (11.13) and (11.14) that

$$g(\pi^1, \pi^{*2}) \le g(\pi^{*1}, \pi^{*2}) \le g(\pi^{*1}, \pi^2);$$

that is, (π^{*1}, π^{*2}) is average optimal. Now, if we choose in (11.13) and (11.14) a pair (π^1, π^2) of average optimal strategies, we obtain

$$h(i, \pi^1, \pi^{*2}) \le h(i, \pi^{*1}, \pi^{*2}) \le h(i, \pi^{*1}, \pi^2) \quad \forall i \in S,$$

which shows that (π^{*1}, π^{*2}) is bias optimal. $\square$

11.4.   A counterexample on bias and overtaking optimality

As was already mentioned, Theorem 11.4 does not establish the existence of a pair of overtaking optimal strategies in $\Pi_s^1 \times \Pi_s^2$. Our next example shows that such optimal strategies might not exist.

Consider the following zero-sum two-person game. The state space is $S = \{0, 1\}$. The action sets are

$$A(0) = \{0\}, \quad A(1) = \{0, 1\}, \quad B(0) = \{0\}, \quad B(1) = \{0, 1\}.$$

The reward/cost rate and the transition rates are

$$r(0, 0, 0) = 4, \quad r(1, 0, 0) = 1, \quad r(1, 0, 1) = -2, \quad r(1, 1, 0) = 0, \quad r(1, 1, 1) = 2,$$

and

$$q_{00}(0, 0) = -2, \quad q_{11}(0, 0) = -1, \quad q_{11}(0, 1) = q_{11}(1, 0) = q_{11}(1, 1) = -2,$$

respectively.

We will denote by π_x^1, for $x \in [0, 1]$, a randomized stationary strategy for player 1, where $\pi_x^1(\cdot|1)$ takes the values 0 and 1 with probabilities x and $1 - x$, respectively. Similarly, we will denote by π_y^2, for $y \in [0, 1]$, the randomized stationary strategy for player 2 such that $\pi_y^2(\cdot|1)$ takes the values 0 and 1 with respective probabilities y and $1 - y$.

Obviously, this game model satisfies Assumptions 10.2, 10.7, 10.9, and 10.10.

Direct calculations show that the payoff rates for the stationary strategies (π_x^1, π_y^2), for $0 \le x, y \le 1$, are

$$r(0, \pi_x^1, \pi_y^2) = 4 \quad \text{and} \quad r(1, \pi_x^1, \pi_y^2) = 5xy - 4x - 2y + 2,$$

while the transition rate matrices are

$$Q(\pi_x^1, \pi_y^2) = \begin{pmatrix} -2 & 2 \\ 2 - xy & xy - 2 \end{pmatrix}.$$

Hence, the invariant probability measures are given by

$$\mu_{\pi_x^1, \pi_y^2}(0) = \frac{2 - xy}{4 - xy} \quad \text{and} \quad \mu_{\pi_x^1, \pi_y^2}(1) = \frac{2}{4 - xy}$$

for arbitrary $0 \le x, y \le 1$. Consequently, the long-run expected average payoff of the stationary strategy (π_x^1, π_y^2) is

$$g(x, y) := g(\pi_x^1, \pi_y^2) = \frac{6xy - 8x - 4y + 12}{4 - xy}.$$

Let us now determine the family of average optimal strategies. For a fixed $x \in [0, 1]$, we have

$$\inf_{0 \le y \le 1} g(x, y) = \begin{cases} g(x, 1) = 2, & \text{for } 0 \le x < 1/2, \\ g(x, y) = 2, & \text{for } x = 1/2 \quad \text{and} \quad 0 \le y \le 1, \\ g(x, 0) = 3 - 2x, & \text{for } 1/2 < x \le 1. \end{cases}$$

On the other hand, for fixed $y \in [0, 1]$,

$$\sup_{0 \le x \le 1} g(x, y) = \begin{cases} g(0, y) = 3 - y, & \text{for } 0 \le y < 1, \\ g(x, y) = 2, & \text{for } y = 1 \quad \text{and} \quad 0 \le x \le 1. \end{cases}$$

Therefore, the value of the game is $g^* = 2$. We also deduce that the family of average optimal stationary strategies is given by the pairs (x^*, y^*) such that

$$g(x^*, y^*) = \inf_{0 \le y \le 1} g(x^*, y) = \sup_{0 \le x \le 1} g(x, y^*) = g^* = 2,$$

is

$$\left\{ (\pi^1_{x^*}, \pi^2_1) : 0 \le x^* \le 1/2 \right\}.$$

Hence, the optimal strategies for player 1 are $\pi^1_{x^*}$, for $0 \le x^* \le 1/2$, while the unique optimal strategy for player 2 is π^2_1.

The average payoff and the bias of the stationary strategy (π^1_x, π^2_1), for $0 \le x \le 1$, are

$$g(x, 1) = 2 \quad \text{and} \quad h(\cdot, x, 1) := h(\cdot, \pi^1_x, \pi^2_1) = \begin{pmatrix} \dfrac{2}{4 - x} \\ \dfrac{x - 2}{4 - x} \end{pmatrix}, \qquad (11.15)$$

respectively. It follows that the unique pair of bias optimal strategies is $(\pi^1_{1/2}, \pi^2_1)$, and the corresponding average payoff and bias are

$$g(1/2, 1) = 2 \quad \text{and} \quad h(\cdot, 1/2, 1) = \begin{pmatrix} 4/7 \\ -3/7 \end{pmatrix}. \qquad (11.16)$$

Observe that, since player 2 has a unique average optimal strategy, the problem of finding bias optimal strategies was reduced to a one-dimensional problem.

Let us now suppose that there exists a pair of overtaking optimal strategies in $\Pi^1_s \times \Pi^2_s$. By Theorem 11.4, this pair of strategies is necessarily $(\pi^1_{1/2}, \pi^2_1)$, the unique pair of bias optimal strategies.

Recalling (11.15), it follows that the respective average payoff and bias of (π^1_1, π^2_1) are

$$g(1, 1) = 2 \quad \text{and} \quad h(\cdot, 1, 1) = \begin{pmatrix} 2/3 \\ -1/3 \end{pmatrix}, \qquad (11.17)$$

and, thus, by (11.16) and (11.17),

$$\lim_{T \to \infty} [J_T(i, \pi^1_{1/2}, \pi^2_1) - J_T(i, \pi^1_1, \pi^2_1)] = -2/21 < 0 \quad \forall i \in S,$$

which contradicts Definition 11.3.

Therefore, for this game model, there does not exist a pair of overtaking optimal strategies in $\Pi_s^1 \times \Pi_s^2$.

The reason is that, even if the optimal strategy π_1^2 of player 2 remains fixed, the available strategies for player 1 are $\pi_{x^*}^1$ for $0 \leq x^* \leq 1/2$. On the contrary, if the game model were interpreted as a control model with a single player (player 1) in which the strategy π_1^2 of player 2 remains fixed, then the available strategies for player 1 would be π_x^1 for $0 \leq x \leq 1$.

This example shows an interesting discrepancy between games and control models, for which there indeed exist overtaking optimal policies, as shown in Theorems 5.3 and 5.4.

11.5. Conclusions

In this chapter we have studied refinements of the average payoff optimality criterion such as bias and overtaking equilibria. We have observed that the analogy between control and game models no longer holds. Indeed, as shown in Sec. 11.4, there might not exist overtaking optimal strategies in the class of stationary strategies (as erroneously claimed in [118, Theorem 5]).

We note that it is precisely the characterization of the set of average optimal strategies as a rectangle (recall Sec. 10.5) that allows us to pose the bias game. This is the reason why, when restricted to the class of average optimal strategies, the action sets of the two players are well defined.

Regarding extensions of bias optimality, we can also define the sensitive n-discount equilibria by considering a sequence of nested average games, as we did for control models in Chapter 6. The fact that, in Theorem 11.4(i), bias optimal strategies are overtaking optimal only in the class of average optimal strategies shows that, in general, n-discount optimal strategies will be n-discount optimal *only* in the class of $(n-1)$-discount optimal strategies (and not in the class of *all* stationary strategies, as happened for control models). Therefore, intuitively, the n-discount optimality criteria for stochastic games seem to be of a lesser interest.

Notation List

$\emptyset$	the empty set
$\mathbf{0}$	the 0 sequence, p. 159
$\mathbf{1}$	the constant function equal to 1, p. 14
$\\|\cdot\\|_1$	the 1-norm, p. 21
$\\|\cdot\\|_w$	the w-norm, pp. 20, 90
$\\|\cdot\\|_{w^2}$	the w^2-norm, p. 21
$\square$	the end of a proof
$\xrightarrow{p}$	convergence in probability, p. 14
$\xrightarrow{w}$	weak convergence, p. 14
$[T]$	the integer part of the real number T, p. 123
$\mathbf{u} \succeq \mathbf{v}$	$\mathbf{u}$ is lexicographically greater than or equal to $\mathbf{v}$, p. 142
$\mathbf{u} \succ \mathbf{v}$	$\mathbf{u}$ is lexicographically greater than $\mathbf{v}$, p. 142
$\sim$	asymptotic equivalence of sequences, p. 14
$\vee$	the maximum of two real numbers, p. 93
$\wedge$	the minimum of two real numbers, p. 35
A	the action space (control models), p. 16
	the action space for player 1 (games), p. 228
$A(i)$	the action sets (control models), p. 16
	the action sets for player 1 (games), p. 228
$A^*(i)$	the set of maxima in the AROE, p. 67
$A^n(i)$	the set of maxima in the average reward optimality equations, p. 140
$A_n(i)$	the action sets of the approximating control models, p. 89
$\overline{A}(i)$	the family of probability measures on $A(i)$ (games), p. 232
$\overline{A}_0(i)$	the set of $\phi \in \overline{A}(i)$ attaining the maximum in the Shapley equation, p. 242
$a(t)$	the action process for player 1 (games), p. 231
α	the discount rate (control models), p. 49
	the discount rate (games), p. 233
AROE	the average reward optimality equation, p. 67

a.s.	almost surely, p. 14
B	the action space for player 2 (games), p. 228
$B(i)$	the action sets for player 2 (games), p. 228
$\overline{B}(i)$	the family of probability measures on $B(i)$ (games), p. 232
$\overline{B}_0(i)$	the set of $\psi \in \overline{B}(i)$ attaining the maximum in the Shapley equation, p. 242
$b(t)$	the action process for player 2 (games), p. 231
$\mathcal{B}_1(S)$	the bounded functions on S, p. 21
$\mathcal{B}_w(S)$	the w-bounded functions on S, p. 20
$\mathcal{B}_w(S_n)$	the w-bounded functions on S_n, p. 90
$\mathcal{B}_w(K)$	the w-bounded measurable functions on K, p. 21
$\mathcal{B}_{w^2}(S)$	the w^2-bounded functions on S, p. 21
$\mathcal{B}_{w^2}(K)$	the w^2-bounded measurable functions on K, p. 21
$\mathcal{B}_{w+w'}(S)$	the $(w+w')$-bounded functions on S, p. 21
$\mathcal{B}_{w+w'}(K)$	the $(w+w')$-bounded measurable functions on K, p. 21
$\mathbb{B}(X)$	the Borel subsets of X, p. 13
$C(1)$	the ball with radius 1 in $\mathcal{B}_1(S)$, p. 32
$C^1[0, \infty)$	the class of continuously differentiable functions on $[0, \infty)$, p. 34
$C_w^1([0, T] \times S)$	the class of functions $v(t, i)$ continuously differentiable in t, p. 47
$\mathcal{C}_w(K)$	the family of continuous functions in $\mathcal{B}_w(K)$, p. 176
CMC	controlled Markov chain, p. 9
$\mathbf{d}(f, f')$	the metric in $\mathbb{F}$, p. 18
δ_{ij}	the Kronecker delta, p. 14
Δ_0	the 0th discrepancy function, p. 144
Δ_n	the nth discrepancy function, p. 144
DROE	the discounted reward optimality equation, p. 53
$E_{s,i}^\varphi$	the expectation operator when using a policy φ starting at $x(s) = i$ at time $s \geq 0$, p. 20
E_i^φ	$\equiv E_{0,i}^\varphi$
$E_{s,i}^{\pi^1, \pi^2}$	the expectation operator of a pair of strategies π^1, π^2 starting at $x(s) = i$ at time $s \geq 0$, p. 231
$E_i^{\pi^1, \pi^2}$	$\equiv E_{0,i}^{\pi^1, \pi^2}$
$\tilde{E}_i^{\pi^1, \pi^2}$	the state-actions expectation operator of a pair of strategies, p. 232
$E_{n,i}^f$	the expectation operator of the policy $f \in \mathbb{F}$ under the control model $\mathcal{M}_n$, p. 100

E_{γ_i}	the expectation operator of the catastrophe size distribution, p. 209
$\mathbb{F}$	the set of deterministic stationary policies, p. 18
$\mathbb{F}_{\text{ao}}$	the set of average optimal policies in $\mathbb{F}$, p. 70
$\mathbb{F}_{\text{ca}}$	the set of canonical policies, p. 70
$\mathbb{F}_n$	the set of deterministic stationary policies of the approximating control model $\mathcal{M}_n$, p. 90
$\mathbb{F}^n$	the policies in $\mathbb{F}$ supported on the $A^{n+1}(i)$, p. 142
$\mathbb{F}^\infty$	the intersection of all the $\mathbb{F}^n$, p. 142
$\mathbb{F}_\lambda^*$	the set of optimal policies for the reward rate v_λ, p. 169
$f_n \to f$	convergence of policies in $\mathbb{F}$, p. 18
$\varphi, \varphi_t(B\|i)$	a Markov policy, p. 17
$\varphi_n \to \varphi$	convergence of policies, p. 18
ϕ	an element of $\overline{A}(i)$ (games), p. 232
Φ	the set of Markov policies, p. 17
Φ_{s}	the set of stationary policies, p. 18
Φ_n	the set of Markov policies of $\mathcal{M}_n$, p. 90
$\Phi^{-1} \equiv \Phi_{\text{s}}$	the family of stationary policies, p. 147
Φ^n	the family of stationary policies supported on $A^n(i)$, p. 147
$\Phi_m^*(i)$	the set of policies with minimal m such that $\mathbf{u}_m(i, \varphi) < 0$, p. 160
$\Phi_\infty^*(i)$	the set of policies such that $\mathbf{u}(i, \varphi) = \mathbf{0}$, p. 160
g^*	the optimal gain (control models), p. 67
	the optimal average payoff (games), p. 239
$g^*, h^0, h^1, \ldots$	the solutions of the average reward optimality equations, p. 141
g_n^*	the optimal gain of the control model $\mathcal{M}_n$, pp. 101, 109
$g(f)$	the gain of a policy in $\mathbb{F}$, p. 61
$g(\varphi)$	the gain of a policy in Φ_{s}, p. 61
$g(\lambda)$	the optimal average reward with the Lagrange multiplier λ, p. 183
$g(\pi^1, \pi^2)$	the average payoff of a pair of stationary strategies, p. 238
g_n	the gain in the nth step of the policy iteration algorithm, p. 86
$g_n(f)$	the gain of the policy $f \in \mathbb{F}$ under the control model $\mathcal{M}_n$, p. 101
$g_u(\varphi)$	the average cost of a policy in Φ_{s}, p. 175
$\mathcal{G}M$	the game model, p. 228
Γ	the family of state-action invariant probability measures, p. 177

$\gamma_i(j)$	the probability distribution of the catastrophe size, p. 209
h_f	the bias of a policy in $\mathbb{F}$, p. 61
h_f^k	the compositions of the bias operator, p. 135
h_φ	the bias of a policy in Φ_s, p. 63
$h(\cdot, \pi^1, \pi^2)$	the bias of a pair of stationary strategies, p. 238
h_n	the bias in the nth step of the policy iteration algorithm, p. 86
$\hat{h}$	the optimal bias function (control models), p. 118
H_f	the bias operator, p. 131
H^*	the optimal bias function (games), p. 246
$\mathbf{I}_B$	the indicator function of the set B, p. 14
$J(i, \varphi)$	the average reward of a policy, p. 59
$J(i, \pi^1, \pi^2)$	the average payoff of a pair of strategies, p. 236
$J_T(i, \varphi)$	the total expected reward on $[0, T]$, p. 46
$J_T(i, \varphi, h)$	the total expected reward with terminal payoff function h, p. 46
$J_T(s, i, \varphi, h)$	the total expected reward on $[s, T]$ with terminal payoff function h, p. 47
$J_T(i, \pi^1, \pi^2)$	the total expected payoff on $[0, T]$ (games), p. 236
$J^0(i, \varphi)$	the long-run pathwise average reward of a policy, p. 73
$J_T^0(i, \varphi)$	the total pathwise reward of a policy, p. 73
$J_u(i, \varphi)$	the average cost of a policy, p. 174
$J_{u,T}(i, \varphi)$	the total expected cost on $[0, T]$, p. 174
$J_u^0(i, \varphi)$	the long-run pathwise average cost of a policy, p. 185
$J_{u,T}^0(i, \varphi)$	the total pathwise cost of a policy, p. 185
$J^*(i)$	the optimal average reward function, p. 59
$J_T^*(i, h)$	the optimal total expected reward with terminal payoff h, p. 46
$J_T^*(s, i, h)$	the optimal total expected reward on $[s, T]$ with terminal payoff h, p. 47
K	the set of state-action pairs (control models), p. 16 the set of state-actions triplets (games), p. 228
K_n	the set of state-action pairs of the approximating control model $\mathcal{M}_n$, p. 89
L	the infinitesimal generator operator, p. 177
L^f, L^φ	the infinitesimal generator of a stationary policy, p. 22
$L^{t,\varphi}$	the infinitesimal generator of a Markov policy, pp. 22, 158
$L(i)$	the lower value function of the average game, p. 237
$L_\alpha(i)$	the lower value function of the discounted game, p. 234

log	the logarithmic function, with $\log(e) = 1$.
$\mathcal{M}$	the control model, p. 16
	the constrained control model, p. 165
	the original control model, pp. 89, 98
$\mathcal{M}_n$	the approximating control models, pp. 89, 96, 98, 109
$\overline{\mathcal{M}}_n$	the finite state and continuous action control models
	for the numerical approximations, pp. 206, 217
$\mathcal{M}_n \to \mathcal{M}$	convergence of control models, pp. 92, 101
$\mathcal{M}_d$	a discrete-time control model, p. 56
MDP	Markov decision process, p. 9
μ_φ, μ_f	the invariant probability measure of a stationary policy,
	p. 25
$\hat{\mu}_\varphi$	the state-action invariant probability measure of a
	stationary policy, p. 176
μ_f^n	the invariant probability measure of $f \in \mathbb{F}$ under the
	control model $\mathcal{M}_n$, p. 100
μ_{π^1, π^2}	the invariant probability measure of a pair of
	stationary strategies, p. 237
$\mu(u)$	the expectation of u with respect to μ, p. 24
$\mu_{\pi^1, \pi^2}(u)$	the expectation of u with respect to μ_{π^1, π^2}, p. 238
$n(i)$	the first n such that $i \in S_n$, p. 92
o, O	the Landau notation, p. 14
Ω	the product space $K^{[0,\infty)}$ (games), p. 231
P-a.s.	almost surely with respect to P, p. 14
$P_{ij}(s,t)$	a nonhomogeneous transition function, p. 19
$P_{s,i}^\varphi$	the state probability measure of a policy φ
	starting at $x(s) = i$ at time $s \geq 0$, p. 20
P_i^φ	$\equiv P_{0,i}^\varphi$
$P_{ij}^\varphi(t)$	the transition function of a stationary policy, p. 24
$P_{ij}^\varphi(s,t)$	the nonhomogeneous transition function of a policy, p. 19
$P_{iC}^\varphi(s,t)$	the sum of the $P_{ij}^\varphi(s,t)$ over $j \in C$, p. 30
$P_{ij}^{n,f}(t)$	the transition function of the policy $f \in \mathbb{F}$ under
	the control model $\mathcal{M}_n$, p. 100
$\mathcal{P}_w(S)$	the family of probability measures on S
	with $\mu(w) < \infty$, p. 176
$\mathcal{P}_w(K)$	the family of probability measures on K
	with $\mu(w) < \infty$, p. 176
$\mathcal{P}_n$	the discretization of the action space, pp. 202, 214
$\pi^1, \pi_t^1(C|i)$	a randomized Markov strategy for player 1, p. 229

$\pi^2, \pi_t^2(C\|i)$	a randomized Markov strategy for player 2, p. 229
Π^1, Π^2	the family of randomized Markov strategies for players 1 and 2, p. 229
Π_s^1, Π_s^2	the family of randomized stationary strategies for players 1 and 2, p. 229
Π_s^{*1}, Π_s^{*2}	the family of average optimal stationary strategies for players 1 and 2, p. 244
$P_{s,i}^{\pi^1,\pi^2}$	the state probability measure of a pair of strategies π^1, π^2 starting at $x(s) = i$ at time $s \geq 0$, p. 231
$P_i^{\pi^1,\pi^2}$	$\equiv P_{0,i}^{\pi^1,\pi^2}$
$\tilde{P}_i^{\pi^1,\pi^2}$	the state-actions probability measure of a pair of strategies, p. 232
$P_{ij}^{\pi^1,\pi^2}(s,t)$	the nonhomogeneous transition function of a pair of strategies, p. 230
ψ	an element of $\overline{B}(i)$ (games), p. 232
$q(i)$	the supremum of $-q_{ii}(a)$ (control models), p. 16
$q_{ij}(a)$	the transition rates, p. 16
$q_{ij}(f)$	the transition rate under $f \in \mathbb{F}$, p. 18
$q_{ij}(\varphi)$	the transition rate under $\varphi \in \Phi_s$, p. 18
$q_{ij}(t,\varphi)$	the transition rate under a Markov policy, p. 17
$q_n(i)$	the supremum of $-q_{ii}^n(a)$, pp. 89, 99
$q_{ij}^n(a)$	the transition rates of $\mathcal{M}_n$, pp. 89, 99
$q(i)$	the supremum of $-q_{ii}(a,b)$ (games), p. 229
$q_{ij}(a,b)$	the transition rates (games), p. 228
$q_{ij}(\pi^1,\pi^2)$	the transition rate under a pair of stationary strategies, p. 230
$q_{ij}(\pi^1,\psi)$	the transition rate under π^1 and ψ, p. 232
$q_{ij}(\phi,\pi^2)$	the transition rate under ϕ and π^2, p. 232
$q_{ij}(\phi,\psi)$	the transition rate under ϕ and ψ, p. 232
$q_{ij}(t,\pi^1,\pi^2)$	the transition rate under a pair of Markov strategies, p. 230
$Q^\varphi(t)$	the Q^φ-matrix of a policy, p. 19
$Q^{\pi^1,\pi^2}(t)$	the Q-matrix of a pair of strategies, p. 230
$r(i,a)$	the reward rate, p. 16
$r(i,f)$	the reward rate of a policy $f \in \mathbb{F}$, p. 46
$r(i,\varphi)$	the reward rate of a policy $\varphi \in \Phi_s$, p. 46
$r(t,i,\varphi)$	the reward rate of a Markov policy, p. 45
$r_n(i,a)$	the reward rates in the control model $\mathcal{M}_n$, pp. 89, 99
$r(i,a,b)$	the reward/cost rate (games), p. 229

$r(i, \pi^1, \pi^2)$	the reward/cost rate of a pair of stationary strategies, p. 232
$r(i, \pi^1, \psi)$	the reward/cost rate of π^1 and ψ, p. 232
$r(i, \phi, \pi^2)$	the reward/cost rate of ϕ and π^2, p. 232
$r(i, \phi, \psi)$	the reward/cost rate of ϕ and ψ, p. 232
$r(t, i, \pi^1, \pi^2)$	the reward/cost rate of a pair of strategies, p. 232
$\mathbb{R}$	the set of real numbers, p. 14
$\mathbb{R}^+$	the set of nonnegative real numbers, p. 14
$R_k(f, u, \alpha)$	the residual of the Laurent series expansion, p. 134
s.t.	subject to, p. 174
S	the state space (control models), p. 16
	the state space (games), p. 228
S_n	the state space of the approximating models $\mathcal{M}_n$, pp. 89, 109
$\sigma^2(f)$	the limiting average variance of a policy in $\mathbb{F}$, p. 122
$\sigma^2(i, f)$	the limiting average variance of a policy in $\mathbb{F}$, p. 121
θ	the constraint constant, p. 168
θ_0	the constraint constant (for average problems), p. 174
θ_α	the constraint constant (for the vanishing discount approach), p. 187
$\theta_{\min}, \theta_{\max}$	the range of the expected average cost, p. 176
$u(i, a)$	the cost rate for constrained problems, p. 165
$u(t, i, \varphi)$	the integral of $u(i, \cdot)$ with respect to $\varphi_t(\cdot|i)$, p. 50
u_α	the optimal discounted reward differences, p. 66
$\mathbf{u}_n(i, \varphi)$	the integral of the expectation of $\Delta_n(t, i, \varphi)$, p. 159
$\mathbf{u}(i, \varphi)$	the sequence of the $\mathbf{u}_n(i, \varphi)$, p. 159
$U(i)$	the upper value function of the average game, p. 237
$U_\alpha(i)$	the upper value function of the discounted game, p. 234
$v_\lambda(i, a)$	the reward rate with the Lagrange multiplier, p. 168
$V(i)$	the value function of the average game, p. 237
$V_\alpha(i)$	the value function of the discounted game, p. 234
$V(\theta)$	the optimal average constrained reward function, p. 179
$V_\alpha(i, \varphi)$	the total expected discounted reward of a policy, p. 49
$V_\alpha^n(i, \varphi)$	the total expected discounted reward of a policy in the control model $\mathcal{M}_n$, p. 91
$V_\alpha(\mu, \varphi)$	the total expected discounted reward of a policy, given the initial distribution μ, p. 172
$V_\alpha(i, \varphi, u)$	the total expected discounted reward of a policy with reward rate u, p. 50

$V_\alpha(\mu, \varphi, u)$	the total expected discounted cost of a policy with cost rate u, given the initial distribution μ, p. 173
$V_\alpha(i, \pi^1, \pi^2)$	the total expected discounted payoff of a pair of strategies, p. 233
$V_u(\lambda)$	the discounted cost of a policy in $\mathbb{F}_\lambda^*$, p. 169
$V_\alpha^*(i)$	the optimal discounted reward, p. 50
$V_\alpha^{*n}(i)$	the optimal discounted reward of $\mathcal{M}_n$, p. 92
$V^*(i, \lambda)$	the optimal discounted reward under v_λ, p. 168
$V^*(i, \theta_0)$	the optimal average constrained reward function, p. 174
w	a Lyapunov function, p. 19
$x^\varphi(t), x(t)$	a controlled Markov chain, p. 20
$x^{\pi^1, \pi^2}(t), x(t)$	the state process (games), p. 231
$x(t), a(t), b(t)$	the state-actions process (games), p. 231
X_i	the random size of the catastrophe, p. 209

Bibliography

[1] Albright, S.C., Winston, W. (1979). A birth-death model of advertising and pricing. *Adv. Appl. Probab.* **11**, pp. 134–152.

[2] Allen, L.J.S. (2011). *An Introduction to Stochastic Processes with Applications to Biology. Second Edition.* CRC Press, Boca Raton, FL.

[3] Altman, E. (1994). Denumerable constrained Markov decision processes and finite approximations. *Math. Oper. Res.* **19**, pp. 169–191.

[4] Altman, E. (1999). *Constrained Markov Decision Processes.* Chapman & Hall/CRC, Boca Raton, FL.

[5] Álvarez-Mena, J., Hernández-Lerma, O. (2002). Convergence of the optimal values of constrained Markov control processes. *Math. Methods Oper. Res.* **55**, pp. 461–484.

[6] Anderson, W.J. (1991). *Continuous-Time Markov Chains.* Springer, New York.

[7] Ardanuy, R., Alcalá, A. (1992). Weak infinitesimal operators and stochastic differential games. *Stochastica* **13**, pp. 5–12.

[8] Bailey, N.T.J. (1975). *The Mathematical Theory of Infectious Diseases and its Applications.* Hafner Press [MacMillan Publishing], New York.

[9] Bartholomew, D.J. (1973). *Stochastic Models for Social Processes.* Wiley, New York.

[10] Bäuerle, N. (2005). Benchmark and mean-variance problems for insurers. *Math. Methods Oper. Res.* **62**, pp. 159–165.

[11] Bellman, R. (1957). *Dynamic Programming.* Princeton University Press, Princeton, NJ.

[12] Berge, C. (1997). *Topological Spaces.* Dover Publications, Mineola, NY.

[13] Bertsekas, D.P. (2001). *Dynamic Programming and Optimal Control. Vol. II.* Athena Scientific, Belmont, MA.

[14] Bewley, T., Kohlberg, E. (1976). The asymptotic theory of stochastic games. *Math. Oper. Res.* **1**, pp. 197–208.

[15] Bhattacharya, R.N. (1982). On the functional central limit theorem and the law of the iterated logarithm for Markov processes. *Z. Wahrsch. Verw. Gebiete* **60**, pp. 185–201.

[16] Bhulai, S., Spieksma, F.M. (2003). On the uniqueness of solutions to the Poisson equations for average cost Markov chains with unbounded cost functions. *Math. Methods Oper. Res.* **58**, pp. 221–236.

[17] Blackwell, D. (1962). Discrete dynamic programming. *Ann. Math. Statist.* **33**, pp. 719–726.

[18] Bojadziev, G. (1989). A cooperative differential game in predator-prey systems. In: *Differential Games and Applications*, Lecture Notes in Control and Inform. Sci. **119**, Springer, Berlin, pp. 170–177.

[19] Brauer, F., Castillo-Chávez, C. (2001). *Mathematical Models in Population Biology and Epidemiology*. Springer, New York.

[20] Carlson, D., Haurie, A., Leizarowitz, A. (1994). Overtaking equilibria for switching regulator and tracking games. In: *Advances in Dynamic Games and Applications*, edited by T. Basar and A. Haurie, *Annals of the International Society of Dynamic Games 1*, Birkhäuser, Boston, MA, pp. 247–268.

[21] Cavazos-Cadena, R., Lasserre, J.B. (1988). Strong 1-optimal stationary policies in denumerable Markov decision processes. *Systems Control Lett.* **11**, pp. 65–71.

[22] Davies, E.B. (1980). *One-Parameter Semigroups*. Academic Press, London.

[23] Dekker, R., Hordijk, A. (1988). Average, sensitive and Blackwell optimal policies in denumerable Markov decision chains with unbounded rewards. *Math. Oper. Res.* **13**, pp. 395–420.

[24] Dekker, R., Hordijk, A. (1992). Recurrence conditions for average and Blackwell optimality in denumerable state Markov decision chains. *Math. Oper. Res.* **17**, pp. 271–289.

[25] Dekker, R., Hordijk, A., Spieksma, F.M. (1994). On the relation between recurrence and ergodicity properties in denumerable Markov decision chains. *Math. Oper. Res.* **19**, pp. 539–559.

[26] Denardo, E.V. (1970). Computing a bias-optimal policy in a discrete-time Markov decision problem. *Oper. Res.* **18**, pp. 279–289.

[27] Denardo, E.V., Veinott, A.F., Jr. (1968). An optimality condition for discrete dynamic programming with no discounting. *Ann. Math. Statist.* **39**, pp. 1220–1227.

[28] Dieudonné, J. (1960). *Foundations of Modern Analysis*. Academic Press, New York.

[29] Doshi, B.T. (1976). Continuous time control of Markov processes on an arbitrary state space: average return criterion. *Stochastic Process. Appl.* **4**, pp. 55–77.

[30] Down, D., Meyn, S.P., Tweedie, R.L. (1995). Exponential and uniform ergodicity of Markov processes. *Ann. Probab.* **23**, pp. 1671–1691.

[31] Durrett, R. (1999). *Essentials of Stochastic Processes*. Springer, New York.

[32] Dynkin, E.B. (1965). *Markov Processes. Vol. I*. Springer, Berlin.

[33] Federgruen, A. (1978). On N-person stochastic games with denumerable state space. *Adv. Appl. Probab.* **10**, pp. 452–471.

[34] Feinberg, E.A., Shwartz, A. (1996). Constrained discounted dynamic programming. *Math. Oper. Res.* **21**, pp. 922–945.

[35] Feller, W. (1940). On the integro-differential equations of purely discontinuous Markoff processes. *Trans. Amer. Math. Soc.* **48**, pp. 488–515.

[36] Fleming, W.H. (1963). Some Markovian optimization problems. *J. Math. Mech.* **12**, pp. 131–140.

[37] Fleming, W.H., Soner, H.M. (2006). *Controlled Markov Processes and Viscosity Solutions.* Springer, New York.

[38] Föllmer, H., Schied, A. (2004). *Stochastic Finance.* Walter De Gruyter, Berlin.

[39] Freedman, H.I. (1980). *Deterministic Mathematical Models in Population Ecology.* Marcel Dekker, New York.

[40] Frenk, J.B.G., Kassay, G., Kolumbán, J. (2004). On equivalent results in minimax theory. *Euro. J. Oper. Res.* **157**, pp. 46–58.

[41] Gale, D. (1967). On optimal development in a multi-sector economy. *Rev. Econom. Studies* **34**, pp. 1–19.

[42] González-Hernández, J., Hernández-Lerma, O. (2005). Extreme points of sets of randomized strategies in constrained optimization and control problems. *SIAM J. Optim.* **15**, pp. 1085–1104.

[43] Guo, X.P., Cao, X.R. (2005). Optimal control of ergodic continuous-time Markov chains with average sample-path rewards. *SIAM J. Control Optim.* **44**, pp. 29–48.

[44] Guo, X.P., Hernández-Lerma, O. (2003). Continuous-time controlled Markov chains with discounted rewards. *Acta Appl. Math.* **79**, pp. 195–216.

[45] Guo, X.P., Hernández-Lerma, O. (2003). Drift and monotonicity conditions for continuous-time controlled Markov chains with an average criterion. *IEEE Trans. Automat. Control* **48**, pp. 236–244.

[46] Guo, X.P., Hernández-Lerma, O. (2003). Constrained continuous-time Markov control processes with discounted criteria. *Stoch. Anal. Appl.* **21**, pp. 379–399.

[47] Guo, X.P., Hernández-Lerma, O. (2003). Zero-sum games for continuous-time Markov chains with unbounded transition and average payoff rates. *J. Appl. Probab.* **40**, pp. 327–345.

[48] Guo, X.P., Hernández-Lerma, O. (2003). Continuous-time controlled Markov chains. *Ann. Appl. Probab.* **13**, pp. 363–388.

[49] Guo, X.P., Hernández-Lerma, O. (2004). Zero-sum games for nonhomogeneous Markov chains with an expected average payoff criterion. *Appl. Comput. Math.* **3**, pp. 10–22.

[50] Guo, X.P., Hernández-Lerma, O. (2005). Nonzero-sum games for continuous-time Markov chains with unbounded discounted payoffs. *Bernoulli* **11**, pp. 327–345.

[51] Guo, X.P., Hernández-Lerma, O. (2007). Zero-sum games for continuous-time jump Markov processes in Polish spaces: discounted payoffs. *Adv. Appl. Probab.* **39**, pp. 645–668.

[52] Guo, X.P., Hernández-Lerma, O. (2009). *Continuous-Time Markov Decision Processes: Theory and Applications.* Springer, New York.

[53] Guo, X.P., Hernández-Lerma, O., Prieto-Rumeau, T. (2006). A survey of recent results on continuous-time Markov decision processes. *Top* **14**, pp. 177–261.

[54] Guo, X.P., Liu, K. (2001). A note on optimality conditions for continuous-time controlled Markov chains. *IEEE Trans. Automat. Control* **46**, pp. 1984–1989.

[55] Guo, X.P., Rieder, U. (2006). Average optimality for continuous-time Markov decision processes in Polish spaces. *Ann. Appl. Probab.* **16**, pp. 730–756.

[56] Guo, X.P., Song, X.Y., Zhang, J.Y. (2009). Bias optimality for multichain continuous-time Markov decision processes. *Oper. Res. Lett.* **37**, pp. 317–321.

[57] Guo, X.P., Zhu, W.P. (2002). Denumerable-state continuous-time Markov decision processes with unbounded transition and reward rates under the discounted criterion. *J. Appl. Probab.* **39**, pp. 233–250.

[58] Guo, X.P., Zhu, W.P. (2002). Denumerable-state continuous-time Markov decision processes with unbounded transition and reward rates under average criterion. *ANZIAM J.* **43**, pp. 541–557.

[59] Hall, P., Heyde, C.C. (1980). *Martingale Limit Theory and its Applications.* Academic Press, New York.

[60] Hamadène, S. (1999). Nonzero sum linear-quadratic stochastic differential games and backward-forward equations. *Stoch. Anal. Appl.* **17**, pp. 117–130.

[61] Haviv, M. (1996). On constrained Markov decision processes. *Oper. Res. Lett.* **19**, pp. 25–28.

[62] Haviv, M., Puterman, M.L. (1998). Bias optimality in controlled queueing systems. *J. Appl. Probab.* **35**, pp. 136–150.

[63] Hernández-Lerma, O. (1989). *Adaptive Markov Control Processes.* Springer, New York.

[64] Hernández-Lerma, O. (1994). *Lectures on Continuous-Time Markov Control Processes.* Sociedad Matemática Mexicana, México.

[65] Hernández-Lerma, O., González-Hernández, J. (2000). Constrained Markov control processes in Borel spaces: the discounted case. *Math. Methods Oper. Res.* **52**, pp. 271–285.

[66] Hernández-Lerma, O., González-Hernández, J., López-Martínez, R.R. (2003). Constrained average cost Markov control processes in Borel spaces. *SIAM J. Control Optim.* **42**, pp. 442–468.

[67] Hernández-Lerma, O., Govindan, T.E. (2001). Nonstationary continuous-time Markov control processes with discounted costs on infinite horizon. *Acta Appl. Math.* **67**, pp. 277–293.

[68] Hernández-Lerma, O., Lasserre, J.B. (1996). *Discrete-Time Markov Control Processes: Basic Optimality Criteria.* Springer, New York.

[69] Hernández-Lerma, O., Lasserre, J.B. (1997). Policy iteration for average cost Markov control processes on Borel spaces. *Acta Appl. Math.* **47**, pp. 125–154.

[70] Hernández-Lerma, O., Lasserre, J.B. (1999). *Further Topics on Discrete-Time Markov Control Processes.* Springer, New York.

[71] Hernández-Lerma, O., Lasserre, J.B. (2001). Zero-sum stochastic games in Borel spaces: average payoff criteria. *SIAM J. Control Optim.* **39**, pp. 1520–1539.

[72] Hernández-Lerma, O., Romera, R. (2004). The scalarization approach to multiobjective Markov control problems: *why* does it work? *Appl. Math. Optim.* **50**, pp. 279–293.

[73] Hernández-Lerma, O., Vega-Amaya, O., Carrasco, G. (1999). Sample-path optimality and variance-minimization of average cost Markov control processes. *SIAM J. Control Optim.* **38**, pp. 79–93.

[74] Hilgert, N., Hernández-Lerma, O. (2003). Bias optimality versus strong 0-discount optimality in Markov control processes with unbounded costs. *Acta Appl. Math.* **77**, pp. 215–235.

[75] Hordijk, A., Yushkevich, A.A. (1999). Blackwell optimality in the class of stationary policies in Markov decision chains with a Borel state and unbounded rewards. *Math. Methods Oper. Res.* **49**, pp. 1–39.

[76] Hordijk, A., Yushkevich, A.A. (1999). Blackwell optimality in the class of all policies in Markov decision chains with a Borel state and unbounded rewards. *Math. Methods Oper. Res.* **50**, pp. 421–448.

[77] Howard, R.A. (1960). *Dynamic Programming and Markov Processes.* Wiley, New York.

[78] Iosifescu, M., Tautu, P. (1973). *Stochastic Processes and Applications in Biology and Medicine. II: Models.* Springer, Berlin.

[79] Jaśkiewicz, A. (2000). On strong 1-optimal policies in Markov control processes with Borel state spaces. *Bull. Pol. Acad. Sci. Math.* **48**, pp. 439–450.

[80] Jaśkiewicz, A. (2009). Zero-sum ergodic semi-Markov games with weakly continuous transition probabilities. *J. Optim. Theory Appl.* **141**, pp. 321–347.

[81] Jaśkiewicz, A., Nowak, A.S. (2001). On the optimality equation for zero-sum ergodic Markov games. *Math. Methods Oper. Res.* **54**, pp. 291–301.

[82] Jasso-Fuentes, H., Hernández-Lerma, O. (2008). Characterizations of overtaking optimality for controlled diffusion processes. *Appl. Math. Optim.* **57**, pp. 349–369.

[83] Jasso-Fuentes, H., Hernández-Lerma, O. (2009). Blackwell optimality for controlled diffusion processes. *J. Appl. Probab.* **49**, pp. 372–391.

[84] Jasso-Fuentes, H., Hernández-Lerma, O. (2009). Ergodic control, bias, and sensitive discount optimality for controlled diffusion processes. *Stoch. Anal. Appl.* **27**, pp. 363–385.

[85] Kadota, Y. (2002). Deviation matrix, Laurent series and Blackwell optimality in countable state Markov decision processes. *Optimization* **51**, pp. 191–202.

[86] Kakumanu, P. (1971). Continuously discounted Markov decision models with countable state and action space. *Ann. Math. Statist.* **42**, pp. 919–926.

[87] Kakumanu, P. (1972). Nondiscounted continuous-time Markov decision processes with countable state and action spaces. *SIAM J. Control* **10**, pp. 210–220.

[88] Kakumanu, P. (1975). Continuous time Markov decision processes with average return criterion. *J. Math. Anal. Appl.* **52**, pp. 173–188.

[89] Kato, T. (1966). *Perturbation Theory for Linear Operators*. Springer, New York.

[90] Kitaev, M.Y., Rykov, V.V. (1995). *Controlled Queueing Systems*. CRC Press, Boca Raton, FL.

[91] Küenle, H.U. (2007). On Markov games with average reward criterion and weakly continuous transition probabilities. *SIAM J. Control Optim.* **45**, pp. 2156–2168.

[92] Kurtz, T.G., Stockbridge, R.H. (1998). Existence of Markov controls and characterization of optimal Markov controls. *SIAM J. Control Optim.* **36**, pp. 609–653.

[93] Lai, H.C., Tanaka, K. (1984). On an N-person noncooperative Markov game with a metric state space. *J. Math. Anal. Appl.* **101**, pp. 78–96.

[94] Lal, A.K., Sinha, S. (1992). Zero-sum two-person semi-Markov games. *J. Appl. Probab.* **29**, pp. 56–72.

[95] Lasserre, J.B. (1988). Conditions for the existence of average and Blackwell optimal stationary policies in denumerable Markov decision processes. *J. Math. Anal. Appl.* **136**, pp. 479–490.

[96] Lasserre, J.B. (1999). Sample-path average optimality for Markov control processes. *IEEE Trans. Automat. Control* **44**, pp. 1966–1971.

[97] Lefèvre, C. (1979). Optimal control of the simple stochastic epidemic with variable recovery rates. *Math. Biosci.* **44**, pp. 209–219.

[98] Lefèvre, C. (1981). Optimal control of a birth and death epidemic process. *Oper. Res.* **29**, pp. 971–982.

[99] Leizarowitz, A. (1996). Overtaking and almost-sure optimality for infinite horizon Markov decision processes. *Math. Oper. Res.* **21**, pp. 158–181.

[100] Leizarowitz, A., Shwartz, A. (2008). Exact finite approximations of average-cost countable Markov decision processes. *Automatica J. IFAC* **44**, pp. 1480–1487.

[101] Lembersky, M.R. (1974). On maximal rewards and ε-optimal policies in continuous time Markov chains. *Ann. Statist.* **2**, pp. 159–169.

[102] Lewis, M.E., Puterman, M.L. (2001). A probabilistic analysis of bias optimality in unichain Markov decision processes. *IEEE Trans. Automat. Control* **46**, pp. 96–100.

[103] Lewis, M.E., Puterman, M.L. (2002). Bias optimality. In: *Handbook of Markov Decision Processes*, edited by E.A. Feinberg and A. Shwartz. *Internat. Ser. Oper. Res. Management Sci.* **40**, Kluwer, Boston, MA, pp. 89–111.

[104] Lippman, S.A. (1975). Applying a new device in the optimization of exponential queuing systems. *Oper. Res.* **23**, pp. 687–710.

[105] Lorenzo, J.M. (2007). *Control of an Epidemic Process: A Markov Decision Processes Approach* (in Spanish). M.Sc. Thesis, UNED, Madrid, Spain.

[106] Lund, R.B., Meyn, S.P., Tweedie, R.L. (1996). Computable exponential convergence rates for stochastically ordered Markov processes. *Ann. Appl. Probab.* **6**, pp. 218–237.

[107] McNeil, D.R. (1972). On the simple stochastic epidemic. *Biometrika* **59**, pp. 494–497.

[108] Mendoza-Pérez, A.F., Hernández-Lerma, O. (2010). Markov control processes with pathwise constraints. *Math. Methods Oper. Res.* **71**, pp. 477–502.

[109] Mertens, J.F., Neyman, A. (1981). Stochastic games. *Internat. J. Game Theory* **10**, pp. 53–66.

[110] Mertens, J.F., Neyman, A. (1982). Stochastic games have a value. *Proc. Nat. Acad. Sci. USA* **79**, pp. 2145–2146.

[111] Meyn, S.P., Tweedie, R.L. (1993). Stability of Markovian processes III: Foster-Lyapunov criteria for continuous-time processes. *Adv. Appl. Probab.* **25**, pp. 518–548.

[112] Miller, B.L. (1968). Finite state continuous time Markov decision processes with an infinite planning horizon. *J. Math. Anal. Appl.* **22**, pp. 552–569.

[113] Miller, B.L. (1968). Finite state continuous time Markov decision processes with a finite planning horizon. *SIAM J. Control* **6**, pp. 266–280.

[114] Miller, B.L., Veinott, A.F., Jr. (1969). Discrete dynamic programming with a small interest rate. *Ann. Math. Statist.* **40**, pp. 366–370.

[115] Muruaga, M.A., Vélez, R. (1996). Asymptotic behavior of continuous stochastic games. *Top* **4**, pp. 187–214.

[116] Nowak, A.S. (1999). A note on strong 1-optimal policies in Markov decision chains with unbounded costs. *Math. Methods Oper. Res.* **49**, pp. 475–482.

[117] Nowak, A.S. (1999). Sensitive equilibria for ergodic stochastic games with countable state spaces. *Math. Methods Oper. Res.* **50**, pp. 65–76.

[118] Nowak, A.S. (1999). Optimal strategies in a class of zero-sum ergodic stochastic games. *Math. Methods Oper. Res.* **50**, pp. 399–419.

[119] Nowak, A.S. (2003). On a new class of nonzero-sum discounted stochastic games having stationary Nash equilibrium points. *Internat. J. Game Theory* **32**, pp. 121–132.

[120] Parthasarathy, T. (1973). Discounted, positive, and noncooperative stochastic games. *Internat. J. Game Theory* **2**, pp. 25–37.

[121] Piunovskiy, A.B. (1997). *Optimal Control of Random Sequences in Problems with Constraints*. Kluwer Academic Publishers, Dordrecht, Netherlands.

[122] Piunovskiy, A.B. (1998). A controlled jump discounted model with constraints. *Theory Probab. Appl.* **42**, pp. 51–72.

[123] Piunovskiy, A.B. (2004). Multicriteria impulsive control of jump Markov processes. *Math. Methods Oper. Res.* **60**, pp. 125–144.

[124] Polowczuk, W. (2000). Nonzero-sum semi-Markov games with countable state spaces. *Appl. Math. (Warsaw)* **27**, pp. 395–402.

[125] Ponstein, J. (1980). *Approaches to the Theory of Optimization*. Cambridge University Press, Cambridge, UK.

[126] Prieto-Rumeau, T. (2006). Blackwell optimality in the class of Markov policies for continuous-time controlled Markov chains. *Acta Appl. Math.* **92**, pp. 77–96.

[127] Prieto-Rumeau, T., Hernández-Lerma, O. (2005). The Laurent series, sensitive discount and Blackwell optimality for continuous-time controlled Markov chains. *Math. Methods Oper. Res.* **61**, pp. 123–145.

[128] Prieto-Rumeau, T., Hernández-Lerma, O. (2005). Bias and overtaking equilibria for zero-sum continuous-time Markov games. *Math. Methods Oper. Res.* **61**, pp. 437–454.

[129] Prieto-Rumeau, T., Hernández-Lerma, O. (2006). Bias optimality for continuous-time controlled Markov chains. *SIAM J. Control Optim.* **45**, pp. 51–73.

[130] Prieto-Rumeau, T., Hernández-Lerma, O. (2006). A unified approach to continuous-time discounted Markov control processes. *Morfismos* **10**, pp. 1–40.

[131] Prieto-Rumeau, T., Hernández-Lerma, O. (2008). Ergodic control of continuous-time Markov chains with pathwise constraints. *SIAM J. Control Optim.* **47**, pp. 1888–1908.

[132] Prieto-Rumeau, T., Hernández-Lerma, O. (2009). Variance minimization and the overtaking optimality approach to continuous-time controlled Markov chains. *Math. Methods Oper. Res.* **70**, pp. 527–540.

[133] Prieto-Rumeau, T., Hernández-Lerma, O. (2010). The vanishing discount approach to constrained continuous-time controlled Markov chains. *Systems Control Lett.* **59**, pp. 504–509.

[134] Prieto-Rumeau, T., Hernández-Lerma, O. (2010). Policy iteration and finite approximations to discounted continuous-time controlled Markov chains. In: *Modern Trends in Controlled Stochastic Processes: Theory and Applications*, edited by A.B. Piunovskiy, Luniver Press, Frome, UK, pp. 84–101.

[135] Prieto-Rumeau, T., Hernández-Lerma, O. (2011). Discounted continuous-time controlled Markov chains: convergence of control models. *Submitted.*

[136] Prieto-Rumeau, T., Lorenzo, J.M. (2010). Approximating ergodic average reward continuous-time controlled Markov chains. *IEEE Trans. Automat. Control* **55**, pp. 201–207.

[137] Puterman, M.L. (1974). Sensitive discount optimality in controlled one dimensional diffusions. *Ann. Probab.* **2**, pp. 408–419.

[138] Puterman, M.L. (1994). *Markov Decision Processes*. Wiley, New York.

[139] Ramachandran, K.M. (1999). A convergence method for stochastic differential games with a small parameter. *Stoch. Anal. Appl.* **17**, pp. 219–252.

[140] Ramsey, F.P. (1928). A mathematical theory of savings. *Econom. J.* **38**, pp. 543–559.

[141] Ross, K.W., Varadarajan, R. (1989). Markov decision processes with sample path constraints: The communicating case. *Oper. Res.* **37**, pp. 780–790.

[142] Ross, K.W., Varadarajan, R. (1991). Multichain Markov decision processes with a sample path constraint: A decomposition approach. *Math. Oper. Res.* **16**, pp. 195–207.

[143] Rubinstein, A. (1979). Equilibria in supergames with the overtaking criterion. *J. Econom. Theory* **21**, pp. 1–9.

[144] Rykov, V.V. (1966). Markov sequential decision processes with finite state and decision space. *Theory Probab. Appl.* **11**, pp. 302–311.

[145] Sennott, L.I. (1991). Constrained discounted Markov decision chains. *Probab. Engrg. Inform. Sci.* **5**, pp. 463–475.

[146] Sennott, L.I. (1994). Zero-sum stochastic games with unbounded cost: discounted and average cost cases. *Z. Oper. Res.* **39**, pp. 209–225.

[147] Sennott, L.I. (1999). *Stochastic Dynamic Programming and the Control of Queueing Systems.* Wiley, New York.

[148] Sennott, L.I. (2001). Computing average optimal constrained policies in stochastic dynamic programming. *Probab. Engrg. Inform. Sci.* **15**, pp. 103–133.

[149] Serfozo, R.F. (1979). An equivalence between continuous and discrete time Markov decision processes. *Oper. Res.* **27**, pp. 616–620.

[150] Shapley, L.S. (1953). Stochastic games. *Proc. Nat. Acad. Sci. USA* **39**, pp. 1095–1100.

[151] Sladký, K. (1978). Sensitive optimality criteria for continuous time Markov processes. *Trans. 8th Prague Conference Inform. Theory B*, pp. 211–225.

[152] Sobel, M.J. (1973). Continuous stochastic games. *J. Appl. Probab.* **10**, pp. 597–604.

[153] Sundaram, R.K. (1996). *A First Course in Optimization Theory.* Cambridge University Press, Cambridge, UK.

[154] Tanaka, K., Homma, H. (1978). Continuous time non-cooperative *n*-person Markov games. *Bull. Math. Statist.* **15**, pp. 93–105.

[155] Tanaka, K., Wakuta, K. (1977). On continuous time Markov games with the expected average reward criterion. *Sci. Rep. Niigata Univ. A* **14**, pp. 15–24.

[156] Tanaka, K., Wakuta, K. (1978). On continuous time Markov games with countable state space. *J. Oper. Res. Soc. Japan* **21**, pp. 17–27.

[157] Taylor, H.M. (1976). A Laurent series for the resolvent of a strongly continuous stochastic semi-group. *Math. Programming Stud.* **6**, pp. 258–263.

[158] Tidball, M.M., Lombardi, A., Pourtallier, O., Altman, E. (2000). Continuity of optimal values and solutions for control of Markov chains with constraints. *SIAM J. Control Optim.* **38** pp. 1204–1222.

[159] Vega-Amaya, O. (2003). Zero-sum average semi-Markov games: fixed-point solutions of the Shapley equation. *SIAM J. Control Optim.* **42**, pp. 1876–1894.

[160] Veinott, A.F., Jr. (1966). On finding optimal policies in discrete dynamic programming with no discounting. *Ann. Math. Statist.* **37**, pp. 1284–1294.

[161] Veinott, A.F., Jr. (1969). Discrete dynamic programming with sensitive discount optimality criteria. *Ann. Math. Statist.* **40**, pp. 1635–1660.

[162] von Weizsäcker, C.C. (1965). Existence of optimal programs of accumulation for an infinite horizon. *Rev. Econom. Stud.* **32**, pp. 85–104.

[163] Wessels, J., van Nunen, J.A.E.E. (1975). Discounted semi-Markov decision processes: linear programming and policy iteration. *Statist. Neerlandica* **29**, pp. 1–7.

[164] Wickwire, K. (1977). Mathematical models for the control of pests and infectious diseases: a survey. *Theoret. Population Biology* **11**, pp. 182–238.

[165] Wu, C.B., Zhang, J.H. (1997). Continuous time discounted Markov decision processes with unbounded rewards. *Chinese J. Appl. Probab. Statist.* **13**, pp. 1–10.

[166] Wu, W., Arapostathis, A., Shakkottai, S. (2006). Optimal power allocation for a time-varying wireless channel under heavy-traffic approximation. *IEEE Trans. Automat. Control* **51**, pp. 580–594.

[167] Ye, L., Guo, X.P., Hernández-Lerma, O. (2008). Existence and regularity of a nonhomogeneous transition matrix under measurability conditions. *J. Theoret. Probab.* **21**, pp. 604–627.

[168] Yosida, K. (1980). *Functional Analysis.* Springer, Berlin.

[169] Yushkevich, A.A. (1994). Blackwell optimal policies in a Markov decision process with a Borel state space. *Math. Methods Oper. Res.* **40**, pp. 253–288.

[170] Yushkevich, A.A. (1997). Blackwell optimality in continuous in action Markov decision processes. *SIAM J. Control Optim.* **35**, pp. 2157–2182.

[171] Zhang, L., Guo, X.P. (2008). Constrained continuous-time Markov decision processes with average criteria. *Math. Methods Oper. Res.* **67**, pp. 323–340.

[172] Zhu, Q.X. (2007). Bias optimality and strong n ($n = -1, 0$) discount optimality for Markov decision processes. *J. Math. Anal. Appl.* **334**, pp. 576–592.

[173] Zhu, Q.X., Guo, X.P. (2006). Another set of conditions for Markov decision processes with average sample-path costs. *J. Math. Anal. Appl.* **322**, pp. 1199–1214.

[174] Zhu, Q.X., Prieto-Rumeau, T. (2008). Bias and overtaking optimality for continuous-time jump Markov decision processes in Polish spaces. *J. Appl. Probab.* **45**, pp. 417–429.

Index